PRESERVING THE ASTRONOMICAL SKY

IAU SYMPOSIUM VOLUME 196

COVER ILLUSTRATION:

Future radio telescopes will have to be located in radio-quiet parts of the world (pp. 199 and 271). The global distribution of radio background emission at 131 MHz shows how few radio-quiet regions there are at low frequencies. The quantity plotted is the median root-mean-square electric field measured by the FORTE satellite (http://forte.lanl.gov/) at 800-840 km altitude, averaged over several months and all local times. The centre frequency is 131 MHz and the bandwidth is 1 MHz. The FORTE satellite is a joint project of the Los Alamos National Laboratory and the Sandia National Laboratory, under the auspices of the United States Department of Energy. Persons interested in more information on radio-frequency backgrounds and other aspects of FORTE data should contact the project leader, Dr. Abram R. Jacobson (ajacobson@lanl.gov).

THE ASTRONOMICAL SOCIETY OF THE PACIFIC
390 Ashton Avenue – San Francisco, California – USA   94112-1722
Phone: (415) 337-1100   E-Mail: catalog@aspsky.org
Fax: (415) 337-5205   Web Site: www.aspsky.org

Publisher

ASP CONFERENCE SERIES - EDITORIAL STAFF
Managing Editor: D. H. McNamara   LaTeX-Computer Consultant: T. J. Mahoney
Associate Managing Editor: J. W. Moody   Production Manager: Enid L. Livingston

PO Box 24453, 211 KMB, Brigham Young University, Provo, Utah, 84602-4463
Phone: (801) 378-2111   Fax: (801) 378-4049   E-Mail: pasp@byu.edu

ASP CONFERENCE SERIES PUBLICATION COMMITTEE:
Alexei V. Filippenko   Geoffrey Marcy
Ray Norris   Donald Terndrup
Frank X. Timmes   C. Megan Urry

A listing of all other ASP Conference Series Volumes and IAU Volumes
published by the ASP is cited at the back of this volume

INTERNATIONAL ASTRONOMICAL UNION
98bis, Bd Arago – F-75014 Paris – France
Tel: +33 1 4325 8358   E-mail: iau@iap.fr
Fax: +33 1 4325 2616   Web Site: www.iau.org

PRESERVING THE ASTRONOMICAL SKY

Proceedings of the 196TH Symposium of the IAU
held in
United Nations Vienna International Conference Centre
in conjunction with UNISPACE III
at Vienna, Austria
12-16 July 1999

9757

Edited by

R. J. Cohen
*University of Manchester, Jodrell Bank Observatory, Macclesfield
Cheshire, United Kingdom*

and

W. T. Sullivan, III
University of Washington, Department of Astronomy, Seattle, Washington, USA

Library of Congress Cataloging in Publication Data
Main entry under title

Card Number: 2001091699
ISBN: 1-58381-078-1

IAU Publications - First Edition

Published on behalf of IAU by Astronomical Society of the Pacific

Printed in United States of America by Sheridan Books, Chelsea, Michigan

Contents

Preface

IAU Symposium 196 addressed the problems of light pollution, radio interference and space debris that hinder astronomers' ability to study the Universe at the exquisitely sensitive levels necessary for current research. The Symposium was a follow up to the very successful 1988 IAU Colloquium 112 in Washington, DC ("Light Pollution, Radio Interference, and Space Debris", ed. D. L. Crawford) and the 1992 Paris workshop sponsored by UNESCO, ICSU, IAU and COSPAR ("The Vanishing Universe: Adverse Environmental Impacts on Astronomy", ed. D. McNally). The Symposium was organized by IAU Commission 50 (Protection of Existing and Potential Observatory Sites), with the support of Commissions 9 (Instrumentation and Techniques), 21 (Light of the Night Sky), 25 (Stellar Photometry and Polarimetry), 40 (Radio Astronomy), 46 (Astronomy Education and Development) and 51 (Bioastronomy: Search for Extraterrestrial Life) .

Increasing levels of "technological fog" are threatening not only the work of professional astronomers but also the ability of the general public to study and appreciate the wonders of the sky. The threats are global in scale and the effects are long-term in nature. The meeting had two primary purposes: to share information on the status of the environment and adverse effects on observations, and (2) to develop technical and political action plans to preserve the astronomical sky. The meeting was held at the United Nations centre in Vienna, as part of the "Technical Forum" of UNISPACE III, the third United Nations conference on the Exploration and Peaceful Uses of Outer Space.

The meeting had 70 participants from 25 nations and 3 Press Officers who helped to get the word out. After three days of talks and a delightful banquet at the old Vienna Observatory, the fourth day was devoted to workshop sessions for brainstorming effective strategies for dealing with the issues. On the final day, the workshop results were debated in plenary and polished further, before being approved by the entire Symposium. Primarily, a statement was prepared which was presented to the national delegations to UNISPACE III, with the intention that this will provide a basis for future UN discussions and international action. The statement was approved at UNISPACE III and published in their Report: it is included as Appendix 1 to these proceedings.

On the optical front, the most interesting new results for light pollution were based on quantitative calibrations newly available from the US Defense Meteorological Satellite Program (DMSP). Qualitative images of the night-time earth showing the glow from urban areas have been available for twenty years, but now one can do proper quantitative measurements. Based on these images, S. Isobe (Japan) presented estimates of the money wasted annually by major cities from street lighting scatterred upwards: e.g., US$3M for London and US$14M for New York City. P. Cinzano (Italy) presented a marvellous new night-time image of Europe based on DMSP data combined with a light-scattering model to show estimated sky brightness (limiting magnitude) at any given location. Back on Earth, M. Smith, Director of Cerro Tololo Inter-American Observatory, Chile, emphasized the challenges and successes which that observatory has had in dealing with local authorities to preserve their telescopes' capabilities.

Radio astronomers were preparing for the next major meeting of the International Telecommunication Union to allocate radio frequences, in May/June 2000. The millimetre-wavebands, little used until now by industry, will be a major focus for radio astronomy in the 21st century, with the development of powerful new facilities such as ALMA. H. Butcher (The Netherlands) told of the strong recommendation of the Working Group on Radio Astronomy of the Megascience Forum of the Organization for Economic Cooperation and Development (OECD) to seek protection for the new major radio observatories in the mm- and cm-bands, now in various stages of technical development and implementation. As a result, in June 1999 the science ministers of the OECD recommended the formation of a high-level task force to develop long-term solutions (such as radio-quiet zones) that may safeguard our radio windows on the Universe while allowing efficient development of commercial telecommunications. The Terms of Reference of this group are given in Appendix 2. R. Ekers (Australia) and R. Fisher (USA) emphasized that radio astronomers need to become better in rejecting radio interference, using techniques that are quite feasible and known in military circles but not yet developed at observatories. Furthermore, radio (and optical) observatories have been remiss in quantitatively documenting the interference they encounter over time.

W. Flury (Germany) reported on the space debris problem. Already, some 100,000 objects larger than 1 cm circle the Earth and menace operations of all spacecraft, including scientific missions. While this is a serious problem, our meeting did not focus specially on it because it featured prominently elsewhere on the UNISPACE agenda and the major space agencies are themselves giving the problem high priority.

In summary, we achieved our goal not only to exchange information among ourselves and develop new strategies, but also to bring the issues of radio interference and light pollution to an entirely new forum, the United Nations. The 'Vienna Declaration' and Final Report of UNISPACE III (now available from bookshops and at http://www.un.or.at/OOSA/) show that our recommendations were heard and, to a gratifying extent, accepted by the ~100 UN Member States attending UNISPACE III. This forms an excellent basis for the further work, through the UN Committee on the Peaceful Uses of Outer Space, UN-COPUOS, to improve international protection of the astronomical sky as the cultural heritage of all humankind and to protect everyone's access to pristine and uncluttered skies.

In conclusion we would like to thank all members of the Scientific Organizing Committee and the Local Organizing Committee for their help in organizing and running the meeting. We also gratefully thank our sponsors the IAU, COSPAR, the UN Office of Outer Space Affairs, the CIE, IUCAF, URSI and the International Dark-Sky Association, without whose financial support the meeting would not have been possible.

Jim Cohen and Woody Sullivan
April 2001

Participants

Johannes Andersen, Copenhagen University, Astronomical Observatory, Juliane Maries Vej 30, Copenhagen, DK-2100, Denmark ⟨ ja@astro.ku.dk ⟩

Zeki Aslan, Akdeniz University and TUBITAK National Observatory, Akdeniz Universitesi Yerleskesi, Antalya, 07058, Turkey ⟨ aslan@pascal.sci.akdeniz.edu.tr ⟩

J. J. (Harry) Blom, Kluwer Academic Publishers, Spuiboulevard 50, P.O. Box 17, Dordrecht 3300 AA, The Netherlands ⟨ harry.blom@wkap.nl ⟩

H. R. Butcher, Netherlands Foundation for Research in Astronomy, P.O. Box 2, 7990 AA Dwingeloo, The Netherlands ⟨ butcher@astron.nl ⟩

Pierantonio Cinzano, Dipartimento di Astronomia, Università di Padova, vicolo dell'Osservatorio 5, Padova I-35122, Italy ⟨ cinzano@pd.astro.it ⟩

R. J. Cohen, University of Manchester, Jodrell Bank Observatory, Macclesfield SK11 9DL, UK ⟨ rjc@jb.man.ac.uk ⟩

James C. Cornell, International Science Writers Association, 65 Mt. Vernon Street #11, Boston, MA 02103-130, USA ⟨ cornelljc@earthlink.net ⟩

David L. Crawford, IDA, 3225 North First Avenue, Tucson, Arizona 85719-2103, USA ⟨ crawford@darksky.org ⟩

Mary Crawford, IDA, 3225 North First Avenue, Tucson, Arizona 85719-2103, USA ⟨ crawford@darksky.org ⟩

Donald Davis, Planetary Science Institute, 620 North Sixth Avenue, Tucson, Arizona 85705, USA ⟨ drd@psi.edu ⟩

Mario Di Sora, Astronomical Observatory Campo-Catino, Via Plebiscito, 65, Frosinone, 03100, Italy ⟨ mario.disora@rtmol.stt.it ⟩

Francisco Javier Diaz Castro, Instituto de Astrofisica de Canarias, Via Lactea S/N, La Laguna, Canary Islands 38200, Spain ⟨ fdc@iac.es ⟩

Yvan Dutil, Special Project in Radiometry Division, ABB Bomem Inc., 585 Charest blvd East, Suite 300, Quebec, G1K 9H4, Canada ⟨ yvan.r.dutil@ca.abb.com ⟩

R. D. Ekers, CSIRO Australia Telescope National Facility, P.O. Box 76, Epping, NSW 1710, Australia ⟨ rekers@atnf.csiro.au ⟩

Pierre Encrenaz, DEMIRM, Observatoire de Paris, 61 Avenue de L'Observatoire, Paris 75014, France ⟨ pierre.encrenaz@obspm.fr ⟩

Istvan Fejes, FOMI Satellite Geodetic Observatory, PF 546, Budapest, Hungary ⟨ fejes@sgo.fomi.hu ⟩

David Finley, NRAO, P.O. Box 0, 1003 Lopezville Road, Socorro, New Mexico 87801, USA ⟨dfinley@nrao.edu⟩

J. Richard Fisher, National Radio Astronomy Observatory, P.O. Box 2, Greenbank, West Virginia 24944, USA ⟨rfisher@nrao.edu⟩

Walter Flury, ESOC, Robert-Bosch Str. 5, Darmstadt D-64293, Germany ⟨walter.flury@esa.int⟩

Tomas Gergely, National Science Foundation, Division of Astronomical Sciences, 4201 Wilson Boulevard, Room 1045, Arlington, Virginia 22230, USA ⟨tgergely@nsf.gov⟩

Marc Gillet, Schreder Group G.I.E., R-Tech, 3b, Rue de Mons, Liège 4000, Belgium ⟨mg@rtech.be⟩

Andreas Hänel, Museum am Schölerberg, Am Schölerberg 8, D-49082 Osnabrück, Germany ⟨ahaenel@rz.uni-osnabrueck.de⟩

Hans J. Haubold, Office for Outer Space Affairs, United Nations, Vienna International Centre, P.O. Box 500, Vienna A-1400, Austria ⟨haubold@kph.tuwien.ac.at⟩

Christine Hermann, CIE - International Commission on Illumination, Kegelgasse 27, Vienna A-1030, Austria ⟨ciecb@ping.at⟩

Anita Heward, National Space Centre, Exploration Drive, Leicester LE4 5NS, UK ⟨anitah@spacecentre.co.uk⟩

Bambang Hidayat, Bosscha Observatory, Lembang 40391, Indonesia ⟨bhidayat99@hotmail.com⟩

Syuzo Isobe, National Astronomical Observatory, 2-21-1- Osawa, Mitaka, Tokyo, Japan ⟨isobesz@cc.nao.ac.jp⟩

Jacob Kanelbaum, IMI Israel, 2 Haimahot Street, Rehovot 76488, Israel ⟨kanelbaum@hotmail.com⟩

Stella Y. Lioubtchenko, Astro Space. Center, Profsoyuznaya 84/32, Moscow 117810 GSP, Russia ⟨stella@tanatos.asc.rssi.ru⟩

Christian B. Luginbuhl, U.S. Naval Observatory, Flagstaff Station, P.O. Box 1149, Flagstaff, Arizona 86002, USA ⟨cbl@nofs.navy.mil⟩

Donald H. Martins, Physics and Astronomy Dept., University of Alaska Anchorage, 3221 Providence Drive, Anchorage, Alaska 99508, USA ⟨afdhm@uaa.alaska.edu⟩

John R. Mattox, Institute for Astrophysical Research, Boston University, 725 Commonwealth Avenue, Boston, Massachusetts 02215, USA ⟨mattox@bu.edu⟩

Derek McNally, University of London Observatory, Mill Hill Park, London NW7 2Q3, UK ⟨dmn@star.herts.ac.uk⟩

Margarita Metaxa, Arsakeio School of Athens, 63, Ethnikis Antistaseos Str, Athens 152 31, Greece ⟨mmetaxa@compulink.gr⟩

Jacqueline Mitton, Royal Astronomical Society, Burlington House, Piccadilly, London W1J 0BQ, UK ⟨jmitton@dial.pipex.com⟩

Stelio Montebugnoli, Instituto di Radioastronomia - Consiglio Nazionale Ricerche, Via Fiorentina, Villafontana Bologna 40060, Italy ⟨ stelio@ira.bo.cnr.it ⟩

Nawar Samir, Institute of Astronomy and geophysics (NRIAG), Helwan, Cairo, Egypt ⟨ wanas@frcu.eun.eg ⟩

Masatoshi Ohishi, National Astronomical Observatory of Japan, 2-21-1, Osawa, Mitaka, Tokyo 181-8588, Japan ⟨ masatoshi.ohishi@nao.ac.jp ⟩

A. I. I. Osman, Institute of Astronomy and Geophysics (NRIAG), Helwan, Cairo, Egypt ⟨ aibrosman@frcu.eun.eg ⟩

Holger Pedersen, Copenhagen University Observatory, Juliane Maries Vej 30, Copenhagen OE DK_2100, Denmark ⟨ holger@ursa.astro.ku.dk ⟩

Bo Peng, Beijing Astronomical Observatory, Datun Rd. A20, Chaoyang District, Beijing 100012, China ⟨ pb@bao.ac.cn ⟩

John R. Percy, University of Toronto, Mississauga, Ontario L5L 1C6, Canada ⟨ jpercy@erin.utoronto.ca ⟩

Lubos Perek, Astronomical Institute, Czech Academy of Science, Bocni LI 1401, CZ 141 31 Praha 4, Czech Republic ⟨ perek@ig.cas.cz ⟩

Tingyi Piao, Beijing Astronomical Observatory, Chinese Academy of Sciences, Datun Rd. A20, Chaoyang District, Beijing 100012, China ⟨ pty@class1.ba0.ac.cn ⟩

Nigel E. Pollard, NEP Lighting Consultancy, 6 Leopold Buildings, Bath BA1 5NY, England, UK ⟨ nep_lighting@compuserve.com ⟩

Francois Rene Querci, Observatoire Midi-Pyreinees, 14 Av. Edward Belin, Toulouse 31400, France ⟨ fquerci@obs-mip.fr ⟩

Brian Robinson, ATNF, P.O. Box 256, Milsons Point, NSW 1565, Australia ⟨ brobinso@ozemail.com.au ⟩

Klaus Ruf, Max-Planck-Institute für Radioastronomie, Auf Dem Hügel 69, D-53121 Bonn, Germany ⟨ kruf@mpifr-bonn.mpg.de ⟩

Aneliese Schnell, Institute for Astronomy, University of Vienna, Tuerkenschanzstrasse 17, A-1180 Vienna, Austria ⟨ schnell@astro.univie.ac.at ⟩

Duco A. Schreuder, Duco Schreuder Consultancies, Spechtlaan 303, Leidschendam 2261 BH, The Netherlands ⟨ d.a.schreuder@wxs.nl ⟩

Thomas Sheasby, National Space Centre, Exploration Drive, Leicester LE4 5NS, UK ⟨ tom.sheasby@astrium-space.com ⟩

Malcom Smith, Cerro Tololo Inter-American Observatory, Casilla 603, La Serena, Chile ⟨ msmith@noao.edu ⟩

C. R. Subrahmanya, Raman Research Institute, C.V. Raman Avenue Sadasivanagar, Bangalore, Karnataka 560 080, India ⟨ crs@rri.res.in ⟩

Woodruff T. Sullivan, III, University of Washington. Dept. of Astronomy, Seattle, Washington 98195, USA ⟨ woody@astro.washington.edu ⟩

Arthur Upgren, Wesleyan University, 349 Science Center, Middletown, Connecticut 06459, USA ⟨ aupgren@wesleyan.edu ⟩

Johan Vandewalle, Bond Beter Leefmilieu Vlaanderen VZW (BBL),
    Tweekerkenstraat 47, Brussels 1000, Belgium
    ⟨ johan.vandewalle1@pandora.be ⟩

Wim van Driel, Paris Observatory. DAEC , Obs. De Meudon, 5 Place Jules
    Janssen, Meudon 92195, France ⟨ wim.vandriel@obspm.fr ⟩

Irina B. Vavilova, Astronomical Observatory, Schevchenko Kyiv National
    University, 3 Observatorna St., Kyiv 04053, Ukraine
    ⟨ vavilova@rcrm.freenet.kiev.ua ⟩

Colin Vincent, Particle Physics and Astronomy Research Council, Polaris
    House, North Star Ave., Swindon SN2 1SZ, UK
    ⟨ colin_vincent@pparc.ac.uk ⟩

Richard West, ESO, Karl Schwarzschildstr 2, Garching D-85748, Germany
    ⟨ rwest@eso.org ⟩

Yihua Yan, Beijing Astronomical Observatory, Chinese Academy of Sciences,
    Datun Road A20, Chaoyang District, Beijing 100012, China
    ⟨ yyh@class1.bao.ac.cn ⟩

Zhang Xizhen, Beijing Astronomical Observatory, Chinese Academy of
    Sciences, Datun Road A20, Chaoyang District, Beijing 100012, China
    ⟨ zxz@ns.bao.ac.cn ⟩

Organizing Committees

Scientific Organizing Committee:

J. Andersen (Denmark), W. Baan (The Netherlands), R. J. Cohen (UK, Co-Chair) D. Crawford (USA, Co-Chair) W. Flury (Germany), S. Isobe (Japan), P. Encrenaz (France), D. McNally (UK), W. T. Sullivan, III (USA, Co-Chair) and G. Swarup (India).

Local Organizing Committee:

D. Crawford (IDA, USA), H. Haubold, (UN Office for Outer Space Affairs, Vienna, Chair), C. Hermann (CIE, Vienna), A. Schnell (University of Vienna) and R. West (ESO, Germany).

Editors:

Part 1
Introduction

Preserving the Astronomical Sky
IAU Symposium, Vol. 196, 2001
R. J. Cohen and W. T. Sullivan, III, eds.

Opening Remarks

Hans J. Haubold

UN Office For Outer Space Affairs, United Nations Office Vienna, Austria

Distinguished Ladies and Gentlemen, dear Colleagues,

On behalf of the United Nations Office for Outer Space Affairs, I am very pleased to welcome all of you at the United Nations Office Vienna for convening the IAU/COSPAR/UN Special Environmental Symposium entitled "Preserving the Astronomical Sky".

The decision of the International Astronomical Union (IAU) to organize this Symposium as an integral part of UNISPACE III [1] was first conveyed to the United Nations Committee on the Peaceful Uses of Outer Space (COPUOS) in February 1998. The idea for holding such a Symposium was born during the deliberations of the XXIIIrd General Assembly of the IAU at Kyoto, Japan, in 1997. Since that time, the Office for Outer Space Affairs of the United Nations has worked very closely with the IAU to develop an appropriate programme for this Symposium and to prepare the necessary logistics for your participation. In this connection, it has been a great pleasure for the Office for Outer Space Affairs to have worked closely on a continuing basis, on all aspects of this Symposium, with the following individuals:

1. Dr. Johannes Andersen, General Secretary of the International Astronomical Union;

2. Dr. David Crawford, Executive Director of the International Dark-Sky Association;

3. Dr. Woodruff T. Sullivan III; Chairman of the IAU Commission 50: Protection of Existing and Potential Observatory Sites; and

4. Dr. Michel Breger, Director of the Institute of Astronomy at the University of Vienna.

The United Nations Office for Outer Space Affairs is very pleased to have worked hand-in-hand with these distinguished individuals.

We offer our special thanks to the many co-organizers of this Symposium and their representatives, who have come to participate in and contribute to the Symposium.

[Applause]

Distinguished Ladies and Gentlemen, Colleagues,

There are at least three remarkable facts related to this Symposium which I would like to address briefly: the first concerning astronomy in general; the second exploring the cooperation between the United Nations and the International Astronomical Union; and the third focusing on Unispace III and IAU Symposium 196.

1.   Astronomy

Education and research in astronomy are international enterprises and the astronomical community has long shown leadership in creating international collaboration, dating back to a network of comet observers established by Newton and Halley in the 17th century. The International Astronomical Union was the first of the modern international scientific unions organized under the Versailles treaty.
    Further

- Astronomy has deep roots in virtually every human culture;

- It helps us to understand humanity's place in the vast scale of the Universe; and

- It increases the knowledge of humanity about its origins and evolution.

    However, as stated in the IAU Press Release for this Symposium [2, 3],

    "astronomy, a science that has been a leading engine of human
    progress since ancient times, now finds itself increasingly at risk from
    a new type of environmental degradation - that of space itself".

The participants of this Symposium, coming from around the world, are now gathering for the first time at the United Nations to discuss the threats of light pollution, radio interference and space debris to their research.
    You are the specialists to undertake an assessment of the adverse environmental impacts on astronomy. This is the prime reason for calling this Symposium. I will not address this matter further at this point in time.

2.   The United Nations and the International Astronomical Union

First contact between the UN Office for Outer Space Affairs and the International Astronomical Union regarding closer cooperation was established in 1989; at that time Dr. Derek McNally was the General Secretary of IAU.
    In 1991 the UN Office for Outer Space Affairs, through its Programme on Space Applications, initiated the organization of annual United Nations/European Space Agency Workshops on Basic Space Science for the benefit of developing countries in all major economic regions on Earth [4]. The term 'basic space science' subsumes astronomy and planetary exploration, which are two agenda items of the UN COPUOS. While these Workshops are designed to provide a

forum for basic space scientists to inform each other of their research findings, a number of those scientists addressed the issue of adverse environmental impacts on astronomy already at the first Workshop in 1991. Participants of these Workshops soon started exploring opportunities to inform Member States of the United Nations about adverse environmental impacts on astronomy.

In 1995 the IAU was granted observer status in COPUOS and in the period of time from 1994 to 1997, Dr. Derek McNally addressed the Member States, represented in the Scientific and Technical Subcommittee of COPUOS, on an annual basis about adverse environmental impacts on astronomy [5-8]. In 1998 and 1999 Dr. Johannes Andersen, presiding General Secretary of IAU, took the floor in the Scientific and Technical Subcommittee of COPUOS to initiate further communication with representatives of UN Member States of COPUOS by bringing the astronomical community to UNISPACE III to elaborate on the pressing issue of light pollution and radio interference in astronomical research [9,10].

Having said this, I am taking the opportunity to recall that one of the distinguished participants of this Symposium, Dr. L. Perek from the Czech Republic, served as General Secretary of the IAU from 1967 to 1970 and as Director of the United Nations Office for Outer Space Affairs from 1975-1980 [11].

3. UNISPACE III and IAU Symposium 196

The International Astronomical Union has chosen to organize this Symposium as part of the Technical Forum of the Third United Nations Conference on the Exploration and Peaceful Uses of Outer Space [12]. This Conference is convened as a special session of COPUOS open to all Member States of the United Nations. The justification for holding this Symposium at UNISPACE III is that it is the first astronomical activity ever that has the opportunity to address the 185 Member States of the United Nations and representatives of space industry on the issue of adverse environmental impacts on astronomy in the most direct way.

In the period of time from 19 to 30 July 1999 representatives of Governments will gather in the core meetings of UNISPACE III: Committee I, Committee II and the Plenary. At that point in time, the General Secretary of the IAU, Dr. Johannes Andersen, will have the opportunity to inform Member States about the "conclusions and/or proposals" emanating from the Symposium "Preserving the Astronomical Sky". The draft "conclusions and/or proposals", prepared by the Secretariat of the IAU for this Symposium, are before you, distinguished participants, and it is now your responsibility to finalize them for transfer to the UNISPACE III core meetings, where Member States will consider them. It is too early to predict how your "conclusions and/or proposals" will be reflected in the final report of UNISPACE III, adopted by Member States. I would like to recall that the Committee on the Peaceful Uses of Outer Space works on the basis of consensus [12]. However, I should also mention that all concerns of the IAU in COPUOS to date have been reflected in United Nations documents issued by this body [5-10, 12].

It is my hope and belief that over the next five days you will get to know one another better, increase your cooperation on the issue of preserving the astronomical sky, and return home ready to use your knowledge and experience, gained at the United Nations, to address this issue in your respective countries and organizations.

Ladies and gentlemen, it is in this context that I take this opportunity to wish you all very successful deliberations in the days ahead. Again, welcome to the United Nations; please feel at home - it is your organization.

Thank you for your attention.

References

[1] Abstract of the Paper of the International Astronomical Union for UNIS-PACE III (A/CONF.184/AB/NGO/1)

[2] Press Release of the IAU: Astronomy at Risk from Space Environment Degradation, http://www.iau.org/sym196pr.html

[3] International Dark-Sky Association, http://www.darksky.org/ida/iau196/

[4] UN/ESA Workshops on Basic Space Science, 1991-1999,
http://www.seas.columbia.edu/ ah297/un-esa

[5] Scientific and Technical Presentations to the Scientific and Technical Subcommittee of COPUOS at its Thirty-first Session, Vienna, 21-22 February 1994 (A/AC.105/574)

[6] Scientific and Technical Presentations to the Scientific and Technical Subcommittee of COPUOS at its Thirty-second Session, Vienna, 6-7 February 1995 (A/AC.105/606)

[7] Scientific and Technical Presentations to the Scientific and Technical Subcommittee of COPUOS, Vienna, 12-13 February 1996 (A/AC.105/638)

[8] Scientific and Technical Presentations to the Scientific and Technical Subcommittee of COPUOS at its Thirty-fourth Session, Vienna, 17-18 February 1997 (A/AC.105/673)

[9] Report of the Scientific and Technical Subcommittee of COPUOS on the Work of its Thirty-fifth Session, Vienna, 9-20 February 1998 (A/AC.105/697)

[10] Report of the Scientific and Technical Subcommittee of COPUOS on the Work of its Thirty-sixth Session, Vienna, 22-26 February 1999 (A/AC.105/719)

[11] A. Blaauw 1994 History of the IAU - The Birth and First Half-Century of the International Astronomical Union, Kluwer Academic Publishers, Dordrecht

[12] Draft Report of the Third United Nations Conference on the Exploration and Peaceful Uses of Outer Space, Vienna, 19-30 July 1999 (A/CONF.184/3)

Preserving the Astronomical Sky
IAU Symposium, Vol. 196, 2001
R. J. Cohen and W. T. Sullivan, III, eds.

Remarks on the Effort to Preserve the Astronomical Sky

Robert P. Kraft[1]

IAU President,
Lick Observatory, University of California, Santa Cruz, CA 95064, USA

The challenge to astronomy of interference from optical and radio pollution has been around for a very long time, even before I got into astronomy almost fifty years ago. In its infancy at that time, radio astronomy was already in conflict with rapidly developing commercial communication interests. The deterioration of prime optical sites was also rapidly advancing. I recall observing runs at Mt Wilson in the 1960s, when in the late summer one sometimes had seeing of one-quarter arcsec - at the 100-inch Coudé you could even resolve the two components of Mira! Yet the explosive growth of Los Angeles had already doomed Mt Wilson as a dark-sky site, in spite of the excellent seeing conditions which are present to this very day.

Sometime later, after I had joined the staff of Lick Observatory, we entered a long and sometimes acrimonious debate with the city of San Jose concerning street lights. It was a battle pitting low-pressure sodium (which we wanted since it concentrated the light in narrow, if bright, D lines) versus high-pressure sodium (which the city supported since high-pressure lamps emit over a broad spectrum, more like natural daylight). Besides, high-pressure lamps were manufactured by General Electric, an American company with "connections" in San Jose. But, allowing for some compromises, we won that one. From that experience, lessons were learned. Some of these may help us in developing guidelines as we confront the challenges that now face us. I'll come back to this later.

There isn't time to review the often heroic efforts to preserve radio frequencies and protect the night sky that have surfaced in the intervening forty years, many of them led by delegates to this symposium and especially by chairs and members of your SOC and LOC. You, who have done the work and fought the battles, know the story far better than I. But developments over the past decade, especially in space, have led to new challenges, and discussion of these will take centre stage at this Symposium. General Secretary Johannes Andersen has summmarized them succinctly:

> ... interference at radio frequencies from telecommunications satellites and their ever-increasing demand for frequency space cloud the future of radio astronomy and communication with scientific satellites; space debris is a growing threat to scientific satellites and inter-

---

[1]This paper was delivered by Woody Sullivan.

feres with ground-based observations; and projects to launch highly luminous objects into space for earth illumination, artistic, celebratory or advertising purposes present a growing danger to observational cosmology ...

The solution to these problems lies in the political domain and requires an international scope. It is thus a project for the IAU. The convening of this Symposium is an important step on the way to finding the international solution we all seek. It would be presumptuous of me to engage in remarks about the technical and scientific issues, which you know far better than I and which you will be discussing here. But I can in good conscience offer a few administrative and political comments that might prove useful. These fall under three headings.

(1) **Follow the "Track".** It will do us little good to wave our hands and gnash our teeth in public displays of dismay. A plan of action is called for that gets to the political heart of the matter. GS Andersen, acting on his considerable knowledge of the United Nations political scene, has developed such a plan and it is outlined in the "Observations and Recommendations of Symposium 196" (see Appendix 1). I call your attention particularly to items 7 and 8, which we hope ultimately will lead to a revision of the UN Space Treaty that will include articles favorable to our concerns. The recognition that a revision of the UN Space Treaty would be needed, and that a direction or "track" had to be developed to engage the right international committees and influential people, is a major contribution of GS Andersen, and I am greatly indebted to him for developing that track in consultation with many of you.

(2) **Recognize Potential Allies.** Space has come to be regarded by some as just another place to do business, never mind the adverse consequences to matters not only scientific, but cultural and even religious. The potential economic value of space advertising alone could dwarf the substantial budgets currently associated with the space sciences. Thus the forces that might be arrayed against us are formidable. Yet there exist allies. Increasing space debris is a concern not only of astronomers but of the space agencies and the telecommunications industry, who fear damage to valuable satellites from potential collisions. Unrestricted growth of inefficient urban lighting is a matter of concern to environmentalists who point out the need for more efficient use of available electric power. In our relatively primitive battle over San Jose lighting, what turned the City authorities in our favour was not in the end appeals to the grandeur of astronomy, but rather a "dollars and cents" issue: for a given expediture of electric power, low-pressure sodium (LPS) gave more lumens per square meter on the ground than did high-pressure sodium (HPS). In short, LPS was cheaper to run than HPS. Do not forget that "astronomy" does not loom so large for others as it does for us: it helps to have allies with economic power and parallel, if different, agendas.

(3) **Persist.** We travel now on a "track" that we hope will take us to our goal: a revised international space treaty that addresses our concerns about the degradation of the astronomical sky. But our train is not the TGV:

getting to the goal will take many years. During the process, member states will have to cooperate via the UN Committee for the Peaceful Uses of Outer Space and its Legal Subcommittee to get the right language into the impending revision of the UN Space Treaties, and the relevant authorities will likely need to be reminded. Personnel may change. Within the IAU, please recall that officers have fixed terms (often three years), and that the Executive Committee also turns over by fifty percent every three years. I have every reason to believe that this matter will remain at or near the top of the agenda at the highest levels of the IAU; nevertheless, continual reminders from activists on the issues will be necessary. Persistence pays. Looking back again over our battle with San Jose, I think the key was that our representatives were present at every hearing, public or private, did their homework on the issues, especially the economic ones, always persisted. Stretching the metaphor a bit more, our train may not be the TGV, but it is a pretty heavy freight train and our momentum is proportional to our persistence. It's a long battle, but we can prevail.

Preserving the Astronomical Sky
IAU Symposium, Vol. 196, 2001
R. J. Cohen and W. T. Sullivan, III, eds.

History, Strategy and Status of IAU Actions

J. Andersen

(IAU General Secretary 1997-2000)
Astronomical Observatory; Niels Bohr Institute for Astronomy,
Physics & Geophysics; University of Copenhagen;
Juliane Maries Vej 30, DK - 2100 Copenhagen, Denmark

Abstract. A brief account is given of how environmental challenges to astronomy have grown and diversified and how the IAU has addressed the problem. In the 1970s and '80s, appeals and Resolutions of increasing urgency were addressed to governments and other authorities. In the 1990s, interdisciplinary organisations such as ICSU and UNESCO were enlisted as allies. The present Symposium marks the beginning of a new phase, where direct collaboration with United Nations Member States is sought through the UN Committee on the Peaceful Uses of Outer Space.

1. Introduction

Observation is the lifeblood of astronomy. Progress in our understanding of the Universe derives from observations of ever greater breadth and depth. Over the past half-century, the breadth has increased by the expansion of the observable wavelength range from traditional visible light to the entire electromagnetic spectrum from γ-rays to long-wavelength radio waves, and immensely greater depth has been reached with larger telescopes and ever more sensitive detectors. The richness of our current picture of the Universe, from the grand design to the detailed physics, was unimaginable a century ago. Yet, the unsolved problems are as challenging as ever and we must push on.

Unfortunately, this is no longer a matter of just improving telescopes, instruments and detectors, or even of placing observatories in space or in dark sites. The limits to what we can do are increasingly set not by our own tools, but by man-made noise at all frequencies and engulfing the globe.

These problems are reviewed in detail in the remainder of this volume. Ground-based light pollution has already driven dark-sky astronomy off more than one continent, to the detriment of astronomy, the environment, wildlife, and budgets alike. Meanwhile, developments in space have not only brought us UV, X-ray and γ-ray astronomy and the *Hubble Space Telescope*, but also a barrage of space debris, satellite trails on astronomical images, Iridium flashes and GLONASS beacons, and ceaseless mobile telephone chat everywhere.

It is a key responsibility of the IAU to gather experience of these problems from all over the world and promote rational solutions, and the IAU has done this for decades. In the beginning, problems were mostly local and a few local authorities, governments and space agencies controlled the scene. Recent years have, however, seen an explosion of activities, particularly in space, fueled by a deliberate policy of privatisation and deregulation of trade.

An effective defence of astronomy in this political environment requires an appropriate update of our strategy. Holding the present Symposium in Vienna at the time of UNISPACE III is part of this revision. After a brief review of the growth of the challenges and the (re)actions of the IAU, I shall outline our current strategy and some tentative directions for the future.

2. How the Problem Evolved and What the IAU Did

Older volumes of the IAU Transactions reveal, often strikingly, how the environmental impacts on astronomy and the reactions of the IAU, have evolved both qualitatively and quantitatively. Very schematically, one can discern three distinct phases in these developments.

2.1. "The Good Old Days"

The proliferation of urbanization and of electric lighting coincided with the increasing recognition of the importance of clear and stable air providing sharp, unobscured views of the heavens. This led to the foundation of some major observatories on mountaintops, e.g. the Lick and Mount Wilson Observatories. But while they still enjoy good seeing, their dark skies are now gone: astronomers are familiar with the pictures of Los Angeles as seen from Mount Wilson early and late in the 20th century, but these are just the most striking examples of a problem that has become global in the meantime - see elsewhere in this volume.

The IAU reaction followed a dual strategy: To identify sites in the world of high potential quality for astronomy for decades; and to help develop measures that would ensure that they remained pristine. For this, IAU Commission 50, *"Identification and Protection of Existing and Potential Observatory Sites"* was created at the XVIth General Assembly in Sydney in 1973. Its first President, Merle F. Walker, described its plan of action thus (Walker 1976):

> The role of the Commission in the protection of existing sites is intended to be three-fold: (1) In collecting and disseminating information regarding site protection measures being considered or that have been adopted. (2) In recommending types of protection actions to be taken. (3) In supporting protection measures for specific sites.

Commission 50 quickly wrote off radio astronomy as a concern, as it was in the good hands of the (IAU-URSI-COSPAR) Inter-Union Committee on Allocation of Radio Frequencies to Astronomy and Space Research (IUCAF). Solar astronomy sites were thought to be in danger 'only' from atmospheric and radio wave pollution and therefore also not an immediate concern - which is probably still largely true. Commission 50 therefore restricted its attention exclusively to sites for optical dark-sky astronomy. It is an amusing sidelight on the development of astronomy that such sites were then referred to as "stellar sites" (!).

Commission 50 also had interesting organisational features, viz., the following quote from Walker (1976):

Owing to the special nature of this Commission, the membership of the Commission consists of: (1) An Organising Committee, consisting of (a) individuals actively working in the field of site investigation and protection and (b) representatives of major national and international observatories. (2) National Representatives, appointed by the National Committees of member countries of the Union, who form the general membership of the Commission. To date, 25 countries have appointed delegates to the Commission.

Perhaps this structure should be revitalised for the future. Good contacts to the world's major astronomical research organizations will certainly be needed to develop technically sound proposals for international measures to protect astronomy. And contacts to national delegates to international organizations such as the United Nations or the International Telecommunication Union (ITU) will be vital for any real action to materialise (see later).

Commission 50 quickly set up a cooperation with lighting engineers as represented by the Commission Internationale d'Eclairage (CIE) - a hallmark also of the present Symposium - and recommendations for controlled lighting near observatories were developed. But the restriction of Commission 50 to "stellar sites" was short-lived. Already the XVIIth General Assembly in Grenoble in 1976 passed its Resolution 9 explicitly in defence of radio astronomy, while the more general Resolution 10 read as follows (Cayrel 1979):

The IAU notes with alarm the increasing levels of interference with astronomical observations resulting from artificial illumination of the night sky, radio emission, atmospheric pollution and operation of aircraft above observatory sites.

The IAU therefore urgently requests that the responsible civil authorities take action to preserve existing and planned observatories from such interference. To this end, the IAU undertakes to provide through Commission 50 information on acceptable levels of interference and possible means of control.

Commission 50 then focused on each of the adverse effects listed in the first paragraph and proceed to develop specific recommendations on each of these. A landmark in the field was the joint IAU/CIE publication *"Guidelines for Minimizing Urban Glow near Astronomical Observatories"* (Cayrel & Smith 1980). These recommendations were heeded, e.g. by Tucson (Arizona, USA), which adopted lighting regulations that not only protected nearby Kitt Peak National Observatory from the full impact of the population growth of Tucson, but also led to better-quality lighting and substantial energy savings.

The second paragraph could no doubt be unanimously endorsed also by the General Assembly in 2000. In hindsight, it is easy to conclude that either these recommendations were not realistic, or the "responsible civil authorities" were not contacted in the right way, or were unable to resist commercial resistance to restrictions. It is harder to translate such insight into advice for the future.

In this first decade, optical and radio observatories were mostly affected by local sources of radiation that could at least in principle be controlled by local or national authorities. The IAU strategy, defined and implemented by the Commission with the backing of the Executive Committee and the General Assembly, was to investigate conditions at observatories worldwide and to systematise and disseminate the data. Stock was taken and recommendations formulated at General Assemblies, often as Resolutions published officially in the IAU Transactions. Supported by these endorsements, the Commission worked to find solutions to existing or impending local problems through information, education and persuasion, often with notable success.

However, with the 80s, a new class of global threats appeared on the horizon: The 1982 report of Commission 50 (Smith 1982) ends:

> A proposal to place a network of very large solar power collectors in orbit round the Earth (the SPS system) has disastrous implications both for optical and for radio astronomy. If the system under study were eventually to be put into operation, reflected sunlight from the satellites, each of which might have 55 km^2 of solar cells, would remove all possibility of dark sky observations over large portions of the sky (Boyce 1980).
>
> The Commission brings this to the attention of the [IAU] General Assembly.

Such plans, if not yet implemented, remain alive and well: UNISPACE III featured a whole Workshop on (sic!) "Clean and Inexhaustible Space Solar Power" (UN 1999, p. 148). "The Good Old Days" were indeed over for good ...

2.2. Two Decades of Proliferating Problems: A Mounting Struggle

The 1980s and 1990s saw accelerating growth of adverse environmental impacts on astronomy, qualitatively, quantitatively and geographically. At the XIXth General Assembly in Delhi in 1985, G. Swarup reported on "Radio Noise Surveys for India's Giant Meter-Wavelength Radio Telescope". Not even in a developing country was it now obvious that radio quiet areas could be found.

The 1988 Commission report (van den Bergh 1988) lists a number of successful actions by the Commission to prevent or reduce interference from light pollution at a number of observatories worldwide. But the centrepiece of the report is a compact "Litany of Horrors" that merits quotation *in extenso*:

> During the period 1985-1987 activities of the Commission centred on dangers posed to all branches of observational astronomy by light pollution, radio interference and "space junk". A proposal to orbit a ring of satellites to celebrate the centenary of the Eiffel Tower was withdrawn following intense pressure by the French and international astronomical communities. Representations were also made to the US Department of Transportation regarding the environmental impact of the proposed launch of cremated human remains into Earth orbit by the Celestis Corporation of Florida. The proposed launch of huge satellites to convert sunlight into electricity for cities and industries on Earth by the USSR is also a source of grave concern. [...]
>
> The principal concern of radio astronomers during this reporting period is related to the transmissions from USSR GLONASS satellites

interfering with observations of the OH spectral line near 1612 MHz. Reports of serious interference have been received from observatories worldwide. At the latest count (June 1987), nine satellites in this system are transmitting at frequencies in the range 1603.125 - 1614.375 MHz, but the system is still evolving. Periodic monitoring of the system status continues. Written enquiries have been made to Soviet officials to get more information on the system and to try and open a dialogue to mitigate some of the problems. To date these inquiries have not been successful.

[An agreement has since been reached to gradually replace the ageing GLONASS satellites with 'cleaner' successors - from about 2006 ...!]

The IAU Transactions through the end of the 1990s contain an unbroken string of Resolutions on the environment. The XXth General Assembly in 1988 passed a particularly poignant set: Resolution A2 recalled the long series of previous resolutions and requested action from all in positions of influence, ICSU (now the International Council for Science) in particular; and Resolutions A5-A7 were urgent calls for protection of the most important frequency bands for radio astronomy, in particular those of the OH lines (McNally 1990).

Commission 50 clearly concluded that Resolutions and contacts to local authorities were not effective enough. It decided to organise an international conference on "Light Pollution, Radio Interference, and Space Debris", in Washington, DC, in August 1988, just after the XXth General Assembly. The Executive Committee approved the meeting as IAU Colloquium 112 (Crawford 1991).

The Colloquium marked a shift in strategy in that *(i)* a dedicated IAU conference on the subject was organised for the first time; *(ii)* it was co-sponsored by our sister Unions CIE, COSPAR (the COmmittee on SPAce Research) and URSI (Union de Radio Science Internationale); and *(iii)* proceedings were published which could serve as a comprehensive reference for further initiatives. In all three respects it set a precedent for the future which is still followed. Other books appeared, notably a report by a Study Group of the NATO Committee on the Challenges of Modern Society (Kovalevsky 1992), covering light pollution, radio interference, pollution by satellites, space debris and aircraft, and - an important first - legal avenues for the protection of observatories.

As a further strategic move, Commission 50 and the Executive Committee decided that even debating and publicising the issues together with other Unions was also not having adequate impact. Accordingly, a high-level Conference on "Adverse Environmental Impacts on Astronomy" was organised jointly by UNESCO, ICSU, the IAU, and COSPAR and held at the UNESCO Headquarters in Paris in July 1992. The Proceedings were published in a beautiful volume entitled *"The Vanishing Universe"* (McNally 1994).

This Conference became a landmark, not only by its high profile and the fine book, but also by defining a set of high-level strategic goals and laying out a specific plan for pursuing these. One possible strategic step envisaged was that some major observatories might be given a status similar to the "World Heritage Sites" which enjoy special national and international protection. Another was to create a Working Group on Adverse Environmental Impacts on Astronomy within the ICSU family. Finally, and in retrospect most importantly, the IAU was advised to apply for Permanent Observer status with the UN Committee on the Peaceful Uses of Outer Space (UN-COPUOS) through which the existing international Space Treaties have been negotiated.

Eventually, the World Heritage Site model proved unsuitable for the problems of astronomy. The ICSU Working Group was created but never given official status or a specific mandate, and it quietly expired in 1997. The IAU did, however, get Permanent Observer status at COPUOS from 1995, a crucial step forward. D. McNally represented the IAU at the Committee and steadily nourished its understanding of and interest in the environmental problems for astronomy until 1998, when the present writer succeeded him.

2.3. Exploding Developments in Space: New Strategy Needed

Regardless of the Paris meeting, a spate of potentially devastating new space projects soon appeared. Already in 1993, Commission 50 reported on the (eventually unsuccessful) test of a 300-square metre solar sail called 'Znamya' ('Banner') from Space Station MIR, intended to illuminate locations on Earth for industry and disaster control (Murdin 1994). A new test in 1999 also failed, but very ambitious plans exist and must be followed with great attention.

Also in 1993, Space Marketing Inc. (Georgia, USA) proposed to launch a "Space Billboard" some 1 kilometre in dimension. Not only would its brightness and size rival the Moon, with obvious consequences for astronomy, but it was estimated that more than 10,000 space debris fragments per day would be created. A similar project was proposed for the 1996 Olympic Games (Isobe 1997).

The most tragi-comical episode of the period was probably the so-called "Star of Tolerance", two very large tethered balloons in low orbit which would be brighter than the brightest planet and beam benevolent messages to nonstop festivities on Earth - all to celebrate the 50th anniversary of UNESCO, the intended sponsor of the project. While veiled in verbose disguises as a humanitarian effort, the project was in reality a space advertising and gadget marketing business. Although appearing only shortly after the 1992 joint meeting, it was apparently given quite serious consideration before being abandoned by UNESCO. It would have been dismaying indeed to see UNESCO championing the commercial pollution of space for raw profit!

These bizarre and potentially damaging projects were, in the end, cancelled after strong protests from the international scientific community, represented by both the IAU and ICSU. With their comical aspects, they still took place in a world where launches were provided exclusively by space agencies under government control, thereby providing some degree of public transparency and accountability.

In the current climate of globalisation, privatisation, and deregulation of business, including the space industry, both these restraints are gone and we should expect neither warning nor means of appeal. Clearly, a proactive defence strategy must be developed, using the channels through which existing rules and treaties have been formulated and negotiated. Thus, while the XXIIIrd General Assembly in Kyoto in 1997 featured yet another general discussion (Isobe & Hirayama 1998), specific marching orders were given in Resolution A1 of the General Assembly, proposed by the Executive Committee (Andersen 1999):

The XXIIIrd General Assembly of the International Astronomical Union,

Considering that

proposals have been made repeatedly to place luminous objects in orbit around the earth to carry messages of various kinds and that the implementation of such proposals would have deleterious effects on astronomical observations,

and that

the night sky is the heritage of all mankind, which should therefore be preserved untouched,

Requests the President

to take steps with the appropriate authorities to ensure that the night sky receive no less protection than has been given to the World Heritage Sites on Earth.

3.   A New Start: COPUOS and UNISPACE III

From this starting point a new strategy had to be developed. To highlight the odds we are up against, recall that the telecommunications industry *alone* plans to launch some 1,700 satellites over the next decade, and forecasts of the total turnover of the space industry in that period hover around 10^{12} US\$. The total investment in astronomy in the wildest dreams of astronomers pales by comparison. In a world that hails free market forces as the best (self-)regulatory mechanism, restrictions on activities in space will clearly not be easily accepted.

But if government officials can be convinced that space is *not* "just another place to do business" but a finite, non-renewable resource that could go the way of the rain forests, the unpolluted atmosphere, or the seas unless "environmental impact assessments" and corresponding international norms are extended also to space, we may have a chance. The place to meet these officials is COPUOS.

The most urgent task of the Scientific and Technical Sub-Committee of COPUOS (S&T for short) in early 1998 was the preparation of the Third United Nations Conference on the Exploration and Peaceful Uses of Outer Space (UNIS-PACE III), a special meeting of COPUOS open to all UN Member States and Observers. This, on the one hand, meant that little time could be spent elaborating on the environmental concerns of astronomers. But in return, UNISPACE III would offer a unique opportunity to bring these concerns directly to the attention of the major governments of the world through their senior officials in space related matters. Further, everybody was urged to help organise topical satellite meetings and workshops for the UNISPACE III "Technical Forum".

The opportunity to address this audience in a format and with a programme of our own choosing, earning goodwill at the same time, was clearly not to be missed. The cooperation of COSPAR, the UN Office of Outer Space Affairs (OOSA), and the Press would be important to strengthen the message. Accordingly, contacts were made with a few key enthusiasts to form the core of the SOC for this meeting. A programme was drafted, suitably weighted towards the space activities which present the greatest long-term dangers to astronomy and the most vital need for internationally concerted action, and which were the focus of UNISPACE III itself. I am grateful to Woody Sullivan, Jim Cohen, and the indefatigable Dave Crawford for rising to this challenge at short notice.

The meeting was proposed as a full Symposium and unanimously approved by the Executive Committee as IAU Symposium 196: "Preserving the Astronomical Sky". Co-sponsorship of COSPAR and UN-OOSA as well as URSI, CIE, IDA, and others was obtained and is gratefully appreciated. Two Press Officers, Richard West (ESO) and David Finley (NRAO) were also recruited and greatly helped to enhance the attention given to the meeting.

Another satellite meeting was also held in parallel: a "Special IAU-COSPAR-UN Workshop on Education in Astronomy and Basic Space Science". The Workshop was very valuable in reviewing our educational activities and drawing general lessons from our experience so far. It also discussed possible opportunities for cooperation between the IAU, COSPAR, and the "Regional Educational Centres for Space Science and Technology" being set up under the auspices of the UN. Most of the papers given there are published elsewhere (Isobe 1999; UN 1999, p. 119).

The Workshop was very useful in its own right, but also helped to portray astronomers as people who not only seek to put restrictions on useful space activities, but also care about one of the greatest concerns of most UN Member States, *Education*. Lack of a scientifically trained workforce is one of the greatest impediments to rapid progress in the space applications which governments consider beneficial for their countries. Astronomy - which interests everyone - may help to recruit more young people into space science.

This volume contains the Proceedings of IAU Symposium 196; hopefully it will remain useful for our colleagues in the coming years. But the Symposium also had another task, unique in the IAU context: to produce a set of concise recommendations to UNISPACE III itself, to be considered by the Conference and hopefully included in its recommendations to the UN General Assembly and, when eventually approved there, to the governments of the world.

The recommendations of the Symposium were issued as a separate paper (A/CONF.184/C.1/L.2). The paper is reprinted in this volume (Appendix A) and also in Annex III of the Final Report of UNISPACE III (UN 1999, p. 111). Also noteworthy in the same publication are the papers from the Workshop on Space Debris (p. 130) and not least the Workshop on Space Law in the Twenty-First Century (p. 122), which presents a remarkable set of detailed, legally well-founded, and strongly-worded statements on the need for international "traffic rules" in space in order to preserve the space environment. The vigorous support of our colleagues in the legal field is as gratifying as it was a revelation to at least the present writer.

4. The Final Report of UNISPACE III and the Follow-Up

The work of IAU Symposium 196 formally ends with its recommendations to UNISPACE III and the publication of the present volume. But this is a continuing process and a never-ending battle. Put mildly, not all recommendations in IAU publications have led promptly to visible progress. The Editors have therefore asked me to add a section on what happened after the Symposium.

One recommendation was to set up a Working Group under IAU Commission 50 to address all scientific, practical, educational, and policy aspects of the problem of light pollution. This WG has been established, chaired by Dr. Malcolm Smith, Director of the AURA observatories in Chile, and is addressing its charge with vigour.

Measures to constrain global, long-term adverse developments in space must be approached by a two-step procedure. First, appropriate recommendations should be made by UNISPACE III to create a formal basis for action. Next, these recommendations must give rise to concrete proposals for action that can be considered, and hopefully adopted, by COPUOS and applied by Member States. The following records what was done, but also illustrates the rules and procedures of the UN system.

As a Permanent Observer, the IAU made a brief statement at the opening of UNISPACE III, of course making strong reference to our recommendations. But to have any force the recommendations must be made *by COPUOS itself*, i.e. be included in its Final Report. This is not straightforward: COPUOS works by consensus, which implies that no text which is voted against by any Member State survives in the Report. Lengthy preparations were therefore necessary in COPUOS and its Sub-Committees for UNISPACE III.

Thus, while it was at first mildly puzzling to an astronomer to find a draft of the Report of UNISPACE III at the COPUOS S&T meeting in February 1998 - 17 months before the conference actually began! - it was obvious by the next meeting in February 1999 that negotiations on the substance of the outcome of UNISPACE III were already in full swing, paragraph by paragraph. It was clearly time to try to have the right things said, using the right of Observers to comment while, of course, being unable to vote. However, when our proposals for inserted or modified text became too explicit and insistent we were politely, but firmly, reminded that "observers may express opinions, but not make proposals" - a subtle, but significant distinction.

There were two lessons: *(i)* to progress in the UN system one must work with the national delegates to convince some of them to adopt a proposal (and others to not veto it), and *(ii)* to do that, one must be known in advance, present when needed, and well prepared. This strategy was followed all through the two weeks of UNISPACE III, and while some cherished recommendations were unable to overcome the resistance to placing any barriers on the commercial development of space, others survived in good health. Some even returned through the back door, apparently looking less suspicious when proposed by lawyers ...

The final document of UNISPACE III is called *The Space Millennium: The Vienna Declaration on Space and Human Development*, and is addressed to the governments of the World through the UN General Assembly (UN 1999, p. 1-4). Its preamble refers to astronomy already in its second paragraph and reaffirms the statement in the original UN Space Treaty that,

> "Outer space should be the province of all humankind, to be utilized for peaceful purposes and in the interests of maintaining international peace and security, and in accordance with international law ...".

Its central recommendations form Chapter I.1(c) (**my emphasis** added here):

"(c) Advancing scientific knowledge of space: action should be taken:

(i) To improve the scientific knowledge of near and outer space by promoting cooperative activities in such areas as astronomy, space biology and medicine, space physics, the study of near-Earth objects and planetary exploration;

(ii) To **improve the protection** of the near-Earth space and outer space environments through further research in **and implementation of** mitigation measures for **space debris**;

(iii) To improve the international coordination of the activities related to near-Earth objects, harmonizing the worldwide efforts directed at identification, follow-up observation and orbit prediction, while at the same time giving consideration to developing a common strategy that would include future activities related to near-Earth objects;

(iv) To protect the near and outer space environments through further research on designs, safety measures and procedures associated with the use of nuclear power sources in outer space.

(v) To ensure that **all users** of space consider the **possible consequences** of their activities, whether ongoing or planned, before **further irreversible actions** are taken affecting future utilization of near-Earth space or outer space, especially in areas such as **astronomy**, Earth observation and remote sensing, as well as global positioning and navigation systems, where **unwanted emissions** have become an issue of concern as they interfere with bands in the electromagnetic spectrum already used for those applications."

We would have preferred stronger and more specific language, but this text in fact recognises that the environment already *is* suffering from the development of space (*"further* irreversible actions"). It also recommends research *and* concrete action against space debris, mentions the need for environmental impact assessments for *all* space activities (even if avoiding the term itself), and mentions the problems for radio astronomy as an international concern. That other disciplines are said to be in danger as well only strengthens our position.

The *Vienna Declaration* has been endorsed by the UN General Assembly and is now UN policy. But the Final Report of UNISPACE III also contains a Chapter II, "Background and Recommendations of the Conference", which was crafted as carefully as the Declaration itself. Paragraphs 57-74 (p. 28-30) deal with astronomy, and via space weather address global climate change. The Chapter then continues with quite graphic descriptions of space debris, satellite flashes, solar reflectors, and space advertising and 'celebrations', and finally recalls that the IAU and COSPAR are strongly opposed to these.

The conclusion is rather meek: "Attention should be given to preserving or restoring astronomical observation conditions to a state as close to natural by any practicable means." Our attempts to strengthen it were in vain, but the tone is clear, and it is followed immediately by the text: "The launch of reflectors for the illumination of parts of the Earth's surface also has a potential negative impact on biological diversity. Research should be undertaken prior to the launch of any such reflectors." Cordial thanks to our biologist allies!

Other key recommendations are in paras. 84-86 (p. 31, **emphasis** added):

"**84.** It was recommended that:

(a) The United Nations continue its work on space debris;

(b) The **entire** international "space-faring" community be invited to apply debris minimization measures **uniformly and consistently**;

(c) Studies be continued on possible solutions to **reduce** the population of in-orbit debris.

85. Member states should continue to cooperate, at the national and regional levels and with industry and through the International Telecommunication Union (ITU), to implement suitable regulations to preserve **quiet frequency bands for radio astronomy** and remote sensing from space and to develop, **as a matter of urgency,** practicable technical solutions to **reduce unwanted radio emissions** and other undesirable side effects from telecommunication satellites.

86. Member states should cooperate to explore new mechanisms to protect selected regions of Earth and space from radio emissions **(radio quiet zones)** and to develop **innovative techniques** that will optimise the conditions for scientific and other space activities to share the radio spectrum and coexist in space."

The Final Report also makes several valuable recommendations on astronomical matters beyond our present scope.

In all, a gratifying number of IAU recommendations have now become part of the policy of the United Nations for future developments in space. With a view to the future it is especially encouraging that this was achieved through the cooperation of several key national delegations who, despite the natural pressure on them from commercial interests, have understood and recognised our arguments. For, as previous history has abundantly shown, words on paper do not by themselves produce action. For any binding agreements to be even thinkable, the cooperation of these delegations is indispensable.

It is most welcome, therefore, that the legal services of ESA have taken the initiative, at the request of the Member States, to begin formulating coordinated proposals which Member and Cooperating States could, in due course, present to COPUOS. The IAU has been asked, in cooperation with COSPAR and other relevant national and international organizations, to assist in formulating internationally agreed, practicable standards for permissible levels of pollution of all kinds. Our reply has, of course, been prompt and enthusiastic.

5. Epilogue

If one compares our past history with that of other major environmental issues - the pollution of the oceans, loss of the tropical rain forest, or growth of fluorocarbons and greenhouse gases in the atmosphere - the timescales for them to become critical are comparable. It is not surprising that solutions will take correspondingly long to implement, let alone become effective. An optimist might hope that the well-known environmental calamities on Earth may have sensitized populations and governments to the fact that some mistakes are irreversible or at least take decades to repair. Not even an optimist could delude himself that we are anywhere near that goal yet; but maybe we are at least under way?

As a final remark, the delay in preparing this contribution which allowed me to include a summary of events since the Symposium serves also to illustrate another aspect of the story. The constant close follow-up of all actions which is needed for progress to occur and opportunities such as UNISPACE III not to be missed, is very time-consuming. When added to the other tasks of a volunteer General Secretary, writing about this work as well as actually doing it comes to rely on marginal resources. If a high profile of the Union in the battle for the environment is to be sustainable in the long term, the IAU Executive Committee will need to review the priorities of the IAU for the use of its human resources and eventually implement the necessary organisational adjustments.

Acknowledgments. On behalf of the IAU, I thank all those colleagues who, over the years, have fought the battle for astronomy to where we stand today. These include, in particular, the past Presidents of IAU Commission 50 and the organisers and Press Officers of this Symposium, but also those in other organisations like ICSU, UNESCO, UN-OOSA, URSI, COSPAR, IUCAF, CIE, and more recently, the International Dark-Sky Association (IDA), who have joined forces with us. Special thanks go to former IAU General Secretaries Derek McNally (1988- 91), whose repeated "Technical Presentations" to COPUOS S&T sensitised it to our subsequent specific proposals, and Lubos Perek (1967- 70!), whose tireless efforts, inside knowledge of the UN system), and present position as a National Delegate to COPUOS have been invaluable to our recent progress. I am also grateful to the IAU Executive Committee for steadfastly supporting my ventures into the unfamiliar terrain of the United Nations system. My personal thanks are due to Dr. Hans Haubold of UN-OOSA for patiently coaching me - through several iterations - in the noble art of achieving tangible progress in the UN system from an Observer's humble position.

Finally, my thankful apologies to the Editors of this volume for waiting for this chapter with more patience than I would likely have shown in their place.

References

Andersen, J. (Ed.) 1999, *IAU Transactions Vol. XXIIIB*, (Kluwer, Dordrecht), p. 31

Boyce, P.B. 1980, Bull. AAS 12, 501

Cayrel, R. 1979, in *IAU Transactions Vol. XVIIA,, Part 1,* Ed. E.A. Müller (Reidel, Dordrecht), p. 215

Cayrel, G., & Smith, F.G. 1980, *Guidelines for Minimizing Urban Glow near Astronomical Observatories,* Publication IAU/CIE No. 1

Crawford, D.L. (Ed.) 1991, *Light Pollution, Radio Interference, and Space Debris, (IAU Colloq. 112),* ASP Conf. Ser. Vol. 17

Isobe, S. 1997, in *IAU Transactions Vol. XXIIIA,* Ed. I. Appenzeller, (Kluwer, Dordrecht), p. 45

Isobe, S. (Ed.) 1999, Proceedings of the IAU-COSPAR-UN Educational Work-shop, in *Teaching of Astronomy in the Asian-Pacific Region,* Bulletin No. 15, 1-4, and volume in press

Isobe, S. & Hirayama, T. (Eds.) 1998, *Preserving the Astronomical Windows,* ASP Conf. Ser., Vol. 139

Kovalevsky, J. (Ed.) 1992, *The Protection of Astronomical and Geophysical Sites,* Ed. Frontières, Gif-sur-Yvette, France.

McNally, D. (Ed.) 1990, *IAU Transactions Vol. XXB,* (Kluwer, Dordrecht), p. 40

McNally, D. 1994, *The Vanishing Universe,* Cambridge Univ. Press, Cambridge.

Murdin, P. 1994, in *IAU Transactions Vol. XXIIA,* Ed. J. Bergeron, (Kluwer, Dordrecht), p. 581

Smith, F. Graham 1982, in *IAU Transactions Vol. XVIIIA,* Ed. P.A. Wayman, (Reidel, Dordrecht), p. 667

United Nations 1999, *Report of the Third United Nations Conference on the Exploration and Peaceful Uses of Outer Space* (UNISPACE III), UN Publication A/CONF.184/6 (Sales No. E.00.I.3), United Nations, New York

van den Bergh, S. 1888, in *IAU Transactions Vol. XXA,* Ed. J.-P. Swings, (Kluwer, Dordrecht), p. 691

Walker, M.F. 1976, in *IAU Transactions Vol. XVIA, Part 1,* Ed. G. Contopoulos (Reidel, Dordrecht), p. 219

2010?

Preserving the Astronomical Sky
IAU Symposium, Vol. 196, 2001
R. J. Cohen and W. T. Sullivan, III, eds.

International Action

D. McNally[1]

University of London Observatory,
Mill Hill Park, London NW7 2QS, UK

Abstract.   The roles of the International Astronomical Union, the other ICSU Unions, UNESCO and the UN are reviewed in order to assess the way forward to resolve a unique situation - the impact of civilisation on the observational procedures of a particular science - astronomy. Action to create a *modus vivendi* to allow the continuance of vigorous astronomical science within a vibrant technological society is suggested.

1.   Introduction

Sections 1 and 2 of "Preparations for the Third United Nations Conference on the Exploration and Peaceful Uses of Outer space" (UN 1998) make reference to the power of astronomy to drive interest in space. That driving force today has suffered considerable dilution by the activities of humankind - also attested in Section 74 of the same document. The ground-based astronomer has to conduct observations of faint cosmic sources with instruments sitting on the ground at the bottom of the Earth's atmosphere. The Earth's atmosphere is, without doubt, the greatest single impediment to successful astronomical observations, given its rapidly varying state in both space and time. Astronomers have invested heavily - time, ingenuity, grit and money - to overcome the worst atmospheric excesses by moving observatories to the highest, driest, darkest or radio-quiet, remote sites this planet can offer. Leaving aside the fact that such paragons among sites often can be tectonically unstable, astronomy must now face activities of humankind which do much to magnify the undesirable consequences of observing from the bottom of the Earth's atmosphere, plus several more without parallel in nature.

The outlook for astronomy is clouded - metaphorically and literally! The technological virtuosity which has permitted the scientific triumphs of the last decades is also the greatest single threat to continued progress in astronomical science. Our widespread exploitation of science has led to a rising tide of electromagnetic noise which threatens to overpower and render faint cosmic signals undetectable. This tide is worst for radio astronomy but affects other wavelength bands as well. Heavy industry and transportation add to background atmospheric pollution and ground vibration. The release of greenhouse gases leads to higher temperatures, greater atmospheric turbulence and the risk of

---

[1]Current address: 17 Greenfield, Hatfield, Herts AL9 5HW, UK

increasing cloudiness - already happening as a result of aircraft contrails. We contend with proposals for solar reflectors in space for earth illumination, artistic, celebratory or advertising purposes. We learn of solar radiation collection projects for major terrestrial power generation purposes which would have disastrous consequences for astronomy.

Yet the science of astronomy has been remarkably successful in understanding the nature of the Universe and the technological virtuosity of this century has created unparalleled, exciting prospects for the next. The progress of astronomy in this century has been dazzling - it could be even more dazzling in the next.

2.   Why should Astronomy have Special Status?

Astronomy is one of the oldest sciences studied by *homo sapiens*. Its roots lie deep in prehistory. Since Galileo have we begun to get to grips with the essential physics of the Universe and the twentieth century has seen explosive progress in that understanding.

The value of astronomy has always lain in its practicality, whether for time-keeping, calendrical regulation, navigation, extending the horizons of laboratory physics or just extending horizons. Often regarded in our own age as being esoteric and "ivory tower", astronomy still has immensely practical value, not only in the above, traditional, areas, but also in obtaining detailed knowledge of the Sun's structure and evolution, solar-terrestrial relations and comparative planetology. The Sun cannot be understood in isolation from other stars in a Galactic context - indeed Section 77 of "Preparations... " (UN 1998) specially mentions the interstellar medium - a response which implies that proper understanding of the interstellar medium will contribute to the understanding of the solar system. Close the windows on the cosmos and our understanding of the Earth becomes incomplete. Where understanding lacks comprehensiveness, error of judgement is never far behind, as a study of astronomy through the ages so clearly demonstrates. Astronomy is an important science. It is one of very few sciences which must share its "laboratory" with the rest of humankind. Humans are making a very thorough mess of the astronomical "laboratory". If astronomy is to remain the creative, vigorous science it now is, it must have agreements in place which will allow astronomical science and a science-based society to flourish together. These agreements must be on an international basis, since the activities of particular groups that can have serious deleterious effects on astronomy are widely spread. We therefore seek a *modus vivendi* that is workable and allows all parties maximum freedom of action.

3.   The Role of the International Astronomical Union

The International Astronomical Union (IAU) is a non-governmental organisation founded in 1919 under the auspices of the League of Nations to promote the development of astronomy internationally and to promote cooperation between astronomers irrespective of nationality, race, religion or ethnic background. In meeting those aims the IAU proved remarkably effective.

Table 1.   Resolutions of the International Astronomical Union which
relate to degradation of observing conditions

Year	Place	Trans B[1]	No.	Title	Source[2,3]
1961	Berkeley	XI	R1	Impact of space projects on optical and radio observations	EC
			R2	Project West Ford	EC
				Allocation of Radio Frequency Space for Astronomy	C40
1964	Hamburg	XII	R3	Support for Radio Frequency Allocation Proposal: greater protection for OH at 1664, 1668 MHz	C40
			R5	Vapour Trails at Total Solar Eclipse	C12
1967	Prague	XIII			
1970	Brighton	XIV	R10	Impact of Use of Space on Astronomy	
1973	Sydney	XV			
1976	Grenoble	XVI	R8	Reduction of interference from bands neighbouring Astronomical Radio Frequency Band Adverse Environmental Impact on Astronomy	
1979	Montreal	XVII	(v)	Extension and Preservation of Astronomical Radio Frequency Bands	
1982	Patras	XVIII	R9	Protection of Radio Frequency Bands	
1985	Delhi	XIX	B3	CCIR (International Radio Consultative Committee) Actions	
			B6	Protection of Observatory Sites	
			B7	Danger of the Contamination of Space	
1988	Baltimore	XX	A2	Adverse Environmental Impacts on Astronomy	
			A5	Cooperation to save Hydroxyl Bands	
			A6	Sharing Hydroxyl Band with Land Mobile Services	
			A7	Revision of Frequency Bands for Astrophysically Significant Lines	
1991	Buenos Aires	XXI	A1	Sharing Hydroxyl Band with Land Mobile Services	EC
			A2	Revision of Frequency Bands for Astrophysically Significant Lines	EC
			A3	Preservation of Radio Frequencies for Radio Astronomy	EC
			A6	Working Group on the Prevention of Interplanetary Pollution	EC
1994	Hague	XXII	B3	Measurement and Mitigation of Adverse Environmental Impacts on Astronomy	C5
			B4	Prohibition of Satellite Systems having potentially adverse impacts on astronomy	C40
			B14	Sharing the Hydroxyl Band with Land Mobile Satellite Services	C40
			B15	Bands to be used for Radiocommunications in the lunar environment	C40/50
1997	Kyoto	XXIII	A1	Protection of the Night Sky	EC

[1] IAU Transaction B; [2]EC = Executive Committee; [3]C = Commission

The impact of the requirements of society on astronomy was first perceived in radio astronomy. It was clear at a very early stage in the development of radio astronomy that the radio spectrum would have to be shared with other users and radio astronomers sought recognition from the International Telecommunication Union (ITU). Their success is a model to be followed but is the subject of another paper (Robinson) in this volume. However, let me recall simply that the three first IAU Resolutions on Adverse Impact, which appeared at the 11th General Assembly in Berkeley in 1961, concerned Radio Astronomy. These resolutions were: two from the Executive Committee on "the Impact of Space Projects on Optical and Radio Observations", "Project West Ford" and one from Commission 40 (Radio Astronomy) on "Allocation of Radio Frequency Space for Astronomy". Clearly we have been thinkng of adverse environmental impact for over 40 years. Table 1 lists the IAU Resolutions since 1961 which relate to the impact of society on astronomy - a significant number of these specifically concern radio astronomy.

The terminology "adverse environmental impact on astronomy" does not appear in IAU Resolutions until 1976 and from 1988 such resolutions have appeared in one form or another at every General Assembly, culminating in the Executive Committee Resolution of 1997 on the Protection of the Night Sky.

The IAU also has Commission 50, whose task is the Protection of Existing and Potential Observatory sites. This Commission has as its mandate, protection of observatory sites at any electromagnetic wavelength and at any site on Earth or in space. The Commission has played a considerable role in seeking protection of observing sites by way of local, regional or national agreements, by giving support to those planning new observatories. C50 (IAU 1978) jointly with CIE (Commission Internationale de l'Eclairage) published a Report and Recommendations on Adverse Environmental Impact on Astronomy and the Protection of Observatory Sites which contains much information of value today. This Report was republished in *The Vanishing Universe* (McNally 1994).

The IAU has sought to maintain awareness of all types of adverse environmental impact that can affect astronomy. Not only are we dealing with electromagnetic pollution at all wavelengths, but we must consider sources of ground vibration, heat input to the atmosphere, changes in weather patterns (both global and local), accumulation of chemical absorbers in the atmosphere, balloons (Sky Stations) for communications relays, laser communications, the effects of satellites trailing astronomical fields under observation, terrestrial pollution of the solar system, and the effects of solar reflectors and other sources of "bright" lighting in space. It is a list which grows with time and the deleterious effect of each of these problems grows with time. Commission 50 has organised two meetings on this topic - IAU Colloquium 112 in 1988 (Crawford 1991) and IAU Symposium 196 in 1999 (the present volume).

IAU Colloquium 112 (Crawford 1991) was a defining marker of astronomical concern. It reviewed electromagnetic pollution and space debris issues. It defines the situation as it existed prior to 1988 - a situation which has, overall, got worse and encompasses more adverse impacts.

There have been other meetings since, e.g., the NATO-sponsored study *The Protection of Astronomical and Geophysical Sites* (Kovalevsky 1992) and the Proceedings of the IAU Joint Discussion in 1997, on *Preserving the Astronomi-*

cal Windows (Isobe and Hirayama 1998). The IAU has put considerable effort into examination of the problems posed by the impact of modern technological civilisation and astronomy. We are very conscious of what these problems are. But there is still a great deal to be done to convince not just the public generally, but also our colleagues, of the seriousness of the problem that exists. The public is unaware of the seriousness of the problem, whereas our colleagues harbour an unreasonable optimism that by some chance the problem will go away. Because the threat to the observational work of the major international observatories is not yet too apparent, the fate of the small observatories closer to civilisation is not a major issue to some. That is a short-sighted view, firstly since much excellent, essential work can only be carried out at small observatories, such is the demand for telescope time at the major international observatories, and secondly because it is only a matter of time before these major, more remote observatories will also be seriously impacted. Simple action can produce worthwhile amelioration, as you have heard, and will hear, many times in this Symposium. Even in radio astronomy - the most seriously affected area of astronomy, as you hear from Brian Robinson - the burden of maintaining reserved frequency space is carried by very few people and the efforts of national and regional bodies like CORF and CRAF, and international bodies such as IUCAF, are the reason why radio astronomy can be conducted at all today. They are to be highly praised for their example and for their dedication.

The IAU can offer support (e.g. IAU/CIE Document 1978) to those attempting to negotiate on behalf of specific types of observation. There is international backing for efforts to obtain local agreements as well as a wealth of experience of such negotiations. It must be recognised that the existence of local agreement, e.g. on outdoor lighting, has consequences for local people which can be restrictive. We astronomers should not forget that our dark skies or low-noise radio observatory comes at a cost to the local neighbourhood. Yet it is remarkable how much goodwill exists and how much local pride is generated by hosting a major observatory.

The threats to astronomy are very far-reaching - particularly those from space. Ground-based threats can often be minimized by retreat or local agreement. Retreat can bring advantages - such as high, dry sites in otherwise inhospitable places. However, there is now no opportunity for further retreat. There is no escape from assaults from space. Signals from space reach the most remote mountain-top observatory as they do the most populous of cities. Bright objects in space cannot avoid passing over the remotest observatory. Radio signals from space are a major threat to the detection of weak cosmic signals by radio telescopes. Bilateral agreements on space activities are of little use, as many of such enterprises are multinational in character, embracing highly diverse groups ranging from governments and multinational companies to small businesses. Therefore there must be international recognition of the problems faced by astronomy. I shall return to this in Section 5.

4. International Partners

It is important to realise that astronomy does not stand alone on the issue of adverse environmental impact. Other sciences also suffer from the activities of a technological civilisation.

(i) Geophysics: Electromagnetic pollution is a problem for geophysics in the study of air-glow, aurora and other geophysical phenomena and indeed the acceptable limit for increase in background levels of illumination for such studies would be less than the 10% increase often quoted for astronomy. The same concern is shared by those studying the zodiacal light and gegenschein. The Earth's magnetic field is a major area of research. However, major electrification schemes give rise to changes in local magnetic fields of the order of two nanotesla at 10 km distance - a variation an order of magnitude greater than the sensitivity of commonly used magnetometers. Ground vibration is as significant an issue for seismometry studies as for astrometric observations. Vibration from trucks, trains, etc. can remain significant at distances of over 3 km. Geophysics is also concerned about the chemistry of the atmosphere and the effect of industrially produced effluents on the balance of the natural atmospheric chemistry.

The geophysics community bring out a further concern - vandalism at automatic recording stations. With a movement towards more robotic telescopes, this is likely to become a greater concern for the astronomy. The IAU therefore cooperates with the IUGG (International Union of Geodesy and Geophysics) and with SCOPE (Standing Committee on the Problems of the Environment) to exchange information and to maintain awareness of common hazards as they arise. Currently the overlap of interest of the IAU with SCOPE is not great, but the contact is valued in maintaining a link with other environmental considerations.

(ii) Meteorology: Astronomers must observe through the Earth's atmosphere - the preserve of meteorologists. Both groups have parallel but not identical interests in changes in atmospheric extinction, the consequences of global warming and changes in global distribution of cloud. Recently, atmospheric scientists have communicated their concerns over space debris, which degrades the quality of observations obtained by northward-looking polar atmospheric radars. The IAU has representation with the WMO (World Meteorological Organisation), but this is an area where more frequent exchanges and better contacts are needed, for example with regard to the meteorological downward-looking radars which are of concern for radio astronomy.

(iii) COSPAR (Committee on Space Research): COSPAR maintains a Panel, PEDAS (Potentially Environmentally Detrimental Activities in Space), to alert COSPAR to proposed space ventures which may prove damaging to the environment. PEDAS and the IAU share the key concerns of increasing levels of radio noise and debris damage to orbiting observatories.

(iv) UNESCO (United Nations Educational, Cultural and Scientific Organization): UNESCO has been a helpful partner in advertising the concerns of the astronomical community regarding adverse environmental impact. UNESCO is the "parent" of the ICSU family of Scientific Unions and Committees. UNESCO supported the efforts of the IAU in mounting an exposition at the UNESCO headquarters in Paris, to advertise the threats to continued high-quality astronomical observation. This exposition resulted in the publication of a book (McNally 1994) which set out in simple terms the range and severity of the

threats facing astronomy. The astronomical community is also very grateful to UNESCO for taking the important, defining decision of turning down an opportunity to mark its first 50 years with a commemorative space project, the "Star of Tolerance".

(v) CIE: The collaboration which has been developed over the years between astronomers and lighting engineers has proved extremely successful, indeed it is another role model. The modern road lighting schemes are a daily reminder that well-considered lighting design gives good visibility and low light-spill.

(vi) UN: The IAU has developed links with the UN COPUOS Scientific and technical Sub-Committee, resulting in an invitation to assume Observer Status. This recognition is highly appreciated by the IAU. One outcome is that this symposium is being held at the UN. The opportunity to have a direct line of communication for the presentation of the serious concerns of the international astronomical community to the UN is of immense value.

The IAU wishes to cooperate and collaborate with all sciences adversely affected by the activities of society, e.g with the biological sciences on the ecological impacts of light pollution. The astronomical community is not alone and must explore with sister sciences the range of measures that urgently need to be taken in order to protect the integrity of observational and other field work. To this end, at first under ICSU auspices and now under IAU auspices, there is a small Working Group which keeps the collective threat under review.

5. International Action

Adverse environmental impact is not the preserve of astronomy alone. Other sciences also suffer. The problems are now so widespread, as pointed out in Section 3, that it is necessary to ask for international action to protect the integrity of the sciences affected.

It is clear that the sciences particularly affected have a close connection with the Earth's environment - astronomy, meteorology and geophysics. It is perhaps novel to consider astronomy in the context of the Earth's environment, except when it is recalled that it is astronomical considerations which are the fundamental determinants of the Earth's environment and that ground-based astronomy is conducted from the bottom of the Earth's atmosphere. Astronomy is irretrievably involved with the Earth's environment. Indeed, because astronomy investigates very faint sources, it is particularly seriously affected by tiny environmental changes.

It was pointed out above that the major sources of adverse environmental impact on astronomy are international. In their severest forms these impacts come from the exploitation of space. Light pollution can be mitigated, terrestrially-generated radio noise can similarly be controlled for specific sites. However, because of the intrinsic nature of space operations, there is no way to escape man-made noise from space. Equally, there are no rational grounds to argue that the science of astronomy should be severely harmed or even nullified in order that the exploitation of space can carry on without paying due regard to consequences for others. The science of astronomy must seek protection of its interests, that is, its sustained capability to carry out significant astronomical observations of faint cosmic sources over a wide range of frequency bands.

Since the exploitation of space is carried out on a multinational basis between cooperating governments, agencies and multinational companies, any protection for astronomy must be agreed at an international level. In this Symposium, very many different environmental impacts on astronomy have been, and will be, laid out. Any single environmental impact can be discussed in isolation, but in the last analysis it is the sum total of environmental impacts that must be addressed.

Because astronomy, unlike most other sciences, is not in sole control of its "laboratory", including the ground-based environment, the actions of others will always affect the quality of that "laboratory". We therefore seek ways in which a *modus vivendi* can be agreed so that astronomy and beneficial exploitation of space can coexist. I urge the UN COPUOS to take up consideration of such protection at their earliest opportunity - otherwise the future for astronomy will be bleak indeed.

6.   Conclusion

The science of astronomy is now threatened on so many fronts, with particularly serious consequences for radio astronomy, that international protection must be urgently sought. The greatest single source of threat is the exploitation of space. However, it is the totality of the threat that must be considered when, for what is probably the first time, a science has requested special regulation to allow it to continue with its essential observations. The situation is potentially so serious that it is all too easy to foresee doomsday scenarios. However, we believe that it is important to the body of all scientific activity that a vigorous astronomical science is maintained, and that ways will be found to establish a *modus vivendi* in which astronomical science can continue to flourish in parallel with a technologically vibrant society.

References

Cohen, R. J. and Sullivan, W. T. (Eds.) 2001, IAU Symposium 196, *Preserving the Astronomical Sky*, Astronomical Society of the Pacific

Crawford, D.L. (Ed.) 1991, IAU Colloquium 112, *Light Pollution, Radio Interference and Space Debris*, Astr. Soc. Pacific Conf. Ser., Vol. 17

IAU/CIE 1978, Report and Recommendations of IAU Commission 50 on Adverse Environmental Impact on Astronomy and Protection of Observatory sites (reprinted in McNally 1994)

Isobe, S., and Hirayama, T. (Eds.) 1998, *Preserving the Astronomical Windows*, Astr. Soc. Pacific Conf. Ser., Vol. 139

Kovalevsky, J. (Ed.) 1992, *The Protection of Astronomical and Geophysical Sites*, Editions Frontières

McNally, D. (Ed.) 1994, *The Vanishing Universe*, Cambridge University Press

Robinson, B. 2001, in these proceedings

UN 1998, Preparations for the Third United Nations Conference on the Exploration and Peaceful Uses of Outer Space, A/CONF.184/

Part 2
Threats to Optical Astronomy

Preserving the Astronomical Sky
IAU Symposium, Vol. 196, 2001
R. J. Cohen and W. T. Sullivan, III, eds.

Light Pollution: Changing the Situation to Everyone's Advantage

David L. Crawford

International Dark-Sky Association, 3225 North First Avenue, Tuscon, AZ 85719-2103, USA

Abstract. There is no question of the need for outdoor lighting to improve the effectiveness of our night-time environment. However, too much of the lighting installed to try to meet this need actually compromises the purpose, and it too often adversely affects the night-time environment, including our view of the stars and of the Universe above us. This urban sky glow severe impacts on all of astronomy, amateur and professional, as well as those of the public who enjoy and profit by the beauty offered by a prime dark sky. In the present paper, I review the issues involved and suggest guidelines to minimize these negative aspects of poor night-time lighting. With good outdoor lighting we all win.

1. Introduction

Outdoor lighting is an essential component of our night-time environment. There are few places in the world today where it is not widely used. The reasons are rather obvious, but they all relate to improved visibility at night. While the eye is a marvelous instrument, and very sensitive over a wide range of lighting levels, we often need to supplement the existing light so as to improve visibility. (It is remarkable, though, how well we can see even by moonlight or less if there are no sources of glare to compromise our vision and the adaptation level of the eye.) The improved visibility we are seeking is for the purpose of moving around safely at night, walking or driving. We can see the ground better, as well as any obstacles such as curbs, unexpected holes, stones, or other objects. In addition, we feel safer, especially in our cities where today the fear of crime can be high. By being able to see better, we expect to be safer, and sometimes we are.

For these reasons we light our streets, walkways, building entrances, parking lots and other areas. The problem comes when we choose, for whatever reason, to do the lighting with a poor choice of lighting fixture or installation. Too much of our night-time lighting worldwide is of poor quality, of poor design, or with no design at all. As such, it then compromises our goals, even to the extent of lowering visibility. Real safety is less, and we ruin the night-time ambience as well. In addition, since we are wasting a good deal of light, we waste expensive energy.

Quality lighting has none of these problems. It exists in many locations, and all good lighting designers and engineers (and most other people when they are aware of the issues) know the difference between good and bad lighting.

While quality lighting might sometimes be more expensive than poor lighting, it is worth any such difference by the increased visibility, safety, security, and energy savings. It usually has a rapid payoff period. Good lighting has great value. There is no excuse for poor lighting. With good lighting, we all win.

2.   Why is Astronomy Interested in Night-Time Lighting?

Think of the adjective *"astronomical"*. What does it mean? There are two definitions: "having to do with astronomy", and "mind-boggling" (overwhelming the mind). We see it used in the media all the time in the latter context. We astronomers deal with mind-boggling things!

Astronomy is the Science of Extremes. Who can wrap their minds around distances of 15 billion light years, with light travelling at 186,000 miles a second. Eight minutes from the Sun, 5 years from the nearest other star. Our Galaxy, the Milky Way, is 100,000 light years in diameter; it is 2 million light years to the nearest galaxy, and over 15 billion light years to the farthest galaxies and quasars that we can observe. And we deal with the astronomically small too: atoms and nuclei. But not much in between. The same holds for temperatures: the highest in the Universe in the centres of some stars, and the lowest, in interstellar space. So it is with densities: some material in stars where an amount the size of my finger tip weighs more than all the people on Earth, and interstellar space where there are only a few atoms per cubic centimeter.

Black holes, quasars, cosmic rays, pulsars, even automobiles and computers get named after these things. Astronomy is an extremely interesting field. The public loves it. Do you know that there are more than 1000 times the number of amateur astronomers as there are professionals? How many amateur physicists or chemists do you know? Or amateur lighting engineers? There is something about astronomy in the press or on television almost every day. The media eats it up. Astronomy conferences have large press rooms and much coverage. How much coverage does street lighting get, unless there is a collision with a utility pole? Astronomy is a small field, but one with high visibility.

In addition to being a science and a frontier technology, astronomy is in many ways also a philosophy and an art, full of beauty and philosophical think-ing. Where did we come from? What does the future hold? What does it all mean? These questions have been in front of humankind for millennia. And they always will be. It is fundamental and exciting stuff, truly mind-boggling. And think also of the potential for intelligent life elsewhere.

There are many beautiful images of things "out there". They regularly show up in the press and in magazines and on television. The Hubble Space Telescope as well as telescopes on the ground such as the VLT and the Keck telescopes have produced many stunning pictures, as well as fundamental research. Even small telescopes, even amateurs' telescopes, can do wonderful things. Essentially no one is so blasé as not to be positively affected by these images and by many of the things going on in astronomy.

One of these mind-boggling things is how faint we work. On a human hand in a typical room there are about 1,000,000,000,000,000,000 photons falling every second. With our telescopes for the faintest objects, we count photons one at a time and are almost always photon-limited in our studies. It is truly a frontier field of research. We are also at the cutting edge of technology, of course, for sensitive detectors and image processing. Much of this technology has applications to everyday life, such as in television and photometric applications.

So in relation to outdoor lighting, Astronomy is really the "Canary in the Mine". We notice the adverse impacts of bad lighting well before almost anyone else, and are strongly affected by such lighting. Two things in particular impact us: urban sky glow and local glaring sources. The latter is a major item for amateurs and the former for both amateurs and professionals. I will discuss both of these in some depth in the following sections of the paper.

Do dark skies have value? I think so. I think almost everyone else does too. Can we afford to lose our view of the stars and of the Universe? Can we toss away this heritage to our children and their children? No we cannot. It is sad to think that the only place most people can see a dark sky today is in a planetarium. The real thing is much better. Light pollution is definitely an environmental issue. Where are the environmentalists? Most have lost sight of the fact that the "day" is more than 12 hours long, that night is a key part of the environment.

In addition, over millenia humans have developed with a day-night cycle. By turning the night into day, we have added a psycho-social stressor to our system. Think about it. Do we need the break of the night? The contemplation of the stars and the Universe that our ancestors had? Too many of us have lost touch with nature, seeing it only on television or in the movies. There are things worth preserving and the dark sky is one of them.

The nice thing is that dark skies and quality lighting are compatible. The sad thing is that there is far too much bad lighting everywhere.

3.   Why the Problem?

Simply put, there is a lot of bad night-time lighting. Far too much of it, everywhere. And such bad lighting is growing rapidly almost everywhere, much faster than the population.

What is bad lighting? We can define it by the following characteristics:

(*i*) **Glare**. Glare never helps visibility, yet is common in most outdoor lighting. Glare is never good. We should never tolerate it. It is not necessary. It can be avoided with good lighting design, in any installation.

(*ii*) **Obtrusive lighting**, or light trespass. This is our neighbor's light bothering us, or the local automobile dealer who has bad lighting, or the local sports complex with bad flood-lighting. There is far too much light trespass, obtrusive lighting. This lighting can even be offensive.

(*iii*) **Clutter and confusion**. This is light that does not add to the night-time ambience. It is a fact that too much of our night lighting actually helps to ruin the night-time environment. Many of us look forward to the time when we can bring back the beauty of the night that existed for many of our ancestors.

(iv) **Wasted light and urban sky glow**. There is far too much up-going light, totally unused light, the major cause of our urban sky glow. In addition, there is the myth of "The More the Better". More light is not always better, no more than more salt is, nor more noise, nor more of almost anything. Certainly there are many locations with inadequate light, but there are also many with too much. The issue of transient adaptation (switching lights on only when they are actually needed, for example using motion sensors) is an important one and we must consider such possibilities in our lighting installations.

(v) **Energy waste**. Lots of energy (and money) is lost by all this wasted light and by inefficient lamps and fixtures. Billions of dollars, literally an astronomical amount, are wasted lighting up the sky and blinding us with glare.

4. Why is There so Much Bad Lighting?

The basic reason for so much bad lighting is that there is little awareness of the problems. Even though night lighting is all around us, most people actually see little of it. The bad stuff has crept up on us, little by little, with little notice, just like a cancer can creep up on us. Once we have begun to notice the problem, it is hard to do anything about it.

Then there is the additional problem of apathy, or perhaps we should call it inertia. Too many people have the attitude that poor lighting is there now and we can't do anything about it, or that it is too hard to change existing standards, or there is not time to be involved in the issues, or many other reasons. None of these reasons is good enough. The International Commission on Illumination (CIE) has addressed these issues, as have many national lighting organizations and their standard practices and recommendations are changing. Good lighting has great value and we must recognize it and market it. We can and should get rid of the old bad stuff and we should use only good lighting for all new installations. It is worth the difference in initial cost, if any. The long-term costs of new quality lighting are always lower than for the bad lighting.

While many push hard for more and brighter lighting as the key to solving the crime and security problem in our cities, we must note that all efforts in this direction so far have failed to solve the problem. In fact, the more we add light, the more crime seems to increase. The correlation between the increase in lighting and the increase in crime in most locales is excellent! But still we add more lighting in an attempt to do better. Maybe we are just adding more bad lighting, not helping. Good lighting can help, I am sure, but there is too little of it. All new recommendations from lighting organizations emphasize *good* lighting, not *more* lighting. Astronomers do not argue for *no* light, but for *good* lighting.

5. What can be Done?

It is impossible to go into the many details here. Let me just review the solutions. They all go a long way to minimize light pollution and preserve dark skies without compromising in any way night-time safety, security, or utility.

1. Use night lighting only when necessary. Turn off lights when they are not needed. Timers can be very effective. Use the correct amount of light for the need, not overkill.

2. Direct the light downward, where it is needed. The use and effective placement of well-designed fixtures achieves excellent lighting control. Whenever possible, retrofit present poor fixtures. In all cases, the goal is to use fixtures that control the light well and minimize glare, light trespass, light pollution and energy usage.

3. Use energy efficient lighting and consider the use of low-pressure sodium (LPS) light sources, especially in the vicinity of major observatories. This is the best possible light source to minimize adverse sky glow effects on professional astronomy. LPS is especially good for street lighting, parking lot lighting, security lighting and any application where colour rendition is not critical. With creative design, it has even been used to illuminate new car dealerships.

4. Establish outdoor lighting ordinances that promote the use of quality lighting. Such controls do not compromise safety and utility. Lighting ordinances have been enacted by many communities worldwide and in several of the states in the USA, all designed to enforce the use of effective night-time lighting and good design standards.

All of these solutions to the problem say: "Do the best possible lighting design for the task. Always consider and minimize all the relevant adverse factors, such as glare, light trespass and urban sky glow." All the solutions needed for protecting astronomy have positive side benefits of maximizing the quality of the lighting, improving visibility and saving energy. We all win!

6. Conclusions

There is a problem.

> The problem is for all of us, not just astronomers.

> The problem is still getting worse almost everywhere.

Why do we tolerate it?

> Lack of awareness. Apathy. Too many laws.

> So bad now that nothing can be done.

> No time to work on it.

None of these reasons is good enough!

None will help us solve the problem or get solutions.

We must do something now!

We know that working solutions exist.

Awareness and education are the keys to getting action.

Quality lighting is the key!

It is an issue in which everyone can win.

Why shouldn't we do it?

The Goals:

- Dark skies.

- Quality lighting, with better visibility at night, hence better safety and security.

- Better night-time ambience.

- Considerable energy savings.

Let's do it!

References

Astronomical Society of the Pacific, The Universe in the Classroom, No.44-Fourth Quarter, ASP, 1998

CIE 1997 "Guidelines for minimzing sky glow", CIE Technical Report 126

Crawford, D. L. (Ed.) 1991, IAU Colloquium No. 112, *Light Pollution, Radio Interference, and Space Debris*, Astronomical Society of the Pacific Conference Series, Vol. 17

Isobe, S. and Hirayama, T. (Eds.) 1998, *Preserving the Astronomical Windows* Astronomical Society of the Pacific Conference Series, Volume 139.

See the IDA Web Page http://www.darksky.org/ ida for much additional information and for links to other resources.

Preserving the Astronomical Sky
IAU Symposium, Vol. 196, 2001
R. J. Cohen and W. T. Sullivan, III, eds.

Controlling Light Pollution in Chile: A Status Report

Malcolm G. Smith

AURA/Cerro Tololo Interamerican Observatory, La Serena, Chile

Abstract. A basic-level summary is provided of work since late 1993 to control light pollution in Chile. The purpose of this article is to stimulate such work inside Chile and to promote good lighting in developing countries in general. Chile is selected as the case study because of its critical importance to optical and radio astronomy, and the related economic and cultural benefits for Chile and the world. Examples are presented in some detail in order to illustrate adjustments that have been made to accommodate local scientific, cultural and economic realities and to show that *it is necessary to anticipate the issues involved in controlling light pollution several decades before it would otherwise become a problem.* It is hoped that international organizations such as the IAU, the IDA and the CIE can soon promote programmes in Chile that can serve as pilot programmes for other parts of the developing world.

1. Introduction

I was invited by the organizers of this Symposium to give a presentation on "The Importance of Ground-Based Optical Astronomy" suitable for lighting engineers and astronomers. The opening remarks by speaker after speaker emphasized the key scientific, economic and cultural aspects of ancient and modern astronomy. Being unwilling to dissociate ground-based from space-based astronomy or optical- from radio- and millimeter-wave astronomy, my talk emphasized the importance of ground-based astronomy within this wider, integrated context.

The *scientific* importance of astrophysics as a laboratory of extreme conditions, e.g. in solar flares, planetary nebulae, supernovae, the nuclei of galaxies, gamma-ray bursters, black holes and the big bang is known to all physicists and astronomers.

Its *cultural* importance in terms of understanding our origins and eventual fate is well known to all. Well-written and authoritative treatments of these matters at the level of a general reader can be found in books on cosmology such as those by Weinberg (1993), Hawking (1988) and Rees (1997).

The *economic* and cultural importance, over the centuries, of knowledge of the seasons, navigation, gravity, nuclear energy and so forth is obvious. Practical spin-offs from modern astronomical technology include such diverse areas as advances in medical x-ray technology, breast-cancer diagnosis, monitoring of the structural integrity of oil rigs and the avoidance of waste light energy from cities.

However, following the various discussions at the Symposium, it became clear that a volume like this would be more useful if it contained some discussion of the issues involved in preserving the astronomical sky - as they arise today in Chile. This is because Chile is at the centre of the greatest construction programme ever undertaken in ground-based astronomy, so the scientific, economic and cultural importance of its astronomical sky is obvious and increasing. Chile also provides examples of issues associated with controlling light pollution in a developing country.

The preservation of the astronomical skies over Chile is a concern for everyone - and an appropriate issue, therefore, for the IAU and other international bodies (such as the UN, UNESCO, the Commission Internationale d'Eclairage and the International Dark-Sky Association), as well as for Chile's own national and regional authorities.

Astronomy has, over the years, found it necessary to run further and further away from industrialized areas in order to be able to work under skies unpolluted by man-made interference. Other, more qualified speakers have addressed the issue of radio interference; with the advent of the Atacama Large Millimeter Array (ALMA), control of such interference is obviously vital to international astronomy and to the economics of that region of Chile. This basic-level article will be focussed on optical interference - primarily the increase over natural sky background produced by light from surrounding communities and industries.

2.   Light Pollution in Chile - Scientific Aspects

Seven telescopes larger than 6 m in diameter are currently being brought into regular scientific use in northern Chile or are nearing their final stages of construction and commissioning. The four 8-m elements of the European Southern Observatory's Very Large Telescope (VLT) on Cerro Paranal, the Carnegie Institute of Washington's twin 6.5-m Magellan telescopes on Cerro Las Campanas and the international 8-m Gemini South telescope on Cerro Pachón will all depend for their success on the skies over northern Chile being preserved in their current pristine condition. New generations of large, wide-field, survey telescopes such as the VLT 2.65 m Survey Telescope (VST), the British 4-m VISTA telescope and the US 8-m Dark-Matter Telescope are being built or proposed specifically to take advantage of the darkest available skies.

Examples of the kinds of science that need such dark skies cover the full range of size and distance in the universe - from the detection of large-scale structure by means of weak gravitational lensing of faint galaxies to the early location and study of near-earth objects (asteriods and comets).

To set a context for the goal of the efforts to preserve these dark astronomical skies, the natural sky background at new moon near the zenith at high ecliptic and galactic latitudes varies by a factor of about 1.7, i.e. by about 0.6 mag, $(21.3 < V < 21.9)$ per square arcsecond over the course of the 11-year solar activity cycle (Krisciunas 1997). This minimum sky brightness is "about 10 million times dimmer than the daylight sky (but easily visible to the dark-adapted eye)" (Benn and Ellison 1999).

Faint galaxies will be observed with telescopes in Chile down to surface brightness levels at or even below 29 mag per square arcsecond, i.e. ~ 700

times fainter than the natural solar-minimum background at V. Our goal must be to ensure that artificial light does not increase the natural sky background by more than about 0.1 magnitudes per square arcsec at these levels of surface brightness, at least in broad passbands (Smith 1979; Krisciunas 1997).

AURA's observatory is the closest major observatory in Chile to city lights (see section 5). It is therefore a natural point from which to lead an effort to study and control light pollution, well before it can become a serious problem at Cerro Pachón and Cerro Tololo.

The measured sky brightness at the zenith at Cerro Tololo is < 0.08 mag per square arcsecond brighter than the natural value, i.e. it is not yet possible to measure *any* artificial light pollution at zenith distances < 45 degrees through broad-band filters (see, for example, Walker and Smith 1999). The sky overhead at Cerro Tololo and Cerro Pachón is still very dark.

Nevertheless, the ability to detect the sodium lights of La Serena spectroscopically by looking low in the sky in the direction of that city is an early warning that action that started in earnest in late 1993 has to continue, especially as such action has implications for astronomy throughout Chile. It may also be possible to detect such radiation in certain directions low over the horizon from other major observatories in Chile. *This should be attempted now at all of the major optical observatories (both in Chile and elsewhere) and monitored regularly. An excellent example of such work in La Palma has been described recently by Benn and Ellison (1999).*

As discussed in section 5.1, it has already taken nearly 6 years to secure a set of government lighting regulations that can now be taken to local municipalities for communication, education, development of municipal ordinances and eventual enforcement. *It is necessary to anticipate the issues involved in controlling light pollution several decades before it would otherwise become a problem.*

3. Light Pollution in Chile - Economic Aspects

The current boom in construction of international astronomical facilities in the north of Chile involves investments already surpassing a billion dollars. Specific plans exist which take this figure to well over US$1.2 billion. The detailed budgets for the observatories involved are, unfortunately, not readily available. Nevertheless, from experience, one can make rough estimates for the level of investment made in the Chilean economy. These lead to averages of about US$60M per annum entering the economy for construction and a steady ramp-up to about US$40M of annual investment in operations.

The Chilean national economy thus receives about US$100M in annual income from astronomy. Chilean astronomers are, in addition, awarded 10% of the observing time on telescopes in Chile. In return, the Chilean government provides a variety of privileges and tax breaks to the international astronomical community, and is generally supportive of astronomical development and the associated benefits arising from this foreign investment.

The AURA Observatory in Chile (Gemini South, SOAR and the telescopes on Cerro Tololo) provides a measure of the economic stimulus that occurs at a regional level. AURA has brought approximately US$150M in construction projects to Region IV of Chile and currently spends about US$4M dollars each year in Chile on operations. From 2002 onwards, with the arrival of Gemini South and SOAR, AURA will be spending about US$7M per annum in Chile. AURA has also stimulated local entrepreneurial activity in astronomical tourism, education and research as described in more detail in section 5.

3.1. DMSP Data as a Monitor of Energy Waste

We have seen that Chile has a strong economic interest in preserving the astronomical sky. Further beneficial economic and environmental impact arises from the potential for more rational use of energy in exterior lighting. As we have heard elsewhere at this conference, the United States shines well over US$1 billion of light annually into the heavens, while each year Japan inadvertently sends about US$200M dollars worth of light into the sky (Isobe and Hamamura 2000).

It will be interesting over the next few years to extend the use by Elvidge and collaborators of unsaturated Defense Meteorological Satellite Program (DMSP) data from the National Geographic Society's databank (Luginbuhl, personal communication; see also Cinzano *et al.*(2001), and Isobe, Hamamura and Elvidge (2001)). Carefully selected subsets of these data should permit measurement and monitoring of the upwardly-directed light from Chilean cities and towns, including those nearest the major observatories.

One may be able to stimulate a positive reaction from authorities in developing countries by providing information on the luminous flux, expressed in terms of wasted money, from their major cities; this information could be accompanied by reasonably simple guidelines, in their own language, about what can be done to reduce the waste. Isobe and Hamamura (2000) provide an example of this idea applied to an industrialized nation. Santiago de Chile would be a good initial test case, given the existing positive attitude of authorities at the national level and a recent awareness of the need to conserve electrical energy (brought on by a drought in the southern lake region which provides most of the country's hydroelectric power).

Early DMSP data were taken at high gain. The data from large cities were saturated. Unsaturated data from the DMSP may, however, be available from the outset for smaller, isolated towns - such as those near major observatories. It will be a challenge to see if sufficient precision and accuracy in calibration can be obtained to monitor changes in the upward flux as remedial lighting programmes come into effect.

3.2. Astronomical Tourism

Chile is beginning to realise the potential for eco-tourism based on its dark skies. The Servicio Nacional de Turismo (SERNATUR) has recently produced a 57-page full-colour booklet on astronomical tourism in the province of Elqui (where AURA has its observatory). In combination with the municipal observatory described in section 4.3, they are generating sufficient funds from publicizing the dark skies in the region to support a growing tourist industry.

4. Light Pollution in Chile - Cultural Aspects

Several speakers at the Symposium mentioned the ancient cultural links in many parts of the world between astronomy, different religions and practical pursuits such as agriculture and navigation. In addition to the cultural impact of the tourism mentioned above, astronomy provides humankind, on our tiny island planet, with a sense of perspective on our place in the universe.

4.1. The Voyage of the Hokule'a

At this time, the Polynesian Voyaging Society's canoe "Hokule'a" (Arcturus) is sailing to Rapa Nui, better known as Easter Island (the most westerly part of Chile). This journey is the culmination of a 25-year-long series of voyages by PVS's canoes across Polynesia; these voyages have led a cultural renaissance in the region. This latest voyage will close the Polynesian Triangle, formed by New Zealand, Hawai'i and Easter Island. The seas (away from the squid fleets - Sullivan 2001) offer dark skies to the navigators.

To quote from the PVS web page, "To get from Hawai'i to Rapa Nui, Hokule'a must travel.., 2820 nautical miles south (from 20 degrees N to 27 degrees S) and 2760 nautical miles east (from 155 degrees W to 109 degrees). The first three destinations (Nukuhiva, Mangareva and Rapa Nui) lie upwind of the departure points, so the canoe will have to struggle to get east against the prevailing winds."

"The sail from Mangareva to Rapa Nui will be the most difficult, as Rapa Nui lies 1450 miles to the east of Magareva...On this leg, Hokule'a will be navigated without instruments by a team of Hawai'i's best navigators, headed by Nainoa Thompson. They will guide the canoe by celestial bodies (sun, moon, planets and stars), ocean swells...."

These bold navigators use astronomy in their endeavors and convey the wisdom of the more successful island peoples about their relationship to the environment. The Earth, too, is a tiny, isolated island; dark starlit skies help remind the whole of humanity of our fragile existence in one, much vaster, universe (Rees 1997).

5. Details of Work on Light Pollution in Chile

5.1. Work at National and Regional Levels: Development of Regulations and Municipal Ordinances in Chile

At about the time of IAU Symposium 196, Chile's President Eduardo Frei Ruiz-Tagle signed the "Norma Luminica" which gives the force of law to a set of lighting regulations covering Regions II, III and IV of (Northern) Chile. Cerro Paranal is in Region II. Cerro La Campanas is in Region III. Cerro La Silla, Cerro Tololo and Cerro Pachón are all in Region IV. This code is currently being translated into English and will be available on the world-wide web by the time this article is published (www.ctio.noao.edu/ctio.html).

The Chilean National Environment Commission, CONAMA, is leading a major programme of outreach and training related to the "Norma Luminica", in conjunction with the major international optical observatories in Chile. A series of light-pollution "events" will be held later this year, one in each of the three astronomically-critical regions.

Each event will include an outreach workshop organized by CONAMA, targeting the general public, local government officials and representatives of companies in the private (commercial) sector. The events will also include a technical training workshop, organized in close consultation with one of the observatories in the relevant region.

CONAMA is also organizing, in consultation with the international observatories, the production of 1,000 copies of an "Application manual" for the "Norma Luminica", along with a special web page with links from the CONAMA home page and links to various observatories and key websites on light pollution.

The first regional CONAMA office of technical support for the "norma" is to be set up in La Serena for a period of two years, with personnel financed by the international observatories in Chile. After 2 years, the value and performance of this office will be assessed. One of its early tasks will be to create a database on sources of light pollution in the three critical regions of Chile.

This set of national regulations provides a framework which allows us now to start serious work with municipalities in Chile, with the aim of mitigating and controlling light pollution. The President of the IVth Chapter of the Chilean Association of Municipalities (which includes those closest to AURA's Observatory) has recently and publicly expressed a wish to help in this effort in specific and significant ways.

5.2. Work at Local Level: 1. Vicuña and its Municipal Observatory at Cerro Mamalluca

Initial efforts five years ago to capture the attention and support of local municipalities did not go so well. The initial reaction was that changing exterior light fixtures would be too expensive, that tourism was flourishing as a result of the bright lights along the beaches and that there were concerns to ensure security at night. Although we used the correct reasoning and examples provided by the IDA, we did not make progress until an employee of the municipality of Vicuña contacted us (Sr. Eduardo Valenzuela, who is also an amateur astronomer). Vicuña is the closest town to Cerro Tololo and Cerro Pachón, with a central population of about 10,000 (excluding outlying regions) at line-of-sight distances of roughly 18 km and 23 km from Tololo and Pachón, respectively. As luck would have it, municipalities in Chile were being offered the chance to apply for a loan from the central government to replace the old, inefficient mercury fixtures with energy-efficient sodium fixtures. The two amateurs asked us to help the municipality design the fixtures in such a way as to minimize running costs (in particular) and minimize the impact on the astronomical research going on nearby! This was the breakthrough we were looking for.

We first studied the situation in the USA, with the help of the IDA materials and Dave Crawford's personal advice. We soon realized that what works in the USA may not satisfy Vicuña's requirements. At that time there was only one supplier of low-pressure sodium lamps in the country. Such fixtures were too expensive for most Chilean municipalities because of losses to vandalism. High-pressure lamps have a wider market and are therefore available in Chile at lower capital cost.

We then made contact with Sr. Enrique Piraino, a lighting engineer from the Universidad Catolica de Valparaiso, who runs a laboratory used to measure and standardize Chilean lighting fixtures available on the Chilean market. Sr. Piraino provided consultant services to the Municipality of Vicuña (at AURA expense) in order to replace and reposition the town's street lights. His recommendations included replacing all the white-light globes which lit the town's main street, the "Avenida de las Delicias". The first step was to draw up an inventory of all existing street lights in the town (exisiting inventories were so inaccurate as to be useless). Powerful (400 W) full-cutoff fixtures were used to illuminate the perimeter of the town square. 150 W fixtures were used for the main streets near the centre of town. Remaining streets were lit with 70 W fixtures, all high-pressure sodium.

A rough calculation shows that the lighting change and redesign saves Vicuña nearly half of its municipal lighting bill. 1534 light fixtures were changed or modified. Illumination increased from 7.8M lumens to 8.9M lumens. Estimated power consumption dropped from 210 kW to 110 kW, with consequent large savings. Estimated power consumed in illuminating the sky dropped from 24 kW to just over 1 kW. The capital cost of the modernization has already been paid off from these savings and the municipality is delighted.

In recognition of Vicuña's seminal co-operation, the US National Science Foundation, at AURA's recommendation, donated a 30-cm commercially-made telescope to Vicuña as the centrepiece of a municipal observatory. The town found the land, installed a 4 km-long road, provided power, water and a building (which did not get funding from the Chilean science foundation, CONICYT - but did get a large grant from the Chilean National Fund for Art and Culture in addition to local, commercial sponsorship). Cerro Tololo provided an obsolete dome that was in good condition.

AURA staff provided the first course of lectures and practical workshops to members of the local community. Graduates of this first course then taught the second course in the series, so that the town now has people qualified to help visitors in various aspects of the observatory.

The new Cerro Mamalluca Municipal Observatory is now making sufficient money from tourism ("come to the darkest, clearest skies in the world") and conferences to be almost self sufficient and, with the help of government grants, will shortly be buying several new telescopes of its own. Phase I, the telescope, building and dome, is in regular operation, every clear night. Phase II, containing a restaurant and a platform for amateur telescopes, is nearly complete. A small-telescope educational and dark-sky research facility is at the proposal stage for Phase III; funds for this third phase are expected to be awarded next year.

Vicunã has worked closely with AURA to become a demonstration community, allowing us to present a way to develop a mutually-beneficial community-wide business approach to the provision of good lighting. Once the cost of international broadband links drops sufficiently, we hope to extend this work to include connection between schools in Chile and those in other countries (initially from the Gemini and SOAR consortium partnerships). Among other plans being discussed are the production of a CD or video, in Spanish, on the subject of light pollution, using illustrations of lighting in the region. This work is likely to be done in close collaboration with M. Metaxas (see paper at this conference).

5.3.   Work at Local Level: 2.   Other Lighting Projects near the AURA Observatory in Chile

The nearest town to the AURA Observatory in Chile is Andacollo, at a distance of only 28 km. from Cerro Tololo and 33.5 km from Cerro Pachón. Although it has only $\sim 12,000$ inhabitants, two international open-pit mines have opened there in the last five years. Fortunately, the lack of vandalism in the controlled area of a mine allows the mine to make economic use of the most energy efficient low-pressure sodium lighting. The mine needs powerful lighting for safe operation; therefore the potential for significant light pollution from such a nearby source, particularly during the early phases of the project, was serious. Fortunately the mine was looking for opportunities to improve its rather tarnished environmental reputation and worked closely with us to direct its lights away from the line of sight to the observatory. We even identified troublesome individual fixtures for them on the telephone, as they kindly switched their lights off and on for us, one by one.

The nearest cities form the coastal conurbation of La Serena and Coquimbo, at a distance of ~ 49 km from Cerro Tololo (and ~ 58 km from Cerro Pachón) and containing about 292,000 inhabitants. Chile's recent economic boom has increased the estimated output of light per inhabitant to levels similar to those in industrialised nations. Fortunately, both La Serena and Coquimbo are largely hidden from direct view from Cerro Tololo and Cerro Pachón by a range of mountains paralleling the coastline just a few kilometers from the shore. The arrival of bright lights, especially those along the beach drive (the "Avenida del Mar") has provided an undeniable stimulus to the economics of local tourism.

AURA has had to proceed very carefully to search for ways in which lighting can be used to mutual benefit (see, e.g., Mendez and Boccas 1999). This proved difficult at first, but efforts to set up a sister-city exchange between the county of Hawai'i and La Serena have allowed us to demonstrate that massive tourism and good lighting can coexist elsewhere and that we should be able to work together in Chile to achieve this. Economic arguments have so far been much more persuasive than cultural and scientific arguments, although even these are now gaining force as the tourist potential of dark skies becomes better understood.

Coquimbo is currently engaged in construction of a US\$6M "Cruz del Tercer Milenio" on a prominent hill overlooking the harbor. This is directly visible from AURA's observatory. The recent publication of the national lighting regulations has undoubtedly helped us in work with the municipality to design downward-looking light fixtures for this 80-meter-high religious structure.

6.   Conclusions - Monitoring the Sky Background

Lighting engineers and others at this symposium with the qualifications and the will to help have repeatedly expressed their frustration that professional astronomers have published so little information on the night sky background that engineers could use to check cause and effect in light pollution. AURA will again be taking the lead on the professional side in trying to contribute to an improvement in this situation.

Tololo and Pachón can be seen as a "canary in the mine" for modern astronomy in the southern hemisphere, just as astronomy itself acts in a similar way to provide early warning to the rest of humankind. Even a worst-case scenario with aggressive population growth and no lighting control measures would predict an increase of ~ 0.1 mag above the natural solar-minimum background over AURA's observatory (21.9V mag per square arcsecond, or 58 nanolamberts, at the zenith) by 2020 (Walker and Smith 1999). Dave Crawford and the IDA have shown, in Tucson, what can be done to contain the effects of such population growth. A more realistic model shows that, even by 2030, light pollution at the zenith will still be only ~ 0.08 mag (to be compared with the 0.6 mag change in natural background during the solar cycle). This, however, does assume some success at light pollution control. We are working hard on such controls in Chile; the canary is not going to die readily.

Acknowledgments. The early efforts of the late Dr. Art Hoag in Tucson, Victor Blanco in Chile and the sustained efforts everywhere of Dave Crawford and the International Dark-Sky Association provided a vital foundation for this work. Sr. Edgardo Boeninger K. (then Minister/Secretary General to the Chilean President) explained to me in late 1993 the need to contact the Comision Nacional del Medio Ambiente (CONAMA), the newly-formed national environment agency. This national-level approach contrasted with the exclusively local- and regional-level emphasis that has proved quite effective in many parts of the USA (Davis 2001).

Sr. Rodrigo Egaña, the National Director of CONAMA, has proved a staunch and constant supporter of light-pollution control (following a visit to Cerro Tololo at night). Sr. Pedro Sanhueza, the director of COREMA (the regional version of CONAMA for Region IV of Chile) and *ex officio* member of the CONAMA, took repeated risks in coming out publically at regional and national level in support of light pollution control. Sra. Cecilia Prats, regional director of SERNATUR, the Chilean national tourism agency, was among the first to press effectively for marketing the dark skies in the region as a non-seasonal tourist attraction. Sr. Enrique Piraino provided technical assessment appropriate to the realities confronting lighting engineers in Chile.

The Cerro Tololo Light Pollution Control Committee was chaired most effectively for much of the last five years by Dr. Mark Phillips. He and his colleagues, Sres. Ricardo Schmidt, Mario Hamuy, Oscar Saa, M. Maxime Boccas and Dr. Rene Mendez (the current committee chair), must take most of the credit for what has been achieved from the astronomers' side in Chile.

This effort would have hardly started without the vision and persistence of the amateur astronomers and municipal authorities in Vicuña, Chile.

References

Benn, C. R. and Ellison, S. L. 1999, submitted .

Cinzano, P., Falchi, F., Elvidge, C. D. and Baugh, K. E. 2001, this conference.

Davis, D. R. 2001, this conference.

Hawking, S.W. 1988, A Brief History of Time, Bantam (Random House) New York.

Isobe, S., Hamamura, S. and Elvidge, C. 2001, this conference.

Isobe, S. and Hamamura, S. 2000, Memorie della Società Astronomica Italiana, 71, No. 1, 131.

Krisciunas, K. 1997, P.A.S.P., 109, 1181.

Mendez, R. and Boccas, M. 1999, NOAO Newsletter No. 59 (Sept.), 23.

Rees, M. 1997, Before the Beginning: Our Universe and Others, Perseus Books.

Sullivan, W. 2001, this conference.

Walker, A and Smith, C. 1999, NOAO Newsletter No. 59 (Sept.), 21.

Weinberg, S. 1993, The First Three Minutes, Basic Books.

CTIO www.ctio.noao.edu (contains links to the NOAO Newsletter, Gemini, SOAR, VLT, Magellan and the Vicuña municipal observatory).

IDA www.darksky.org

PVS leahi.kcc.hawaii.edu/org/pvs

Preserving the Astronomical Sky
IAU Symposium, Vol. 196, 2001
R. J. Cohen and W. T. Sullivan, III, eds.

Light Pollution: How High-Performance Luminaires Can Reduce It

Christian Remande[1], Chairman of the "Exterior Lighting" Committee
of the "Syndicat de l'Eclairage Franais" and the members of the
Lighting Applications Department, R-Tech Company

R-Tech, Rue de Mons, 3 - 4000 Liège, Belgium

Abstract. In 1984 the International Commission on Lighting (CIE)
set up a Technical Committee TC 4.21 to study interference by artificial
light in astronomical observation. The resulting document "Guidelines
for minimizing sky glow" (CIE No. 126) complements previous CIE documents from 1978 and the joint CIE-IAU (International Astronomical
Union) publication of 1984. In this paper, we demonstrate how high
performance luminaires based on more global and pertinent criteria can
significantly reduce sky glow.

1. Introduction

A detailed analysis of the phenomenon 'sky glow' produced by street lighting luminaires clearly shows that not only the direct but also the indirect (or reflected)
contribution has a strong impact on the total upward-emitted light. Street lighting design calculations were made for several typical light distributions, several
types of light sources, the families of road reflection surfaces (R1 to R4), and
also for various types of installations (single- and double-sided). Taking into account the variation of all these parameters made it possible to calculate several
criteria among which are:

- the 'classical' Upward Waste Light Ratio (CIE criterion)

- the total needed flux for a 'one km installation' to reach a given common
  value of luminance (reference level)

- the total direct emitted flux from the installation

- the total reflected flux from the installation (road plus surroundings)

These calculations clearly show that the reflected contribution of light can reach
5 to 8 times larger values than the direct upward-emitted light. If, for the
same luminance level, a given installation can lead to much lower values of installed lm/km (sum of the flux of all the light sources), then the global (direct
plus reflected) portion of light that will be upward-emitted can also be largely

---

[1]This paper was presented by Marc Gillet.

reduced. This can be achieved by high-performance optical systems. This approach enabled us to show clearly that the existing CIE criteria are much too restrictive and that there is an urgent need to take into account more global criteria incorporating not only the direct but also the reflected portion of light.

2. Artificial Lighting

More and more human activity is carried out during the hours of darkness, either as an extension of daylight activities (public works, police, transport, maintenance, etc.), or activities of a specifically nocturnal nature (parties, walks, theatre performances, sporting events, security patrols). This is only possible thanks to the existence of artificial light, which is an essential part of our lives. It is important to evaluate the environmental impact of this proliferation of exterior lighting in order to keep in check certain harmful effects which might exceed permissible limits (unwarranted intrusion into homes, glare, interference with observation of the night sky). The happy medium is yet to be found.

It is particularly in the urban environment that we encounter all the existing forms of lighting:

- functional and decorative public lighting for roads and security

- ambient lighting for pedestrian areas, footpaths, residential areas

- floodlighting of buildings

- floodlighting with high masts, road and rail complexes and some car parks

This is because amazing progress has been made in lighting techniques in the last forty years: efficiency of lighting equipment has doubled, expected life of light sources has doubled then quadrupled, and there is an immense range of lights of all levels of power, allowing a wide choice for aesthetic effects in surroundings where colour temperature and requirements for exact reproduction of the desired colours are of prime importance. The photometric outputs of the lights have been adapted for multiple uses, including decorative, functional, sporting and road lighting.

Artificial lighting has become an indispensable tool for improving the appearance of towns and an essential element in the quality of life of the inhabitants; it lights up the achievements of all the other technologies but, as in all technical progress, success engenders abuses and errors and attracts unqualified people who only seek financial gain. For about fifteen years, certain groups have been worried about the proliferation of exterior lighting and have felt acute concern about the increasing pollution of the night sky, which interferes with astronomical observation.

In the same way, certain lighting installations which are unsuited to the environment where they are sited cause problems at night by:

- extension of their light beyond the area it is intended to illuminate

- producing glare which is noticeable to residents

- casting light on facades and interior of private houses

Lighting professionals must pay attention to these grievances, even if they are questionable.

Recommendations aimed at limiting the various types of light pollution have already been put forward in several countries of the European Union, e.g. the United Kingdom and Italy.

In 1984 the International Commission on Lighting (CIE) set up within Division 4 a technical committee TC 4.21, to study interference by artificial light in astronomical observation. The resulting document "Guidelines for minimizing sky glow" (CIE No. 126) complements previous CIE documents from 1978 and the joint CIE-IAU (International Astronomical Union) publication of 1984.

The other aspects of light obtrusion are covered in detail by another CIE technical committee, TC 5.12 "obtrusive light", which drew up a guide to evaluation of the environmental effects of light and recommendations for limit values which should not be exceeded in new installations; essentially these apply to undesirable effects on

- near neighbours

- road users

- astronomical observation

Light is analysed in terms of quantity, direction, and spectrum, for harmful effects in terms of "discomfort" or "reduction of visual capacity".

3. Sky Glow

Within the framework of this article, we will limit ourselves to the effects of functional public road luminaires and decorative urban luminaires on the luminous halo called "sky glow" in the night sky. But what are we talking about exactly? Sky glow corresponds to the extensive, diffuse band of light in the night sky which is visible over towns, airports and industrial and sport complexes. It results from radiation diffused by the constituents of the atmosphere (molecules, aerosols, pollutant particles).

We can distinguish:

- the natural glow due to radiation from celestial sources and the luminescence of the upper atmosphere, and

- the artificial halo due to:

    1. direct radiation towards the sky from lamps and lights, and

    2. radiation reflected by lighted surfaces.

The artificial glow depends not only to the design of the installations, but also on atmospheric conditions (humidity, clouds, haze, pollution).

To understand the effect of the diffuse sky glow, we must realise that any observation of a light-emitting object (either visual, photographic or electronic) amounts to an appreciation of the contrast C between the luminance of the object and the luminance of the background against which the object is observed:

$$C = \frac{L_{object} - L_{background}}{L_{background}}.$$

The overall stray light causes a light veil that extends over the field of observation. This veil has a luminance as well, that will be called L_v. The veiling luminance has to be added to all luminances in the field of observation. All contrasts will be reduced as follows:

$$C' = \frac{(L_{object} + L_v) - (L_{background} + L_v)}{L_{background} + L_v}$$

$$= \frac{L_{object} - L_{background}}{L_{background} + L_v}$$

C' is always $< C$.

When the luminance of the halo increases, the observed object can disappear from the field of vision, especially when we are dealing with a star which does not have a sharp image when seen through the instrument of observation. With the naked eye, the value of L_{object} is very weak.

To limit the effect due to artificial glow, CIE committee TC 4.21 limits the percentage of a lamp's flux which can be directed above the horizontal plane passing through the light sources in their operating positions. The Upward Waste Light Ratio (UWLR) is the proportion of the flux of a luminaire that is emitted above the horizontal when the luminaire is mounted in its installed position.

$$UWLR(\%) = \frac{ULOR(\%)}{ULOR\% + DLOR\%} = \frac{\%lampflux \nearrow}{\%lampflux \nearrow + \%lampflux \searrow}$$

where the Upward Light Output Ratio (ULOR) is the proportion of the flux of the lamps of a luminaire that is emitted above the horizontal when the luminaire is mounted in its normal, designed position, and the Downward Light Output Ratio (DLOR) is the proportion of the flux of the lamps of a luminaire that is emitted below the horizontal.

The UWLR has limit values which differ according to the environment. The limits recommended are based on:

- the level of lighting existing in the zone

- the types of technologies available for the task to light

- the hours of operation

- the measurable and contractual photometric data

4. Zoning

The concept of zoning has arisen from studies of the potential conflict between the photometric requirements for night-time activity, quality of life and the integrity of the environment. When a pollutant activity can not be completely suppressed, the harmful consequences are not identical for all the sites. Zoning does not stop pollution but it does constitute a frame of reference. Zoning must be complemented by specifying the distance in km between the zone limits and the point of reference (Tables 1, 2 and 3).

Table 1.   Environmental Zones according to the CIE 126 Zoning System

Zone rating	Description
E1	Areas with intrinsically dark landscapes: National Parks, areas of outstanding natural beauty (where roads usually are unlit)
E2	Areas of "low district brightness": generally outer urban and rural residential areas (where roads are lit to residential road standard)
E3	Areas of "middle district brightness": generally urban residential areas (where roads are lit to traffic route standard)
E4	Areas of "high district brightness": generally urban areas having mixed residential and commercial land use with high night-time activity

Table 2.   Recommendations for Limitation of Sky-Glow (CIE 126)

Zone rating	UWLR (%)	Astronomical activities
E1	0	Observatories of (inter)national standing
E2	0 - 5	Postgraduate and academic studies
E3	0 - 15	Undergraduate studies, amateur observations
E4	0 - 25	Casual sky viewing

Table 3.   Minimum Distance between Zone Borderlines and the Reference Point according to the CIE 126 System

Zone rating of reference point	Zone rating of surrounding zones Distance (km) to borderline of surrounding zones		
	E1-E2	E2-E3	E3-E4
E1	1	10	100
E2		1	10
E3			1
E4	No limits		

5.   Another Approach

Knowledge of ULOR or UWLR of a light in its position of operation alone does not seem sufficient to us and certainly not sufficiently selective. To expand on this statement, we calculated distributions of light on a 1-km section of an actual road lit successively by 7 different models of functional luminaires and 3 different models of decorative luminaires (Figure 1).

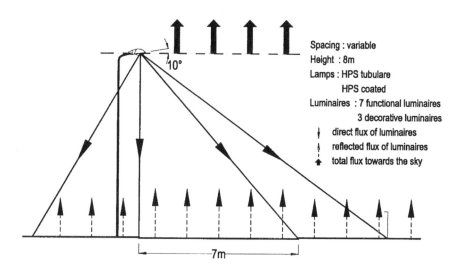

Figure 1.   Upwards Emission of Light.

The functional luminaires are differentiated by their type of protector (flat, rounded, curved, deep, refracting (Figure 2) and by the choice of source (HPS, tubular or coated bulb). All the functional luminaires are inclined between 0°-10° in unilateral arrangement. Therefore the spacing and the flux of the luminaires was calculated so that in all cases, the road surface (R3,Q0 = 0.07) was lit at the start of operation with an average luminance L_{av} of 1 cd/m^2, general uniformity $Ug > 60\%$ and longitudinal uniformity $> 70\%$.

In Tables 4 to 7 we present for each case the calculated total flux per unit length (lumens/km) for (1) the lamps, (2) the light emitted directly upwards, and (3) the light reflected by the ground.

For the total flux/km (lumens) reflected by the ground (3), we calculated separately:

1. the flux reflected by the road surface alone, with a coefficient of reflection of 18% in the case of asphalt (31% in the case of concrete);

2. the flux reflected by all of the ground except the road surface, with a coefficient of reflection of 5%.

In Table 8 we compare for each case (1) UWLR, (2)total flux reflected by the lit surfaces, and (3) total flux lumens/km of lamps.

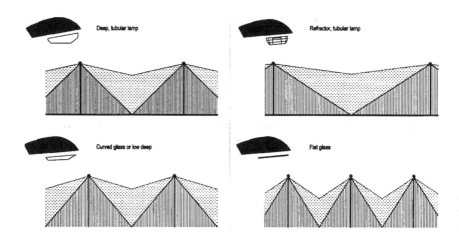

Figure 2.   Light Distribution according to Type of Projector.

Table 4.   Functional Road Lighting Luminaires: Lumens/km for $L_{av} = 1$ cd/m^2

Tubular HPS lamps   R3,Q0 = 0.07   UI > 70%

Protector bowl	Lamps (1)	Direct (2)	Indirect (3)	Total
Flat glass	360 000	0	30 000	30 000
Lowdeep glass	294 000	1200 (0.4%)	26 000	27 200
Curved glass	294 000	1200	27 000	28 200
Deep bowl	270 000	5000 (1.8%)	24 000	29 000
Refractor bowl	220 000	6200	19 000	25 200
Axial deep bowl	200 000	2000	19 000	21 000
		0 to 2.8%	± 9%	± 10%

Table 5.   Functional Road Lighting Luminaires: Lumens/km for $L_{av} = 1$ cd/m^2 – Influence of the Road Surface

Tubular HPS lamps   R1,Q0 = 0.10   UI > 60%

Protector bowl	Lamps (1)	Direct (2)	Indirect (3)	Total
Flat glass	220 000	0	29 000	29 000
Deep bowl	195 000	4000	22 000	26 000
Refractor bowl	175 000	5200	19 000	24 200

Table 6.   Functional Road Lighting Luminaires: Lumens/km for $L_{av} = 1$ cd/m^2

Coated HPS lamps   R3,Q0 = 0.07 $U_L > 70\%$

Protector bowl	Lamp (1)	Direct (2)	Indirect (3)	Total
Clear deep	420 000	6 300	29 000	35 300

Table 7.   Decorative Luminaires: Lumens/km for $L_{av} = 1$ cd/m^2

R3,Q0 = 0.07   UI > 70%

Lamp	Construction	Lamps (1)	Direct (2)	Indirect (3)	Total
Coated HPS	Opal diffuser	585 000	234 000	24 500	258 000 (50%)
Compact fluor.	Clear protector +reflector+roof	430 000	53 000 (12%)	25 000	78 000 (18%)
Ceramic	Clear protector +reflector+roof Indirect light source	480 000	9 000 (1.8%)	24 000	33 000 (7%)

Table 8.   Reflected Flux (Lumens/km of R3/asphalt Road)

For 1 cd/m^2	Zone	UWLR (%) (1)	Total lumens reflected (2)	Flux of lamps (lumens) (3)
Flat glass	E1?	0.5	30 000	360 000
Low deep/curved glass	E1?	0.5	27 500	294 000
Deep bowl (tubular lamp)	E2	2.9	29 000	270 000
Deep bowl (coated lamp)	E2	2.7	35 300	420 000
Refractor (tubular lamp)	E2	3.8	25 200	220 000
Axial deep bowl	E2	1.2	21 300	201 000
Opal diffuser	-	45.7	258 500	585 000
Clear bowl reflector/roof	E4	16	78 000	430 000
Clear bowl reflector/roof/indirect	E2	3.1	33 000	480 000

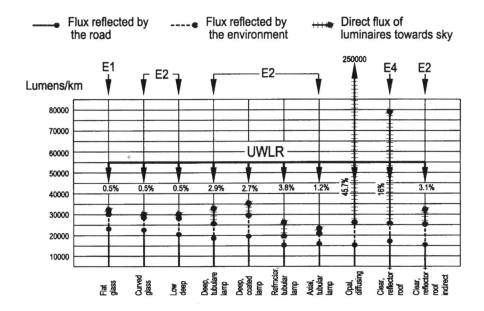

Figure 3.   Upwards Emission of Light.

Figure 3 shows for each installation:

- the flux reflected by the road

- the flux reflected by the surrounding areas

- the flux emitted directly towards the sky by the lights

- the total value in lumens/km of the pollutant flux

- the UWLR (%) and the corresponding zone

5.1. Findings for the Same Level of Performance with R3/Asphalt

1. In tubular high pressure sodium lamps, the required flux/km for the lamps can vary from 220 000 lumens to 360 000 lumens per km (+ 63%) according to the choice of bowls without the UWLR exceeding 3.8% (that is to say qualifying for zone E2, which requires < 5%) (see Tables 4 to 6 and Figure 3).

2. With tubular high-pressure sodium lamps, the use of high-performance optics (flat glass, refractor, deep bowl) limits the flux/km to lower values of both lamp flux and of the total flux reflected or emitted towards the sky (see Tables 4 to 6).

3. The deep bowls associated with HPS diffused bulb lamps are the most pollutant in terms of reflected flux, although their UWLR (2.7%) is satisfactory (see Figure 3).

4. The UWLR calculated from the ULOR measured in the laboratory for functional tubular HPS flat glass shows a minimum of 0.5%, when the reality is 0%. This is due to the "black flux" measured in the laboratory and to interference caused by reflections from the ground and the laboratory walls (Figure 3).

5. The direct flux emitted upwards by functional luminaires is very weak, of the order of 1% of the flux of the lamps, while the reflected flux is of the order of 9% (see Tables 4 to 6).

6. Tubular HPS axial luminaires have high-performance results in all fields, which shows that the maximum efficiency obtained by installation and optical performance results is perfectly compatible with minimum pollution (see Figure 3).

7. In decorative luminaires, the disastrous performance of total diffusers (i.e. spheres) is confirmed in all aspects (reflected flux and UWLR) (see Figure 3).

8. Decorative luminaires with optics and a non-visible lamp have an UWLR of 3.1% and thus can be used in zone E2. The total reflected flux is of the same order ($\pm$ 30 000 lumens/km) as the best functional luminaires. When the lamp is visible, these luminaires are compatible only with zone E4 (very urbanised) and within the limit of zone E3 (UWLR = 16%) (see Figure 3).

5.2. Findings for the Same Level of Performance with R1/Concrete

Complementing the previous findings, we conclude that :

1. As the brightness of the road surface increases, the lighting designer needs less flux to obtain the same level of luminance. This decrease in installed flux will be compensated by an augmentation of the reflection coefficient of the road. Consequently, as a good first approximation, we can say that the total road-reflected flux will remain constant as the brightness of the road varies.

2. Nevertheless, as the brightness of the road surface increases, the lighting designer needs less flux to obtain the same level of luminance and consequently the direct-emitted flux will decrease.

3. On the whole, we can conclude that an increase in road-brightness will lead to a reduced value of the total upward-emitted flux.

6. Conclusions

1. Studying light pollution is more complex than simply evaluating the UWLR ratio.

2. Not all refractors (European models in tubular HPS) are polluting since they can produce results among the best for minimum reflected flux in spite of an UWLR of 3.8% (see Figure 3).

3. The higher the performance of the luminaires within their recommended type of installation, the less light pollution there is.

4. The UWLR is admittedly indicative, but it is hardly selective and shows certain contradictions (Table 8). This is due to the fact that the direct flux upwards (ULOR) is very weak in general in relation to the direct flux downwards (DLOR).

5. The more the optics are hermetically sealed, the less need there will be for primary energy to install them and the less light there will be reflected towards the sky.

6. Lighting specialists need to take on board these new ideas about light pollution. Their resistance can be overcome by the most efficient and cheapest materials compatible with the performances currently proposed.

7. The idea that all kinds of light pollution are associated with the harmful effects created by public and private lighting installations (glare from lights, signs, advertisements, and objects whose lighting has no collective advantage) could be a reason for conflict between astronomers and lighting specialists. In reality, however, in spite of appearances, both camps actually have the same views and objectives. If notice is taken of some of the astronomers' comments, it should be possible to develop very high-performance luminaires, at least for road lighting and urban functional lighting, that satisfy these stricter new criteria.

Preserving the Astronomical Sky
IAU Symposium, Vol. 196, 2001
R. J. Cohen and W. T. Sullivan, III, eds.

The International Commission on Illumination - CIE: What It Is and How It Works

Christine Hermann

CIE General Secretary
CIE Central Bureau, Kegelgasse 27,A-1030 Vienna, Austria

Abstract. This paper explains the structure and organization of the International Commission on Illumination (CIE), shows how it works, and describes its objectives and achievements. It gives details on the technical work done in the CIE Divisions and Technical Committees, and mentions some examples of the work in progress, emphasising those in connection with the subject of this conference. Quality lighting is the key to reducing light pollution. CIE is aware of the problem and takes it very seriously. CIE Divisions are working on problems such as glare, quality lighting, light trespass, light pollution and energy efficiency. In this endeavour CIE is also co-sponsoring this conference.

1. Background

As its name implies, the International Commission on Illumination - abbreviated as CIE from its French title "Commission Internationale de l'Eclairage" - is an organization devoted to international cooperation and exchange of information among its member countries on all matters relating to the science and art of light and lighting. The CIE is an autonomous organization. It was not appointed by any other organization (political or otherwise) but has grown out of the interests of individuals working in illumination.

Its predecessor, the "International Commission on Photometry", was founded in 1900 as the first international body concerned with light measurement (of incandescent gas lamps at that time). As the new technology of illuminating engineering developed rapidly, the measurement of light sources became only part of a much wider activity: the study of how to best use the light which these sources provided. Thus a widening of the Commission's scope was suggested, and an enlarged Commission, the present CIE, was founded in 1913.

Since then, the CIE has gained a worldwide reputation for developing sound guides for lighting practice. These guides are not arrived at lightly, but take into account the vast knowledge of how light and lighting influence visibility, visual and human performance, ease of seeing, visual comfort, safety and security. In addition, they also include consideration of the available light sources, luminaires, lighting techniques and economics.

Since its inception, the CIE has been accepted as representing the best authority on the subject, and as such it is today recognized by the International Organization for Standardization ISO and the International Electrotechnical Commission IEC as an international standardization body. Its standards are submitted to ISO or IEC for direct endorsement and issued as a joint double-logo standard. In those areas where the domains overlap or complement each other, standards are jointly developed. Recently an agreement on technical cooperation between CIE and CEN was signed, with the aim to enhance collaboration and avoid duplication of work. Thus worldwide standardization on fundamentals and metrology of lighting and signalling and related applications of light and colour rely greatly on the CIE.

2. Objectives of the CIE

The CIE is a technical, scientific and cultural non-profit organization whose objectives are:

1. To provide an international forum for the discussion of all matters relating to the science, technology and art in the fields of light and lighting and for the interchange of information in these fields between countries.

2. To develop basic standards and procedures of metrology in the fields of light and lighting.

3. To provide guidance in the application of principles and procedures in the development of international and national standards in the fields of light and lighting.

4. To prepare and publish standards, reports and other publications concerned with all matters relating to science, technology and art in the fields of light and lighting.

5. To maintain liaison and technical interaction with other international organizations concerned with matters related to science, technology, standardization and art in the fields of light and lighting.

It is important to note that in these objectives light and lighting embraces such fundamental subjects as vision, photometry and colorimetry, involving natural and man-made radiations over the ultraviolet, visible and infrared regions of the spectrum. Furthermore, the applications cover all usage of light, indoors and out, including environmental and aesthetic effects, as well as means for the production and control of light and radiation. Recently the CIE created a new Division in the growing field of Image Technology. At the request of international and regional organizations, CIE is leading the efforts to standardize image technology.

3. How CIE is Organized

The affairs of the CIE are vested by the General Assembly, consisting of the Presidents of National Committees which have the responsibility for decisions on all matters relating to the organization (Figure 1). The composition of the National Committees varies from country to country, but each is required to represent and have the cooperation of all organizations having an interest in light and lighting. At the present time the CIE comprises 40 member countries:

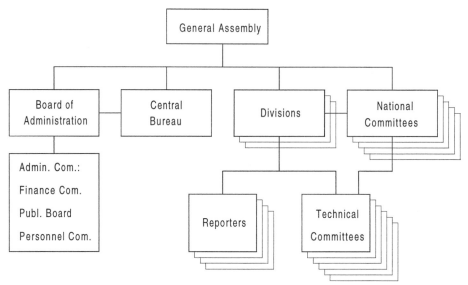

Figure 1.   Organization of the CIE

Argentina	Denmark	Israel	Russian Federation
Australia	Estonia	Italy	Slovak Republic
Austria	Finland	Japan	Slovenia
Belgium	France	Moldova	South Africa
Brazil	Germany	Netherlands	Spain
Bulgaria	Great Britain	New Zealand	Sweden
Canada	Hong Kong	Norway	Switzerland
China	Hungary	Pakistan	Thailand
Croatia	Iceland	Poland	Turkey
Czech Republic	India	Romania	USA

In addition, persons (or organisations) from countries where a National Committee has not yet been established may join as Associate Members and participate in the technical work of the CIE.

Modifications to the CIE Statutes were approved at our recent Session last month (June 1999), to add associate National Committee and supporting membership categories, with the goal of expanding CIE membership to encompass more regions in the world and to obtain financial support from other related organizations.

The affairs of the CIE are discussed and decided by the General Assembly (Figure 1), consisting of the Presidents of the National Committees. The General Assembly meets every two years. Decisions concerning admission and expulsion of members, approval of the budget and publication of Standards are taken by the National Committees. The National Committees delegate the conduct of affairs concerning technical issues to a Board of Administration.

The success of an organization such as the CIE depends upon the effectiveness of its technical work. Indeed, the objectives of the CIE could not be attained without a suitable and active committee structure which draws upon the expertise of people from all the member countries. Each major subject of interest to the CIE was assigned to one of seven Divisions.

Each National Committee is entitled to have one voting member in each Division who keeps the information flow going and assures that information on the status of the technical work becomes known in the member countries.

Technical Committees (TCs), consisting of small international groups of experts, are established in each Division to work on a single subject. The intent is that such committees concentrate on one specific topic and render a report to the Division within a reasonable period of time, after which the committee is dissolved. TC Chairmen are nominated by the Division, and appoint the members of their TC. National Committees may also appoint a representative.

4.   How CIE Works

The work programme can be divided broadly into two areas: fundamental research (covering such topics as vision, colour and measurement) and lighting applications. There is, of course, considerable interaction between these areas of activity.

Division 1: Vision and Colour

Terms of Reference: To study visual responses to light and to establish standards of response functions, models and procedures of specification relevant to photometry, colorimetry, colour rendering, visual performance and visual assessment of light and lighting.

Division 1 works on disability and discomfort glare, covering both fundamentals and specific applications such as discomfort glare experienced by elderly people. This work is much needed by the application divisions. In the colour field this Division is working on colour appearance models and colour rendering. One of the Technical Committees is working on the establishment of a colorimetric system based on fundamental visual functions. This work will eventually lead to a supplementary system of colorimetry.

Reports and standards have been published on: Colorimetry, CIE standard illuminants for colorimetry, CIE standard colorimetric observers, low vision, disability glare, industrial colour difference evaluation, CIE 1997 interim colour appearance model, etc.

The CIE work on colour, which led to the development of the colour rendering index, and on vision with the eye's response curve $V(\lambda)$, which allows meaningful values of lamp lumen output to be calculated for all light sources, provides the basis of every calculation carried out by lighting engineers throughout the world.

Division 2: Measurement of Light and Radiation

Terms of Reference: To study standard procedures for the evaluation of ultraviolet, visible and infrared radiation, global radiation, and optical properties of materials and luminaires, as well as the optical properties and performance of physical detectors and other devices required for their evaluation.

The Division's work ranges from fundamentals in the science of metrology to standards like the $V(\lambda)$ curve and a user guide for selection of illuminance and luminance meters. Reports have been published on: LED measurements, physical photometry, measurement of reflectance and transmittance, and other topics.

Division 3: Interior Environment and Lighting Design

Terms of Reference: To study and evaluate visual factors which influence the satisfaction of the occupants of a building with their environment, and their interaction with thermal and acoustical aspects, and to provide guidance on relevant design criteria for both natural and man-made lighting; as well as to study design techniques, including relevant calculations, for the interior lighting of buildings; incorporating these findings and those of other CIE Divisions into lighting guides for interiors in general, for particular types of interiors and for specific problems in interior lighting practice.

The Technical Committees can be divided into two groups dealing with either electric light or daylight. They work primarily on general aspects of light and light quality. Psychological aspects and post-occupancy evaluation and energy aspects are covered as well.

In the field of daylighting, a major undertaking of the CIE is in daylight data evaluation. Together with the WMO, CIE is conducting the International Daylight Measurement Program (IDMP). More than 50 IDMP stations are in operation around the world, collecting meteorological and light intensity data. The database is maintained on a webserver (http://idmp.entpe.fr).

Another area to mention is discomfort glare. A report on a Unified Glare Rating System (UGR) recommended by the CIE was published as CIE 117-1995.

Other TCs are dealing with questions such as museum lighting, lighting and architecture, and sky luminance models. Reports have been published on such topics as interior lighting, discomfort glare, daylight measurement, standard on overcast sky and clear sky. Last year the Division organised a very successful symposium on lighting quality.

Division 4: Lighting and Signalling for Transport

Terms of Reference: To study lighting and visual signalling and information requirements of transport and traffic, such as road and vehicle lighting, delineation, signing and signalling for all types of public roads and all kinds of users and vehicles, and visual aids for modes other than road transport.

TCs are working on road and tunnel lighting, colour of signal lights, crime and road lighting, etc. Reports have been published on road lighting, standards on road traffic lights, lighting in urban areas, daytime running lights, variable message signs, lighting of road tunnels and underpasses, and other topics.

TC 4-21 is working to produce an international framework to enable national or local regulations or recommendations to be produced to restrict interference by light with astronomical observations (see the paper by D. Schreuder in this volume). Three CIE publications deal specifically with light pollution around astronomical sites:

- CIE 1-1980: Guidelines for minimizing urgan sky glow near astronomical observatories (jointly with IAU)

- CIE x008-1994: Urban sky glow, a worry for astronomy (Proceedings of a Symposium that brought together international experts on astronomy, lighting architecture, lighting engineering and instrumentation)

- CIE 126-1997: Guidelines for minimizing sky glow

This last Technical Report gives general guidance for lighting designers and policy makers on the reduction of sky glow, discussing briefly theoretical aspects of sky glow and giving recommendations about maximum permissible values for lighting installations in relation to the needs of astronomical observations.

Division 5: Exterior Lighting and Other Applications

Terms of Reference: To study procedures and prepare guides for the design of lighting for exterior working areas, security lighting, flood lighting, pedestrian and other urban areas without motorized traffic, areas for sports and recreation, and for mine lighting.

Division 5 also deals with light trespass. Based on a workshop on spill light, TC 5-12 (Obtrusive light) was established with the task to study the effects of obtrusive light from exterior lighting to residents and traffic, and to prepare a technical report that gives measures to describe the effects of light trespass, recommends measuring methods, and recommends acceptable levels. Reports have been published on all kinds of sports lighting, floodlighting, and lighting of exterior work areas.

Division 6: Photobiology and Photochemistry

Terms of Reference: To study and evaluate the effects of optical radiation on biological and photochemical systems (exclusive of vision).

TCs are working on standardization of sunscreen testing, photobiological lamp safety, effects of UVA radiation, standardization of action spectra, etc. Reports have been published on sunscreen testing, standardized erythema dose, and other topics. Last year the Division organised a workshop on Measurement of Optical Radiation Hazards.

Division 7

Division 7 "Generals Aspects of Lighting" was recently dissolved.

Division 8: Image Technology

Terms of Reference: To study procedures and prepare guides and standards for the optical, visual and metrological aspects of the communication, processing, and reproduction of images, using all types of analogue and digital imaging devices, storage media and imaging media.

TCs are working on gamut mapping, colour appearance models for colour management applications, communication of colour information, etc. The Division works in close liaison with other international standards bodies.

Finally, as a project concerning all Divisions, CIE is working on the revision of the International Lighting Vocabulary (joint publication with the International Electrotechnical Commission (IEC), Publ. CIE 17.4-1987).

5.   Results of CIE Work

5.1.   Publications

The CIE produces a range of publications:

- Standards,

- Technical Reports and other technical publications,

- Proceedings of CIE Sessions and Symposia,

- disks (photometric and colorimetric tables, calculation software),

- a CD-ROM with all CIE Technical Reports, as well as individual CD-ROMs (on Low Vision, Proceedings of the last Session). More of these will be published in the near future.

More than 100 such publications have been issued.

Since 1991 when the agreement with ISO was ratified, three dual CIE/ISO standards have been published: one on Colorimetric illuminants (recently revised), one on Colorimetric observers, and one on Overcast sky and clear sky. Several drafts of standards are now in the pipeline.

The official languages of the CIE are English, French and German. Standards are published in all three languages, whereas other publications are written in English, with a summary in French and German.

In addition to Technical Publications, a newsbulletin, *CIE News*, is published quarterly. This gives regular updates on the progress of the technical programme, press releases of new CIE publications and forthcoming meetings, together with news of administrative matters and topics of general interest in the world of light and lighting. It is circulated free of charge to member bodies and is also included on our web home page.

The most up-to-date CIE information can be obtained via the Internet. The CIE home page is

http://www.cie.co.at/cie/

There you will find information on organizational structure, addresses of all our National Committees, a complete list of CIE publications together with abstract information, the contents of the most recent issues of *CIE News*, information on upcoming symposia, as well as some other information of interest to those concerned with light and lighting.

Recently CIE has established a moderated E-mail Lighting List, for the exchange of research results and discussion on open questions. To join, just send an e-mail to

lsg-request@knt.vein.hu

with *subscribe* in the subject line. In addition, Division 3 has its own moderated discussion list:

lighting-request@garnet.nist.gov

5.2. Meetings and Symposia

Every four years the CIE holds a Session, hosted by one of the member countries, which serves the purpose of bringing together all the representatives of the National Committees who are interested in the technical activities of the organization. The General Assembly meets to review and discuss the administrative and technical affairs of the CIE, make plans for the future, and elect the officers for the next quadrennium. The majority of the time, however, is taken up by an international lighting conference and meetings of the Divisions and Technical Committees where the programme for the past term is reviewed and the work for the next term is set up. The 24th Session was held in June 1999 in Warsaw, Poland. The next one is planned for San Diego, California, in 2003.

The first objective of the CIE, as stated in our Statutes, is "to provide an international forum for the discussion of all matters related to the science, technology and art of lighting". Seminars organised on various topical subjects are proving to be one effective way of achieving the objective.

Annual CIE expert seminars were started in 1990, and these have now become a regular event and have proved to be a great success:

- 1990: Light measurement

- 1992: Computer programs for light and lighting

- 1993: Advanced colorimetry

- 1994: Advanced photometry (Lighting quality and energy conservation)

- 1996: Colour standards for image technology

- 1997: Standard methods for specifying and measuring LED characteristics (Symposium and Tutorial Workshop)

- 1997: Colour standards for imaging technology

- 1998: Lighting quality

- 1998: Measurements of optical radiation hazards

- 1999: A symposium on 75 years of CIE photometry will be held in Budapest, Hungary, from 30 September to 2 October 1999.

6. CIE Central Bureau

The CIE headquarters are at:

> Kegelgasse 27, A-1030 Vienna, Austria
> Phone: +43 1 - 714 31 870
> Fax: +43 1 - 713 08 3818
> E-Mail: ciecb@ping.at

We will be pleased to establish contact with our most qualified experts, for any lighting question you might have.

7.   Conclusion

Over the years the administrative and technical organization of the CIE has been modified to meet changing needs and conditions. However, its basic objectives remain the same. It is not unusual for CIE to respond to changing needs and conditions. New subject areas arise, while others lose importance. New technologies are developing, many of which involve aspects of light and lighting in disciplines not traditionally connected with CIE.

CIE is involved in a multidisciplinary field of endeavour. CIE combines the expertise of architects, designers, engineers, physicists, biologists, physiologists, psychologists, eye and health care professionals and many others in its technical work.

This Commission has grown through the engagement and enthusiasm of individuals. The number of member countries has grown from 9 to 40 since its foundation. Some two-thirds of the earth's land surface is now represented in the organization, by either National Committees or individual members.

In view of its authority and the global network it represents, the CIE is in an ideal position to play a key role in the continued progress towards a better-lit environment.

Preserving the Astronomical Sky
IAU Symposium, Vol. 196, 2001
R. J. Cohen and W. T. Sullivan, III, eds.

Recent CIE Activities on Minimizing Interference to Optical Observations

Duco A. Schreuder

Duco Schreuder Consultancies, Spechtlaan 303, 2261 BH Leidschendam, The Netherlands

Abstract. The paper describes the activities of CIE in the field of fighting sky glow. The most recent CIE publications are briefly discussed, as well as suggestions for further work. Modern developments in road lighting as well as alternatives to it are briefly discussed.

1. Introduction

The International Lighting Commission (Commission Internationale de l'Eclairage, or CIE) is a multi-national, professional, non-governmental, non-profit organization that monitors the whole area of lighting and visibility on a worldwide base. The structure and its way of operation are discussed by Hermann (2001) in this volume.

The relationship between CIE and IAU has a long standing. It resulted in two important documents: the CIE Document 'Statement concerning protection of sites for astronomical observatories' (CIE 1978) and the joint IAU/CIE publication 'Guide lines for minimizing urban sky glow near astronomical observatories' (IAU/CIE 1984). Again in close cooperation with IAU, more recently the 'Guidelines for minimizing sky glow; A CIE Technical Report' have been published (CIE 1997). At present, the CIE Technical Committee TC 4-21 is working at a number of additions to this Guide. A more or less parallel CIE activity involves the restriction of the more general 'obtrusive light' (CIE 1995; Pollard 2001). See also the volume edited by Isobe & Hirayama (1998).

2. Sky Glow and its Consequences

Sky glow presents itself as a background luminance over the sky, against which astronomical objects must be observed. Interference with astronomical observations is caused by the resulting reduction in luminance contrast. The glow is caused by non-directional scatter of light by particles in space and in the atmosphere. Part of the sky glow is natural and part is man-made (Levasseur-Regourd, 1994; Leinert & Mattila 1998). It is customary to express the man-made component of the sky glow as a percentage of the natural background luminance.

The sky glow is the result of light that is projected upwards and then reflected and scattered back to the surface of the earth. Even when the light beams are - as is good lighting design practice - downward, part of the light is scattered upwards after being 'used'. So uplight cannot be completely avoided.

The costs of light pollution are considerable. Although the costs per operating hour of an astronomical observatory cannot be defined precisely, a simple estimate suggests that, if one could use the observatory one extra hour per year, the savings in capital costs would be around US$ 15,000 and the running costs another US$ 1,500 or more. This money could be used to improve the lighting around the observatory (Schreuder 1991).

There is another approach to this. Light pollution degrades the visibility. In theory, this may be compensated by making the telescope bigger - and more expensive. The costs of a telescope increase with the third power of the lens or mirror diameter (Murdin 1997). When the light pollution reduces the effective diameter of the mirror by 5% (which seems a reasonable estimate even for weak sky glow), the economic losses are 15%. A big telescope costs easily about US$ 100 million, so fighting light pollution might mean a 'profit' of US$ 15 million. If we assume that there are some 30 such telescopes around the world and another 500 that are a factor 10 smaller, this would lead to a total 'hardware' investment of US$ 8,000 million. Light pollution thus requires an extra, non-profitable investment of US$ 1,200 million. Over 10 years, one might save world-wide some US$ 120 million - a fair percentage of the US$ 4,000 to 5,000 million that are spent on astronomy annually (Woltjer 1998).

Finally, in many ways light may be considered on judicial grounds as a 'pollutant', just as noise, because it is a hazard to safety (i.e. for motorists); it produces ecological disturbances; it may produce environmental and visual nuisance; and it leads to energy wastage (Gray 1993). So, there are legal routes as well to combat (excessive) light pollution. See e.g. Anon (1984); Crawford (1991); IAU/CIE (1980); Murdin (1992). One way may be to include the most important observational sites in the list of World Cultural Heritage sites (see McNally 1994).

3.   Lighting Design Elements

3.1.   Zoning

The two key concepts in the CIE work are the zoning principle and the curfew idea. Zoning is a well-established practice to establish a base for environmental regulations. It is widely in use for describing and limiting noise, air and water pollution, vibration etc. The basic idea is that, in case the polluting activity cannot be avoided altogether, the environmental consequences of the pollution are not equally detrimental for all locations. The zones and the zone requirements are set up in relation to the (human and non-human) activities in these zones. Zoning does not stop environmental pollution, but it may serve as a frame of reference for anti-pollution legislation and regulation (Brouwer & Leroy 1995; Schreuder 1998). CIE has adopted four zones (CIE 1995, 1997). In some cases, when more details are needed, the ALCOR-classification for zones is used (Murdin 1997). The 'curfew' idea is that later in the night (say after midnight) the requirements may be more severe than during the evening (CIE 1995, 1997).

The quantification of the natural background radiation depends on the precise way to convert 'astronomical photometry' into 'lighting engineering photometry'. The TC 4-21 has decided to use the conversion proposed by Crawford (1997) where 26.33 mag/arcsec^2 equals 3.2×10^{-6} cd/m^2. It was also decided to use as a standard geometry for the measurements of sky glow as follows the average luminance in a cone of angular diameter 10° around the zenith, supplemented by six adjacent cone measurements around the zenith. The measurements must be made in photopic terms and the accuracy must be within 15%.

3.2. Lighting Requirements

CIE has given recommendations for the limitation of sky glow (CIE 1997). They are given as the maximum permissible value of ULOR$_{inst}$ (the Upward Light Output Ratio - installed, expressed as a percentage of the luminous flux of the luminaire) for each of the four Environmental Zones, ranging from 0 to 25%. This limit holds for each individual luminaire in that zone. For the actual lighting design, the total ULOR$_{inst}$ per zone, or the average ULOR$_{inst}$ per zone is more relevant.

Lighting designers require more than the upward flux per luminaire. They need to know what is permissible for the different zones as a whole. It is proposed to use the 'average upward flux' as a criterion. This is assessed as follows. For each zone, the maximum illuminated area as a percentage of the total area in the zone is given (roads, parking lots, industrial sites etc.). Furthermore, an 'average area illuminance' is given, which is based on the current CIE recommendations for road lighting (CIE 1995a). The proposal, currently under discussion in CIE TC 4-21, gives maximum values from 0.02 lm/m^2 for the most 'dark' zone to 150 lm/m^2 for city centres. After curfew, the values are 0 and 30 lm/m^2 respectively.

4. Environmentally-Friendly Lighting Design

4.1. Lighting Levels in Road Lighting

Although road lighting is not the only, and even not the most important contributing factor to sky glow, it is good to consider its function as an example of how lighting influences the environment. The function of road lighting is derived from the visual task of the traffic participants (the "driving task"; reaching the destination of the trip and avoiding collisions while doing so; see e.g. Schreuder 1991a, 1998). Large scale studies have established that "good" road lighting as compared with "poor" lighting may result in a reduction of about 30% in night-time injury accidents for major urban roads and for rural motorways - although the quantification of "good" and "poor" turned out not to be easy (CIE 1992; OECD 1972). For other road types, for pedestrian crossings and for tunnels, similar results have been found. Similar figures, although often much higher, were found for road lighting as a crime countermeasure (Schreuder 1994a; Painter 1998).

4.2. Sky Glow Conscious Lighting Design

Generally speaking, 'sky-glow conscious lighting design' for security and road lighting should be an easy matter, and many would simply point to the many local ordinances in the USA that simply demand full cut-off low pressure sodium luminaires. However, a low light pollution luminaire is easier to design for a high pressure discharge lamp than for a low pressure sodium lamp. It could therefore be suggested that while in the vicinity of an observatory, a low pressure sodium lamp should be used, whereas elsewhere in general the smaller and more compact lamp is better.

It has been suggested to introduce light pollution ratings, an "eco label", for luminaires, based on the actual upward flux. In order to arrive at practical limits for such an idea, data have been collected from many luminaire manufacturers of well over 100 luminaires of all different types. Whereas light pollution ratings would be of considerable help in reducing sky glow, there is one word of caution. In some cases, a small number of 'more polluting' luminaires would, over an entire scheme, be less polluting and more economical than a larger number of 'less polluting' luminaires. It must therefore be emphasised how important it is to consider the actual upward light flux of the whole installation.

Floodlighting and lighting for sports facilities usually cause more light pollution than ordinary road lighting. There are, however, national and international standards and recommendations available, such as CIE (1989, 1993). The well-known ILE Guidance Notes are published as an Annex to CIE (1997). Lighting designers usually do not like to have to use standard solutions, but for long-term environmental benefits the designers must learn to contain themselves within sensible guide lines.

This section is based on the paper given by Nigel Pollard at the CIE- Symposium "Urban sky glow, a worry for astronomy", held on 3 April 1993 in Edinburgh (see Pollard 1993).

4.3.   Specific Remedial Measures

Remedial measures follow directly from the characteristic of the disturbance: restricting outdoor lighting in time or location (curfew and zoning); photometric measures (light control, reduction of reflection); colorimetric measures (monochromatic light, filtering the light). Additionally, educational measures must be mentioned. These various measures are discussed in detail in Schreuder (1993, 1994a). See also CIE (1997); Crawford, ed. (1991) and McNally, ed. (1994). It is not possible to give a general rule that is always most effective. Not all countermeasures are directly related to the lighting itself. See NSVV (1997) and Schreuder (1998).

5.   Road Lighting and its Alternatives

Our society is a complicated structure with elements that function closely together while being placed far apart. A complicated network of exchange of persons, goods and information is needed, which also has to function at night. Road lighting is essentially functional, its primary function being to permit societal activities to proceed also during the night and to do so with a high degree of road safety, subjective security and amenity.

A major practical problem is to find out when a road (a road section) needs to be lighted ('Warrants for lighting'). In the past, lighting engineers left the answer to this question to politicians. Nowadays, cost/benefit considerations are used to find answers to this question. CIE has organized a number of meetings on this subject (CIE 1997a, 1999, 1999a). Actually, this question consists of two distinct aspects. The first is that the decision to light or not to light is not a permanent one to be taken once and forever; the second is that there is really no strict dichotomy between 'to light or not to light'.

In the Netherlands, the two different aspects are studied separately and in great depth. The first deals with 'smart road lighting', the second with alternatives to road lighting.

5.1. Smart Road Lighting

Recently, a large experiment has been completed in The Netherlands related to 'smart road lighting'. The system is called 'dynamical lighting' or DYNO for short. Details are given in AVV (1999). On a very busy, multilane highway the light level can be adjusted according to the momentary values of traffic volume and weather. The road is a 14 km stretch of the A12 between the Hague and Utrecht. The speed limit is 120 km/h; the traffic volume was (in 1995) about 210,000 vehicles per day. Before 1995 the road was unlit. The lighting layout is conventional. The two carriage-ways are lit from the sides by means of SON-T lamps of 400 W with cut-off luminaires on 15 m high masts at a spacing of 50 m. The lighting level can be set at one of three levels: 0.2 cd/m^2; 1 cd/m^2 and 2 cd/m^2. The middle value agrees with the current road lighting recommendations for the Netherlands (NSVV 1990) as well as with those of CIE (1995).

The light level can be adjusted according to the following criteria: traffic volume, rain or fog. Under favorable conditions the lowest level (0.2 cd/m^2) is employed. When the traffic volume rises above about 1,400 vehicles per hour per lane, the second, or 'standard', light level (1 cd/m^2) is used. The same is done during rain and fog (visibility under 140 m). Under especially hazardous conditions (road works, snow or ice on the road, accidents) the highest level (2 cd/m^2) may be used. Over the two years of the experiment, the road lighting was used for 50% of the day on average (normally in the Netherlands this is 45%). During darkness, the low level was used 59% of the time, the normal level 37% and the high level 4% of the time. Obviously, the lowest light level causes the lowest light pollution.

The analysis showed that traffic behaviour (speed, headway, etc.) did not depend on the light level, nor did the accident pattern. From this it was concluded that the 'high' level (2 cd/m^2) really did not bring anything extra, so that further analysis and the consequent recommendations are based on an installation that has only two levels: 'low' and 'standard'. From the two years of experimentation, it has been concluded that energy consumption is 35% lower than for a traditional installation. This results, among other things, in 650 tonne less CO_2 emitted per year. In spite of the more expensive installation, the running costs are considerably lower; it is to be expected that maintenance costs will be lower as well. The conclusions are generally speaking very favorable. Similar installations will be used on many more motorways in the Netherlands, particularly in 'areas sensitive to light pollution'.

5.2. Alternatives to Road Lighting

When discussing the ways to fight light pollution, emphasis is on the zone that includes 'Areas with intrinsically dark landscapes: National Parks, Areas of outstanding natural beauty'. Although one tries to let the roads be unlit, this is not always possible. However, especially here, the dichotomy we mentioned earlier is not relevant. We mentioned earlier that the way the function is fulfilled is a matter of adequate visibility of objects. For busy high speed roads, the objects are primarily other (moving) cars, but for rural roads the main part of the visual task is to follow the run of the road (Schreuder 1998). In terms of the driving task, the first is mainly to keep centred in the traffic lane and to avoid collisions, the second to keep on the road. In lighting terms, this means (for the first) seeing road markings and detecting obstacles and (for the second) receiving optical guidance.

For the detection of obstacles, road lighting is essential as soon as lighting by means of vehicle headlamps in not sufficient any more. For road markings and optical guidance, 'full' road lighting is not required at all. It is along these lines that a number of experiments are under way in the Netherlands. As these tests are in the beginning stage, there are not yet any results that can be reported. The major experiments are:

- small non-cutoff fittings at 7 to 10 m high. Usually these are equipped with 7 W compact fluorescent lamps and are mounted halfway up conventional lighting columns that work in dense traffic. The small lamps operate when the traffic is so light that 'full' road lighting is not required;

- a similar design with small non-cutoff fittings on separate masts of 5 to 8 m high equipped with an array of Light Emitting Diodes (LEDs). These lamps operate throughout the night;

- road markings that are equipped with small lamps, usually LEDs, instead of, or supplementary to, retroreflectors. These LEDs may be powered via the electrical net or by stand-alone photovoltaic cells;

- post markers at the road side, equipped with LEDs or with fibre-optic light points;

- light guides (light tubes) along the guard rails at the road side.

All these proposals, and several more, are tested in field experiments that are currently under way. The results seem to be very promising, but it is too early to give any definitive opinion.

References

Anon 1984. *La protection des observatoires astronomiques et geophysiques.* Rapport du Groupe du Travail. Institut de France, Academie des Sciences, Grasse, 1984

AVV 1999. Dynamische openbare verlichting DYNO (Dynamic public lighting DYNO). Ministerie van Verkeer en Waterstaat, Adviesdienst voor Verkeer en Vervoer. Rotterdam, 1999

Brouwer, K. & Leroy, P., eds. 1995. Milieu en ruimte; Analyse en beleid (Environment and space; Analysis and policy making). Amsterdam, Boom

CIE 1978. Statement concerning protection of sites for astronomical observatories. CIE, Paris

CIE 1989. Guide for the lighting of sports events for colour television and film systems. Publication No. 83. CIE, Paris

CIE 1992. Road lighting and accidents. Publication No. 95. CIE, Vienna

CIE 1993. Urban sky glow, A worry for astronomy. Publication No. X008. CIE, Vienna

CIE 1995. Guide on the limitation of the effects of o btrusive light from outdoor lighting installations; Third draft - August 1995; CIE TC 5-12: Obtrusive light

CIE 1995a. Recommendations for the lighting of roads for motor and pedestrian traffic. Technical Report. Publication No. 115-1995. Vienna, CIE

CIE 1997. Guidelines for minimizing sky glow. Publication No. 126-1997. CIE, Vienna

CIE 1997a. Symposium on 'Road Lighting for Developing Countries' held at the SANCI-CIE International Conference on 'Lighting for Developing Countries' Durban, South Africa, 1-3 September 1997

CIE 1999. Proceedings of Workshop "Warrants for road lighting", Saturday, 24 October 1998, Bath, UK. Vienna, CIE

CIE 1999a. Workshop on "Criteria for Road Lighting." held on 24 June 1999 at the CIE 24th Session in Warsaw, Poland, 24 - 30 June 1999 (proceedings to be published)

Crawford, D.L. 1991. Light pollution: a problem for all of us. pp. 7 - 10 in: Crawford, ed., 1991

Crawford, D.L. 1997, Photometry: Terminology and units in the lighting and astronomical sciences. The Observatory, 117, 14-18

Crawford, D.L., ed. 1991. *Light pollution, radio interference and space debris* (Proceedings of the International Astronomical Union colloquium 112, held 13 to 16 August, 1989, Washington DC). Astronomical Society of the Pacific Conference Series Volume 17. San Francisco

Gray, I. 1993. Light pollution and the environmental agenda. In: European colloquium on light pollution. British Astronomical Association. Reading, Saturday, 3rd July, 1993

Hermann, C. 2001. The CIE (Commission Internationale de l'Eclairage): what it is and how it works. In *"Preserving the Astronomical Sky"*, (Proceedings of IAU Symposium No. 196, July 12-16, 1999, Vienna, Austria), eds. Cohen R. J. & Sullivan W. T., ASP Conf. Ser.

IAU/CIE 1984. Guide lines for minimizing urban sky glow near astronomical observatories. Publication of IAU and CIE No. 1. CIE, Paris

Isobe, S. & Hirayama, T. eds. 1998. *Preserving of the astronomical windows* (Proceedings of Joint Discussion 5. XXIIIrd General Assembly International Astronomical Union, 18-30 August 1997, Kyoto, Japan). Astronomical Society of the Pacific, Conference Series, Volume 139, San Francisco, California

Kovalevsky, J., ed. 1992. *The protection of astronomical and geophysical sites.* NATO-committee on the challenges of modern society, Pilot Study No. 189. Paris, Editions Frontieres

Leinert, Ch. & Mattila, K. 1998. Natural optical sky background. p 17-20 in: Isobe & Hirayama, eds., 1998

Levasseur-Regourd, A.C. 1994. Natural background radiation, the light from the night sky. In: McNally, ed., 1994

McNally, D. ed., 1994. *Adverse environmental impacts on astronomy: An exposition* (Proceedings of IAU/ICSU/UNESCO Meeting, 30 June - 2 July, 1992, Paris), Cambridge University Press

Murdin, P. 1992. Protection of observatories; The legal avenues. Chapter 7 in: Kovalevsky, ed. 1992

Murdin, P. 1997, ALCoRs: Astronomical lighting control regions for optical observations, The Observatory, 117, 34-361

NSVV 1990. Aanbevelingen voor openbare verlichting; Deel I. (Recommendations for public lighting; Part I). NSVV, Arnhem

NSVV 1997. Richtlijnen openbare verlichting natuurgebieden. Publicatie 112 (Directives for public lighting in areas of natural beauty. Publication 112). Driebergen, CROW

OECD 1972. Lighting, visibility and accidents. OECD, Paris

Painter, K. 1998. Road lighting and crime prevention. Paper presented at CIE Workshop "Warrants for road lighting", held on Saturday, 24 October 1998 in Bath, UK

Pollard, N.E. 1993. Sky glow conscious lighting design. In: CIE, 1993

Pollard, N.E. 2001. Activities of the CIE Committee on Obtrusive Lighting. IAU Symposium No. 196, "Preserving the Astronomical Sky", July 12-16, 1999, Vienna, Austria

Schreuder, D.A. 1991. Lighting near astronomical observatories. In: CIE, Proceedings 22th Session, Melbourne, Australia, July 1991. Publication No. 91. Paris, CIE, 1992

Schreuder, D.A. 1991a. Visibility aspects of the driving task: Foresight in driving. A theoretical note. R-91-71. SWOV Institute for Road Safety Research, The Netherlands, Leidschendam

Schreuder, D.A. 1993. The assessment of urban sky glow. Paper presented at the ILE Annual Conference, Bournemouth September 1993

Schreuder, D.A. 1994a. Comments on CIE work on sky pollution. Paper presented at 1994 SANCI Congress, South African National Committee on Illumination, 7 - 9 November 1994, Capetown, South Africa

Schreuder, D.A. 1998. Road lighting for security. London, Thomas Telford. (translation of "Openbare verlichting voor verkeer en veiligheid", Kluwer Techniek, Deventer

Woltjer, I. 1998. Economic consequences of the deterioration of the astronomical environment. p. 243 in: Isobe & Hirayama, eds., 1998

Preserving the Astronomical Sky
IAU Symposium, Vol. 196, 2001
R. J. Cohen and W. T. Sullivan, III, eds.

Guide on the Limitation of the Effects of Obtrusive Light from Outdoor Lighting Installations

Nigel Pollard

NEP Lighting Consultancy, Bath, UK

Abstract. Division 5 of the CIE deals with the subject of "Exterior and other Lighting Applications" and has since 1991 had a Technical Committee (TC5.12) working on a "Guide on the limitation of the effects of obtrusive light from outdoor lighting installations". This Committee did much of its work under the chairmanship of Dr. Alec Fisher of Australia, who has now retired and passed on the work to the current author. The present report outlines the content of the proposed Guide and details its relevance to help preserve the astronomical sky. It also highlights the close links with CIE Committee TC4.21, which works directly on the problems of sky glow.

1. Contents of the Guide

The document is currently in draft form, and it is hoped to publish it in 2000. It has the following contents list:

1. Scope

2. Potential Obtrusive Effects and Associated Light Technical Parameters

3. Design, Installation, Operation and Maintenance

4. Documentation of Lighting Design

5. Calculation of Light Technical Parameters

6. Measurement of Light Technical Parameters

Appendices:

A. Investigations into the Obtrusive Effects of Outdoor Lighting

B. Illustration of Floodlight Classifications.

The document is concerned with a wide range of observers as identified in the following list, taken from Section 2.6 of the document:

This leads into the design objectives for which a check list is given in Figure 3.1 of the Guide, "Checks of Potential Obtrusiveness which should be Undertaken in the Design of Outdoor Lighting". The check list includes the following actions to be taken concerning potential obtrusive effects of a lighting installation on astronomical observations:

- Check effects on astronomical observations

- Identify locations of community or scientific optical observatories

- Check existence of planning regulations related to observations

- Check that the illuminances proposed are not excessive in relation to those recommended for the activity

- Assess installation for compliance with recommended limits of *UWLR* in CIE Publication 126

Further design help is given in Table 3.1 of the Guide, which gives the possible effects on spill light from changes to the installation parameters.

All the above are brought together in Section 3.3 of the document, Design Guidelines, which also suggests a definition of three luminaire types as below:

Type A   Symmetrical

Type B   Asymmetrical

Type C   Double-Asymmetrical

These three types of floodlight luminaires and their illumination patterns are illustrated in Figures 1 and 2 of the present paper. The CIE Document then goes into more details on calculation and measurement methods.

Copies of the 4th Draft of the document are available from the present author, who is the Committee Chair. The document will be published on the CIE website

www.cie.co.at/cie

in due course.

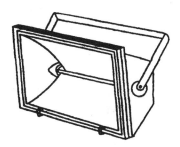

**(a) Type A floodlight
giving a symmetrical beam**

**(b) Type B floodlight
giving a fan-shaped beam**

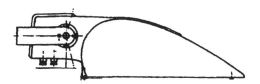

**(c) Type C floodlight giving a fan-shaped beam
with asymmetric distribution in the vertical plane**

Figure 1.   General types of Floodlight Luminaires. From Section 5 of the CIE "Guide on the limitation of the effects of obtrusive light from outdoor lighting installations".

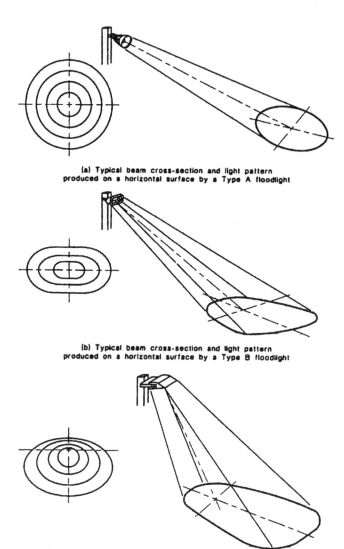

(a) Typical beam cross-section and light pattern
produced on a horizontal surface by a Type A floodlight

(b) Typical beam cross-section and light pattern
produced on a horizontal surface by a Type B floodlight

(c) Typical beam cross-section and light pattern produced on a horizontal surface
by a Type C floodlight

Figure 2. Typical floodlight distributions shown diagramatically. From Section 5 of the CIE "Guide on the limitation of the effects of obtrusive light from outdoor lighting installations".

Preserving the Astronomical Sky
IA U Symposium, Vol. 196, 2001
R. J. Cohen and W. T. Sullivan, III, eds.

Why Astronomy Needs Low-Pressure Sodium Lighting

Christian B. Luginbuhl

United States Naval Observatory, Flagstaff Station, PO Box 1149, Flagstaff AZ 86002 USA

Abstract. The damage to the dark-sky resource upon which ground-based astronomical observatories depend is substantial and increasing, even at what are considered premier dark-sky sites. Due to its nearly monochromatic output at the Na I resonance doublet near 589 nm, extensive use of low-pressure sodium (LPS) lighting in areas near astronomical observatories offers the potential preservation and even restoration of dark skies at other wavelengths, as well as minimal impact on the lighting needs of surrounding communities. Flagstaff, Arizona, with extensive use of LPS for general area lighting, has shown that this goal can be achieved.

1. Introduction

In the last 50 years, the worldwide astronomical community has focussed its efforts on the establishment of observatory facilities located at carefully evaluated sites of the highest astronomical quality. These efforts have been driven by two factors - the value of good site conditions to the quality of the data obtained, and increasing light pollution at existing observatory sites.

But today we find nearly all of these premier astronomical sites are either suffering degradation in their dark-sky conditions, or are threatened with such degradation. Despite their remote locations, the areas near the sites attract settlement because of favourable weather and climate conditions. Ironically, the very qualities that make them astronomically valuable lead to above average population and lighting increases. Further, there is a general trend of increased *per capita* use of outdoor lighting despite energy costs and concerns about global warming.

Arizona, site of Kitt Peak, Mount Hopkins, Mount Graham, and other quality astronomical sites, is experiencing a sustained population growth in excess of three percent per year, two-thirds arising from migration; this should be compared to the average U.S. growth of less than 1% per year for the past twenty years. The Big Island of Hawai'i, home to Mauna Kea, is experiencing growth in excess of two percent per year. The Kailua-Kona coast of the Big Island, where generally clear conditions allow lighting to have the greatest effect on the observatories, is the area where most of this growth is concentrated, and thus the effective growth rate is even higher than the average for the whole island.

2. Lighting Codes

The first line of defence for preserving dark sky over these observatories has already been as fully effected as it can be - namely, the observatories have been located where population densities are low and where high-altitude atmospheric aerosols that back-scatter artificial light are minimal. One of the principal criteria in choosing observatory sites has been the darkness of the sky. Though other conditions affecting the quality of the site such as cloud cover and airflow can not be affected by civil regulation, the sky will not remain dark at these sites without proactive involvement of the astronomical community in lighting use policies of nearby communities.

To minimize brightening of the night sky over observatory sites, civil regulations restricting wasteful and unnecessarily polluting lighting practices must be pursued. Outdoor lighting codes, pioneered by Arizona communities such as Tucson and Flagstaff but in place now in many localities in the U.S., have been actively pursued in many locations. Their effectiveness at reducing the rate of increase of sky brightness is likely to be substantial, although uncertain.

Limited data indicate that the actual effectiveness of lighting codes in limiting sky brightness increases is uncertain. One reason is that there are problems in enforcing lighting codes, and the degree of compliance has not been determined. Pilachowski *et al.* (1989) have presented B and V measures of zenith sky brightness over Kitt Peak for 1987, finding an increase over "natural dark" of only 0.07 and 0.08 mag/arcsec^2 at V and B respectively (Table 1). By comparing these measures to Garstang's "standard model" prediction (Garstang 1989) of 0.11 (V) and 0.07 (B) mag/arcsec^2, they offer evidence that the Tucson and Pima County outdoor lighting codes are having a measureable effect. But measures at nearby Mount Hopkins during the 1996 solar minimum apparently show the opposite, with measured values two to three times higher than Garstang's prediction.

Table 1.   Predicted and Observed B and V Zenith Sky Brightness Increases (relative to a natural sky) at Selected U.S. Observatory Sites

		ΔV		ΔB	
Site	Established	Pre[a]	Obs[b]	Pre[a]	Obs[b]
Kitt Peak	1958	0.13	0.07	0.09	0.08
Mt Hopkins / MMT	1968	0.21	0.49	0.18	0.63
Mt Graham	1995	0.06		0.05	
Lowell / Anderson Mesa	1963	0.27		0.28	
Naval Obs. Flagstaff	1954	0.39	0.45	0.42	0.13
McDonald	1969	0.01		0.01	
Mauna Kea	1970	0.03		0.03	

[a]Predicted values for 1995 from Garstang (1989), except for 1987 for Kitt Peak

[b]Measured values for Kitt Peak from 1987 (Pilachowski *et al.* 1989); Mt Hopkins for 1996 (Caldwell *et al.* 1999); and Naval Obs., Flagstaff, for 1996 (by the author)

The Tucson and Pima County lighting code focuses attention on the fraction of uplight permitted from lighting fixtures, requiring that most lighting fixtures above a certain lumen output project no light above the horizontal plane. There is no overall limitation on the amount of lighting permitted.

Recent efforts, pioneered in Flagstaff and Coconino County, Arizona in 1989 updates to their lighting codes, and now on the verge of adoption in Tucson, include in addition overall lighting caps expressed in lumens per acre or hectare. Again, the lumens per acre caps must be having an effect from simple comparison of uncapped *vs.* capped lighting practices, but the overall effects have not been measured (but see Luginbuhl 2001).

Though these efforts can decrease uplight *per capita*, sometimes by large factors, the inexorable increase in population near observatory sites means that the sky will continue to brighten even in areas with up-to-date and effectively implemented lighting codes. Though with slow replacement of old non-conforming lighting systems there would seem some hope to actually decrease sky brightness, the limited effect of any replacement is swamped by population growth and increased amounts of lighting *per capita*. Near premier astronomical facilities, a further approach offers an avenue to substantially decrease the impact of lighting at most wavelengths - namely, the specification of low-pressure sodium lighting.

3.   Low-Pressure Sodium Lighting

Low-pressure sodium (LPS) lighting produces light by passing an electrical arc through sodium vapour at low pressure, typically 5×10^{-3} torr. The preponderance of LPS lamp emission is at the resonance doublet of Na I at 589.0 nm and 589.6 nm, making LPS lamps nearly monochromatic.

3.1.   Astronomical Value

Extensive use of LPS lighting in areas near astronomical facilities offers the prospect of concentrating the sky-brightening impacts to the sodium D lines, decreasing the brightening at other wavelengths. The upper two panels of Figure 1 show the emission spectra of the two lamp types most commonly used for outdoor lighting, high-pressure sodium (HPS) and metal halide (MH). Both HPS and MH exhibit complex spectra with continuum and emission line components spread across the entire visible spectrum.

In the broad- and intermediate-band photometric systems indicated in the lower panels of Figure 1, HPS and MH contaminate all passbands, whereas LPS lamps can be seen to contaminate none (Strömgren), one (SDSS) or two (Johnson-Cousins) of the passbands. The implications for observations in these systems is obvious. Recent measures of sky brightness at the Flagstaff Station (see Table 1) show that while the V sky brightness increase is approximately as predicted by Garstang (1989) based on population and non-LPS lighting, in B it is less than one-third the predicted value, verifying the practical effectiveness of the Flagstaff LPS requirements.

For spectroscopic observation, the implications are more severe, since the contaminating spectrum of HPS and especially MH are not only broad-spectrum but also extremely complex functions of wavelength. Accurate flux calibration

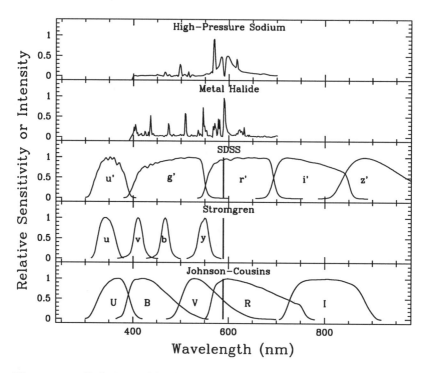

Figure 1. Relation of high-pressure sodium, metal halide and low-pressure sodium emission (vertical bar at 589 nm in the lower three panels) to three photometric filter systems. Only a portion of the emission spectrum of HPS and MH lamps is shown.

against such complex backgrounds is difficult, and slight but common errors will lead to spurious spectral features.

3.2. Issues in LPS Use

The astronomical community has much to gain from increased use of LPS lighting. However, LPS suffers a poor image, particularly within the American lighting industry. This lack of support stems principally from the poor colour rendition and the impression among some that LPS light provides for inferior visibility compared to broad-spectrum sources. There has also been considerable misinformation promulgated about LPS lighting, and the damage done to its image has been substantial. As for many other aspects of outdoor lighting and issues of visibility, safety and utility, more research is needed in some aspects of LPS lighting, such as lamp lifetimes, optimal operating conditions, and system operating costs. But the advantages of LPS are sufficiently well documented to justify its more widespread use in both astronomical and non-astronomical communities. Two of the issues concerning LPS are briefly discussed below.

Colour Perception. Since LPS is a nearly monochromatic source, the eye cannot discriminate colours under pure LPS lighting. But colour perception is not necessary in many outdoor situations, or may be provided by other types of lighting. Many lighting codes distinguish so-called "general illumination" or "Class 2" applications where colour perception is not required. The principal Class 2 examples are roadway and parking lot lighting, and the proportion of outdoor lighting devoted to such uses is large. In Flagstaff, where LPS is strongly encouraged for Class 2 lighting, a sample of five recent development projects shows an average of 82% of outdoor lighting as LPS.

In many cases small amounts of broad-spectrum light will be available either intentionally or accidentally, such as from automobile headlights on roadways. Boynton and Purl (1989) have demonstrated that the addition of as little as 10% white light to otherwise pure LPS illumination can render essentially normal colour perception.

Efficiency and Energy Use. Since the wavelength of the principal LPS emission (589 nm) is so near the eye's peak photopic sensitivity (at 555 nm), LPS has a great natural efficiency advantage. The advantage is further extended by the maintenance of this output during the operating life of the lamps. Whereas MH and HPS lamps dim considerably during use (to about 64% and 73% of initial outputs, respectively), LPS loses little or none of its initial output during its operating life. Some of this advantage is lost in practice, through slightly higher ballast losses and slightly lower efficiencies in some applications, but in most roadway and large area illumination the LPS efficiency advantage remains. Table 2 shows maintained effective lumens per watt for MH, HPS and LPS systems optimized for roadway and parking lot illumination. These figures were calculated by the author using software provided by lighting manufacturers, and use actual luminaires and other factors as summarized in the caption. For the narrowest roadway considered, LPS is 7% more efficient than HPS, and 61% more efficient than MH. The figures for parking lots are actually upper limits, assuming no edge losses, but here the efficiency of LPS towers over HPS and MH systems by 48% and 119% respectively.

In recent years the lighting profession has become aware of the scotopic or "dark-adapted" response of the eye. At low light levels the eye's peak sensitivity shifts toward the blue, with completely scotopic response peaking at about 510 nm. If this sensitivity is the correct one to apply to outdoor lighting conditions, then the effective output of LPS is lowered because the emission is considerably removed from the peak sensitivity. The best work currently available to address the question of which sensitivity is most appropriate under typical outdoor lighting conditions (Adrian 1997) indicates that this effect is relatively minor, but much remains to be learned concerning appropriate metrics of visibility, including peripheral vision questions. For now the indication is that LPS maintains its advantage over MH and HPS under even the lowest levels of lighting recommended by the Illuminating Engineering Society of North America.

Table 2.   Application Efficiencies[a](in effective maintained lumens per watt) for metal halide (MH), high-pressure Sodium (HPS), and low-pressure sodium (LPS) lighting systems evaluated by the author.

Lamp Type	Application Efficiency (lm/W)		
	38' Road	102' Road	Parking
Metal Halide	18	27	31
High Pressure Sodium	27	41	46
Low Pressure Sodium	29	50	68

[a]Figures include circuit losses, light loss factors (MH: 0.54; HPS: 0.62; LPS: 0.95), and application Coefficients of Utilization (MH&HPS: 0.45/0.67/0.75, LPS: 0.31/0.54/0.73 for 38'/102'/PL applications) for comparable *maintained* illumination level (using 175W MH, 150W HPS and 90W LPS). Luminaires used were MH&HPS: GE M250-A2/GE M250-A2/GE Dimension, LPS: Gardco LPSA/LPSA/LPSA.

4. Conclusions

With continued growth in areas near astronomical facilities, prospects for the long-term preservation of dark skies at these sites are dim without involvement of the astronomical community in the effort to enact and enforce strict lighting ordinances. Though lighting codes with shielding standards and lumens per hectacre limits can subtantially slow the deterioration , they are unlikely to stop it. Use of low-pressure sodium (LPS) lighting offers a realistic prospect of nearly dark skies at most wavelengths, even in proximity to substantial population centers. LPS can provide effective lighting for the lowest energy expenditure of all available lighting sources, and it should be more widely used. In Flagstaff, Arizona, lighting ordinances requiring extensive use of LPS have been in effect for over ten years. While recent measures of the sky brightness at V indicate brightening in accord with predictions based on population and non-LPS lighting types, the increase at B is less than one third the predicted value, indicating that such ordinances can be effective.

References

Adrian, W. 1997, paper given at the International Lighting Conference, Durban, South Africa.

Boynton, R. M. and Purl, K. F. 1989, Lighting Res. Technol., 21, 23

Caldwell, N., *et al.* 1999, personal communication.

Garstang, R. H. 1989, Ann. Rev. Astron. Astrophys., 27, 19

Luginbuhl, C. B. 2001, these proceedings

Pilachowski, C. A., *et al.* 1989, PASP, 101, 707

Preserving the Astronomical Sky
IAU Symposium, Vol. 196, 2001
R. J. Cohen and W. T. Sullivan, III, eds.

Methods and Results of Estimating Light Pollution in the Flemish Region of Belgium

J. Vandewalle, Dirk Knapen and Tim Polfliet

Bond Beter Leefmilieu Vlaanderen vzw, Federation of the Flemish environmental movement, Tweekerkenstraat 47, B 1000 Brussels, Belgium
e-mail: johan.vandewalle@bblv.be

H. Dejonghe

Vakgroep wiskundige natuurkunde en sterrenkunde, Universiteit Gent, B 9000 Gent, Belgium
e-mail: Herwig.Dejonghe@rug.ac.be

Abstract. In the Flemish region of Belgium, yearly environmental reports (MIRA) are made on the status and evolution of the environment and its policy. Since 1996 light pollution and light impediment are among the topics considered. By using indicators which can be compared in the different reports it is possible to estimate the evolution of light pollution and light impediment. Suggested changes in lighting policies could save the Flemish region as much as 18 million EURO per year.

1. Introduction

The first Report on the environment and nature in the Flemish region (MIRA-1) was published in summer 1994. It provided a comprehensive but above all qualitative survey of the state of the environment in the Flemish region. The second Report (MIRA-2) was published in autumn 1996. In line with the development of similar reports in other countries, the Vlaamse Milieu Maatschappij (VMM) endeavoured to survey the extent of environmental pollution in quantitative terms. The following elements were considered: (1) social and primarily economic developments which place pressure on the environment; (2) the resulting impact on the quality of the environment; and (3) the effect of the quality of the environment on humans, nature and the economy. Since MIRA-2, light nuisance has been one of the topics.

MIRA-T 1998 and MIRA-T 1999 were the first thematic or T reports. These reports present a concise yet comprehensive picture of the quality of the Flemish environment in 22 chapters. Moreover, the aspiration of MIRA-2 to provide as far as possible a concise and quantitative account of the themes is further developed. To a large degree this approach has been applied to the theme of light nuisance. However, some of the key variables are as yet incomplete, because of lack of data or because they are still relatively recent and for the time being can serve only as a set of observations (Verbruggen 1998).

This paper is mainly based on the light nuisance chapter in the aforementioned MIRA-T reports (Claeys et al. 1998; Vandewalle et al. 1999). More detailed information can be found in these reports and in the MIRA background report available from the World Wide Web (Vandewalle 1999).

2. Definitions

Light pollution is the increased luminosity at nighttime created by excessive use of artificial light (IDA 1997).

Light nuisance is the disruption due to artificial light experienced by humans when executing evening and night-time activities, causing either discomfort or direct blinding. Animals also suffer from light nuisance that can fragment and otherwise affect their habitats. Plants are light sensitive and therefore can change their day-night or seasonal rhythm as a result of light nuisance (IDA 1997).

The link between emissions (illumination and lighting) and immissions (pollution) is not a direct one and is influenced by the direction of the light source, the degree of cover available, reflection of the light on surfaces and the diffusion of light under certain atmospheric conditions. Since there are no exact data available on all light sources, we use extrapolation and estimations to roughly map light pollution and light nuisance in a quantitative way.

3. Pressure: Social Developments

A new key variable to measure the impact of social activities on light pollution and light nuisance is introduced in MIRA-T 1999. For the target group *transport* this variable gives the fraction of lighted road length, expressed by the formula:

$$\frac{\sum d_i * r_i}{d_{tot}}$$

with d_i the length of public roads with lighting type i, r_i the fraction of the night that roads of lighting type i are lighted and d_{tot} the total public road length. The key variable varies between 0 (all roads permanently not lighted) and 1 (all roads permanently lighted). There are different types of road lighting: permanently (from sunset till dawn), evening (from sunset till 11 pm, midnight or 1 am), evening and morning (from sunset till 11 pm, midnight or 1 am and from 5, 5.30, 6 or 6.30 am till dawn). Furthermore road lighting can be uniform (e.g. one lamp every 40 m), only occasional (e.g. only at crossings or dangerous curves ...), or none. The data came from a survey of local and regional authorities as well as energy-distributors.

The total road length in the Flemish region is 59,071 km, of which 7,552 km is unlighted. 47,727 km of local and regional roads with uniform lighting are on average lighted for 94% of the night. We estimate that 2,959 km with occasional lighting are 10% lighted during the entire night. Some highways (190 km) are permanently lighted, some (100 km) are not lighted at all, while for the rest (543 km) lights go out from 0.30 am till 5.30 am. Given these data, on average 45,666 km of the total 59,071 km was lighted in 1997. This results in a key variable value for transport of 0.773. Because of the small sample size (we received data about only 7.4% of the roads) the extrapolated data should be interpreted carefully.

For the target group *agriculture* we measure this key variable by taking the fraction of greenhouse surface that is lighted, corrected by the fraction of the night that the lights are on.

In the Flemish region there are 193.8 ha (4,786 acres) of greenhouses, of which 8.2 ha (202 acres) have assimilation lighting. The total number of nightly hours per annum is 4,380, of which 1,482 (33.8%) are lighted. The key variable value for agriculture is therefore 0.014.

There is no available information about an equavalent key variable for target groups other than transport and agriculture.

Both key variables were calculated for the first time in MIRA-T 1999, based on 1997 data. By measuring the same key variables in the next MIRA reports we should be able to evaluate changes in lighting.

4. Pressure: Use and Emissions

For the target groups *transport* and *agriculture* we use the number of lamps and the installed power (both lamp-power and used power) as a key variable for the amount of outdoor lighting. The data on public lighting (e.g. street lighting and accent lighting of monuments) also came from the previously mentioned survey. We received data from 63% of the local communities and extrapolated these to total road length in the Flemish region. We also received complete data on regional roads which are lighted by the Flemish region. We collected data on the total number of installed lamps, the total power of these lamps ("lamp power") and the total energy-use of the installed fixtures ("net power"). For agriculture we calculated the number of lamps and power-use for assimilation lighting in greenhouses. The results are shown in Table 1.

Table 1. Number of lamps, lamp power and net power in different lamp types in public lighting and agriculture (Flemish region, 1997). Source: Bond Beter Leefmilieu 1999; Knapen 1997.

	Extrapolation local authorities			Flemish Region (1997)			Total		
	Number of lamps	Lamp-power (kW)	Net power (kW)	Number of Lamps	Lamp-power (kW)	Net power (kW)	Number of lamps	Lamp-power (kW)	Net power (kW)
Transport	1,067,190	91,685	120,146	198,000	25,413	35,035	1,265,190	117,098	155,181
HgHP	260,832	36,465	45,526	2,500	881	1,100	263,332	37,346	46,626
HgLP	194,967	3,225	5,030	26,000	1,170	1,824	220,967	4,395	6,854
NaHP	397,757	38,474	51,033	11,500	3,162	4,194	409,257	41,636	55,228
NaLP	203,370	12,526	17,311	158,000	20,200	27,916	361,370	32,726	45,227
Other	10,264	994	1,244	0	0	0	10,264	994	1,245
Agriculture							68,250	27,300	36,210

Not all installed lamps are switched on during the entire night. In Figure 1 we show how the total installed lamp power is used during the night. 11.7% of the public lighting is turned off at night (14.4% of the installed power), and 7.7% of the lamps (9.9% of the installed power) are switched on again during the morning. The average number of lighting hours was 4,112 in 1997, from a total of 4,380 nightly hours. This is an increase of 3.8% compared to 1996.

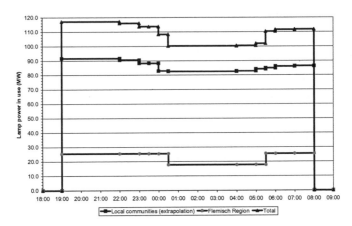

Figure 1.   Use of public lighting during night. Lamp power (MW) switched on in the Flemish region in 1997. Source: Bond Beter Leefmilieu 1999.

Table 2.   Number of lamps and power in different applications in 1997. Source: Bond Beter Leefmilieu 1999.

		Flemish region	
	lamps (n)	Lamp power (MW)	Net power (MW)
Transport			
street lighting	1,208,581	115.5	152.4
accent lighting	56,609	1.6	2.7
total	1,265,190	117.1	155.1
Agriculture			
assimilation lighting	68,250	27.3	36.2

In 1997, 4.5% of the lamps (1.4% of the lamp power and 1.7% of net power) in public lighting were used for accent lighting. In agriculture, 27.3 MW of lamp power (36.2 MW net power) is installed for assimilation lighting in greenhouses. Data about photoperiodic lighting are not available.

The amount of upward light flux is proportional to the power of the upward lighting at a certain moment and also depends on other factors such as weather and air pollution. We define a key variable ULF for the amount of upward lighting flux with the formula:

$$ULF = 0.58KV + 0.224SV + 0.09AV + 0.30RV \, (in \, MW).$$

with KV = net power of installed Accent Lighting,
SV = net power of installed Street Lighting,
AV = net power of installed Assimilation Lighting,
RV = net power of installed Rest Lighting (all other outdoor lighting).

The upward light power of Accent Lighting is estimated as 40% of the installed power directly and 30% of the rest after reflection, giving 58% of the installed power. The same applies for Street Lighting with 3% of the installed power directly and 20% of the rest after reflection, giving 22.4% of the installed power. For Assimilation Lighting in greenhouses we estimate that 5% of the installed power is lost through the greenhouse roof (25% has no roof cover, roofs form 84% of the greenhouse surface and the reflection is about 25%) and 4% is lost through the greenhouse sides (95% has no side cover, sides form 16% of the greenhouse surface and the reflection is about 25%). For the Rest Lighting we estimate that 30% of the installed power is spilt towards the sky (both directly and after reflection). The key variable is shown in Table 3.

Table 3.   Upward light flux from the Flemish region 1996 - 1997. Source: Bond Beter Leefmilieu 1999.

Target group	Upward Light Flux	
	1996 (MW)	1997 (MW)
Transport		
street lighting	33.8	32.1 - 34.1[a]
accent lighting	0.8	1.5 - 1.6[a]
Agriculture		
assimilation lighting		3.1 - 3.4[b]
Rest	18.9 - 61.5[c]	19.2 - 60.6[c]
Total	53.5 - 96.1	55.9 - 99.7

[a] dispersion due to different data sets (lowest = identical to 1996, highest data from inquiry)
[b] dispersion due to uncertainty about greenhouse area
[c] dispersion due to wide margins in estimation of proportion of outdoor lighting in total energy use for the different target groups.

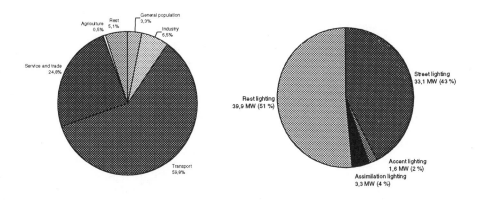

Figure 2.   Contribution of target groups to *Left*: the amount of outdoor lighting and *Right*: the upward lighting, for the Flemish region, 1997. Source: Bond Beter Leefmilieu 1999.

The contribution of the different target groups in the amount of outdoor lighting is estimated by their energy use. Furthermore, we calculated the contribution of the target groups to the upward lighting flux. Results are shown in Figure 2.

We conclude that target groups (summed in the rest term) other than street lighting form an important contribution to the total upward lighting flux. More detailed information is needed to improve our rough estimates and to determine the amount of the different target groups.

Assimilation Lighting is responsible for twice as much upward lighting flux compared with Accent Lighting. Despite the greater fractional loss of light in Accent Lighting (2.5 times that lost in Street Lighting with the same power use) it accounts overall only a few percent of the upward lighting flux compared to Street Lighting.

5.   State: Immissions

Satellites of the Defense Meteorological Satellite Program (DMSP) of the US Air Force circle round the Earth at about 830 km height. They have several instruments on board such as the Operational Linescan System (OLS) The DMSP-OSL system is sensitive enough to detect light sources of urban, industrial and other anthropogenic origin, even at full moon.

We used the technique developed by Imhoff et al. (1997) to make a light pollution map of the Flemish region. Figure 3 was created out of 231 images from October 1994 till March 1995 from the F-10 and F-12 satellites taken at 9.30 pm local time. We calculated the proportion of times a pixel is recognised as a light-emitting source out of the cloud free images. By doing this we only observe stable light sources and not temporary sources such as fires or lightning. Because each pixel shows an area of 2.7 km × 2.7 km, sources such as roads aren't visible. Urban agglomerations on the other hand are clearly visible, as are the industrial complexes near the major waterways and the main greenhouse area. Information from these images is relevant for astronomers who search for suitable locations to observe the skies.

6.   Proposed Measures

We calculated the impact and costs of three measures to restrict light pollution. The first one is replacing all mercury vapour lamps in public lighting with high-pressure sodium vapour lamps. The second one is to change both the lamps and the fixtures and the third is to put all public lighting out between midnight and 6 am.

Changing all mercury vapour lamps would save an energy use of about 103 GWh a year or 17% of the yearly energy use for public lighting. This measure does save a lot of energy, but does not reduce light pollution.

If, on the other hand, one changed not only the lamps but also the old (and bad) fixtures of these same lamps, one would save another 23 GWh a year or 3.9% of the yearly energy use in public lighting and one would significantly reduce light pollution (no exact figure could be calculated). The changing of

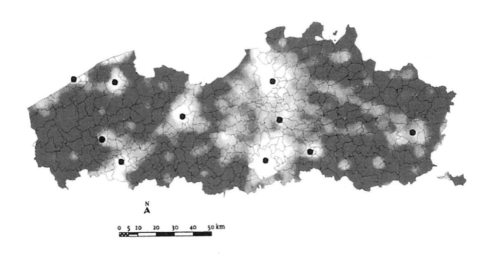

Figure 3.   Light pollution in the Flemish region as seen from the sky
(October 1994 till March 1995). Source: VITO 1998, NOAA/NGDC
1994-1995.

fixtures could also be extended to current old (or bad) fixtures that already use
sodium vapour lamps. We cannot calculate the possible savings of this extended
measure because we don't know how many of the current fixtures with sodium
vapour lamps are bad.

Extinguishing all public lighting between midnight and 6 am gives a saving
potential of 296 GWh a year or 49% of the yearly energy use for public lighting. If
authorities wish to consider extinguishing all lights they should take into account
the positive social effects of continuous public lighting such as ensuring (traffic)
safety. It could be preferable to decrease the public lighting selectively (e.g.
50%), so that crossings with high nightly traffic density, dangerous curves, etc.
would remain lit. This option implies extra costs such as in different scheduling,
distance control, etc. One should balance the pros and cons against each other.

Taking the three measures together gives a potential saving of 362 GWh
per year, or 60% of the present yearly energy use in public lighting.

Despite the low energy price local authorities pay (0.034 EURO per kWh
between 10 pm and 6 am and 0.096 EURO per kWh between 6 am and 10 pm) the
yearly financial saving for all authorities (local and Flemish) is 18×10^6 EURO.
To achieve this saving, 484,299 fixtures with mercury vapour lamps would need
to be changed. The installation of a new (efficient) fixture with sodium vapour
lamps costs 211 EURO. The total cost is therefore 102×10^9 EURO, and in less
than 6 years the costs for this investment would be recovered.

References

Claeys, S., Vandewalle, J. & Dejonghe, H. 1998, Lichthinder. In: Verbruggen, A. (ed.) (1998) Milieu- en natuurrapport Vlaanderen: thema's MIRA-T 1998. Eerste druk, D/1998/5779/32, SIBN 90-5350-726-4, NUGI 825/693, Vlaamse Milieumaatschappij en Garant Uitgevers nv.

International Dark-Sky Association 1997, Glossary of Basic Terms and Definitions. Information Sheet 9, April 1997.

Imhoff, M.L. et al. 1997, A Technique for Using Composite DMSP/OLS "City Lights" Satellite Data to Map Urban Area. Remote Sens. Environ. 61:361-370.

Knapen, J. 1997, Written announcement. BVV (Belgische Vereniging voor Verlichtingskunde), 22/12/1997.

Vandewalle, J. 1999, Lichthinder. In: Vandeweerd, V. (ed.) (1999) Milieu- en natuurrapport: Achtergronddocument.
http://www.vmm.be/milieu/mil_mira_ag.html

Vandewalle, J., Knapen, D., Polfliet, T. & Dejonghe, H 1999. Lichthinder. In: Vandeweerd, V. (ed.) (1999) Milieu- en natuurrapport Vlaanderen: thema's MIRA-T 1999. Eerste druk, D/1999/5779/40, ISBN 90-5350-870-8, NUGI 825/693, Vlaamse Milieumaatschappij en Garant Uitgevers nv.

Vandeweerd, V. (ed.) 1999, Report on the environment and nature in Flanders: themes MIRA-T 1999. First Edition, D/1999/5779/40, ISBN 90-5350-870-8, NUGI 825/693, Vlaamse Milieumaatschappij and Garant Uitgevers nv.

Verbruggen, A. (ed.) 1998, Report on the environment and nature in Flanders: themes MIRA-T 1998 summary. First Edition, D/1998/5779/32, SIBN 90-5350-726-4 NUGI 825/693, Vlaamse Milieumaatschappij and Garant Uitgevers nv.

Preserving the Astronomical Sky
IAU Symposium, Vol. 196, 2001
R. J. Cohen and W. T. Sullivan, III, eds.

The Artificial Sky Brightness in Europe Derived from DMSP Satellite Data.

P. Cinzano[1] and F. Falchi

Dipartimento di Astronomia, Università di Padova, vicolo dell'Osservatorio 5, I-35122 Padova, Italy

C. D. Elvidge and K. E. Baugh

Solar-Terrestrial Physics Division, NOAA National Geophysical Data Center, 3100 Marine Street, Boulder CO 80303

Abstract. We present maps of the artificial sky brightness in Europe in V band with a resolution of ~ 1 km. The aim is to understand the state of night sky pollution in Europe, to quantify the present situation and to allow future monitoring of trends. For each terrestrial site the artificial sky brightness in a given direction on the sky is obtained by integrating the contributions from each surface area in the surroundings, using detailed models of the propagation in the atmosphere of the upward light flux emitted by the area. The top-of-atmosphere light flux is measured by the Operational Linescan System of the Defence Meteorological Satellite Program (DMSP) satellites. The modelling technique, which was introduced and developed by Garstang, takes into account the extinction along light paths, double scattering of light from atmospheric molecules and aerosols, and Earth curvature. Use of this technique allows us to assess the aerosol content of the atmosphere.

1. Introduction

An effective battle against light pollution requires knowledge of the situation of the night sky in large territories, recognition of the worst affected areas, determination of the growth trends and the identification of the more polluting cities. Therefore a method of mapping the artificial sky brightness over large territories is required. This is also useful in order to recognize less polluted areas and potential astronomical sites.

The DMSP satellites provide direct information on the upward light emission from almost all countries around the World (Sullivan 1989; Elvidge et al. 1997a, 1997b, 1997c, 1999; Isobe & Hamamura 1998). We present the outlines of a method to map the artificial sky brightness in large territories using the upward flux measured directly in DMSP satellite night-time images. The method avoids errors which arise when population data are used to estimate upward

---

[1]email: cinzano@pd.astro.it

flux. The effects of light pollution are computed using a detailed model of light pollution propagation in the atmosphere. Details are extensively discussed by Cinzano et al. (2000), where maps for B band are also presented.

2.　Satellite Data

U.S. Air Force Defense Meteorological Satellite Program (DMSP) satellites are in low altitude (830 km) sun/synchronous polar orbits with an orbital period of 101 minutes. With 14 orbits per day they generate a global nightime and daytime coverage of the Earth every 24 hours. The Operational Linescan System (OLS) is an oscillating scan radiometer with low-light visible and thermal infrared imaging capabilities. At night the instrument for visible imagery is a Photo Multiplier Tube (PMT) sensitive to radiation from 485 nm to 765 nm (FWHM), with the highest sensitivity at 550-650 nm where the most widely used lamps for external night-time lighting have their strongest emission. Most of the data received by the National Oceanic and Atmospheric Administration (NOAA) National Geophysics Data Center (NGDC), which has archived DMSP data since 1992, are smoothed by on-board averaging of 5×5 adjacent detector pixels and have a nominal space resolution of 2.8 km.

In three observational runs made during the darkest portions of the lunar cycles during March 1996 and January and February of 1997, NGDC acquired OLS data at reduced gain settings in order to avoid saturation produced in normal-gain operations in a large number of pixels inside cities due to the high OLS-PMT sensivity. Three different gain settings were used on alternating nights to overcome the dynamic range limitations of the OLS. With these data a cloud-free radiance calibrated composite image of the Earth has been obtained (Elvidge et al. 1999). The temporal compositing makes it possible to remove noise and lights from ephemeral events such as fire and lightning. The main steps in generating the night-time lights product are: 1) establishment of a reference grid with finer spatial resolution than the input imagery; 2) identification of the cloud-free section of each orbit based on OLS thermal band data; 3) identification of lights and removal of noise and solar glare; 4) projection of the lights from cloud-free areas from each orbit into the reference grid, with calibration to radiance units; 5) tallying of the total number of light detections in each grid cell and calculation of the average radiance value; 6) filtering images based on frequency of detection to remove ephemeral events. The final image was transformed in a latitude/longitude projection with $30'' \times 30''$ pixel size. The map of Europe was obtained with a portion of 4800×4800 pixels of this final image, starting at longitude 10°30′ west and latitude 72°north.

3.　Mapping Technique

Scattering from atmospheric particles and molecules spreads the light emitted upward by the sources. If $e(x, y)$ is the upward emission per unit area in (x, y), the total artificial sky brightness in a given direction of the sky in a site in (x', y') is:

$$b(x', y') = \int \int e(x,y) f((x,y),(x',y')) \, dx \, dy \qquad (1)$$

where $f((x, y), (x', y'))$ gives the artificial sky brightness per unit of upward light emission produced by unit area in (x, y) in the site in (x', y'). The light pollution propagation function f depends in general on the geometrical disposition (altitude of the site and the area, and their mutual distance), on the atmospheric distribution of molecules and aerosols and their optical characteristics in the choosen photometric band, on the shape of the emission function of the source and on the direction of the sky observed. In some works this function has been approximated with a variety of semi-empirical propagation laws: the Treanor Law (Treanor 1973; Falchi 1998; Falchi & Cinzano 2000), the Walker Law (Walker 1973), the Berry Law (Berry 1976) and the Garstang Law (Garstang 1991b). However, all of them ignore the effects of Earth curvature, that cannot be neglected in accurate mapping of large and non-isolated territories.

We obtained the propagation function $f((x, y), (x', y'))$ for each pair of points (x, y) and (x', y') using detailed models for the light propagation in the atmosphere based on the modelling technique introduced and developed by Garstang (1986, 1987, 1988, 1989a, 1989b, 1991a, 1991b, 1991c, 2000) and also applied by Cinzano (2000a,b,c). The models assume Rayleigh scattering by molecules and Mie scattering by aerosols and take into account extinction along the light path and Earth curvature. They allow association of the predictions with well-defined parameters related to the aerosol content, so the atmospheric conditions to which predictions refer can be well known. The resolution of the maps depends on the results of integrating over a large zone: it is better than the resolution of the original images and is generally of the order of the distance between two pixel centers (less than 1 km). However where sky brightness is dominated by contributions of the nearest land areas, effects of the resolution of the original image could become relevant.

We assumed the atmosphere to be in hydrostatic equilibrium under gravity, with an exponential decrease of number density with height for the atmospheric haze aerosols. Measurements show that for the first 10 km this is a reasonable approximation. We are interested in average or typical atmospheric conditions, not particular conditions of a given night, so a detailed modelling of the local aerosol distribution at a given night is beyond the scope of this work. We neglected the presence of sporadic denser aerosol layers at various heights or at ground level, the effects of the Ozone layer and the presence of volcanic dust. We take into account changes in aerosol content following Garstang (1986), introducing a parameter K which measures the relative importance of aerosol and molecules for scattering light. The adopted modelling technique allows an assessment of the atmospheric conditions for which a map is computed, giving observable quantities like the vertical extinction at sea level in magnitudes. More detailed atmospheric models could be used whenever available.

The angular scattering function for atmospheric haze aerosols can be measured easily with a number of well known remote-sensing techniques. Being interested in a typical average function, we adopted the same function used by Garstang (1991a) and we neglected geographical gradients. The normalized emission function of each area gives the relative upward flux per unit solid angle in each direction. It is the sum of the direct emission from fixtures and the reflected emission from lighted surfaces, normalized to its integral and is not known. In this paper we assumed that all land areas have the same average normalized emission function. This is equivalent to assuming that lighting habits are similar on average in each land area and that differences from the average are randomly distributed in the territory. We choose to assume this function and check its consistency with satellite measurements, rather than directly determine it from satellite measurements because at very low elevation angles the spread is much too large to constrain adequately the function shape. We adopted for the average normalized emission function the normalized city emission function from Garstang (1986).

4.   Results

Figures 1-6 show the maps of the artificial sky brightness in Europe at sea level in V band. The maps have been computed for a clean atmosphere with an aerosol clarity $K = 1$, corresponding to a vertical extinction of $\Delta m = 0.33$ mag in V band, horizontal visibility $\Delta x = 26$ km and optical depth $\tau = 0.36$. Gray levels from black to white correspond to ratios between the artificial sky brightness and the natural sky brightness of: <0.11, 0.11-0.33, 0.33-1.00, 1-3, 3-9 and >9. We limited our computations to zenith sky brightness even if our method allows determination of brightness in other directions. This would be useful to predict visibility in large territories of particular astronomical phenomena. A complete mapping of the artificial brightness of the sky of a site, like Cinzano (2000a), but using satellite data instead of population data, is possible (Cinzano & Elvidge 2000, in prep.).

We are more interested in understanding and comparing light pollution distributions than in predicting the effective sky brightness for observational purposes, so we computed everywhere the artificial sky brightness at sea level, in order to avoid introducing altitude effects in our maps. We will take account of altitudes in a forthcoming paper devoted to mapping the limiting magnitude and naked-eye star visibility, which requires the computation of star-light extinction and natural sky brightness for the altitude of each land area. We neglected the presence of mountains which might shield the light emitted from the sources from a fraction of the atmospheric particles along the line-of-sight of the observer. Given the vertical extent of the atmosphere with respect to the height of the mountains, the shielding is non-negligible only when the source is very near the mountain and both are quite far from the site (Garstang 1989, see also Cinzano 2000a,b). Earth curvature emphasizes this behaviour.

We calibrated the maps on the basis of both (i) accurate measurements of sky brightness together with extinction from the earth-surface and (ii) pre-flight radiance calibration of OLS-PMT. Map calibration based on pre-flight irradiance calibration of OLS PMT requires knowledge, for each land area, of (a) the average vertical extinction Δm during satellite observations and (b) the relation between the radiance in the choosen photometrical band and the radiance measured in the PMT spectral sensitivity range, which depends on the emission spectra. The result of this calibration is well inside the errors of the Earth-based calibration in spite of the large uncertainties both in the extinction and in the average emission spectra. As soon as a large number of sky brightness measurements are available, better calibration will be possible.

We are preparing a World Atlas (Cinzano et al., in prep.).

Figure 1. Artificial sky brightness at sea level in Europe in V band for aerosol content parameter $K = 1$.

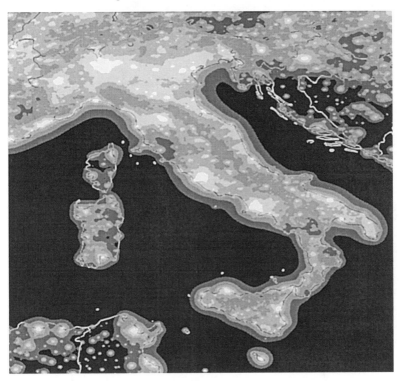

Figure 2. Artificial sky brightness at sea level in Italy in V band for aerosol content parameter $K = 1$.

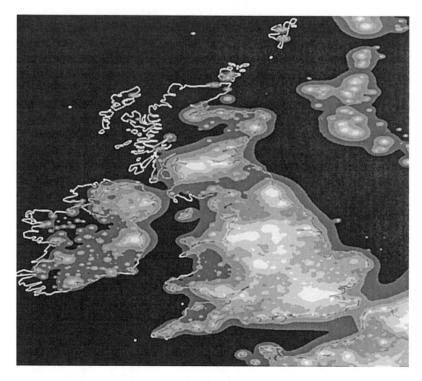

Figure 3.   Artificial sky brightness at sea level in Great Britain and Ireland in V band for aerosol content parameter $K = 1$.

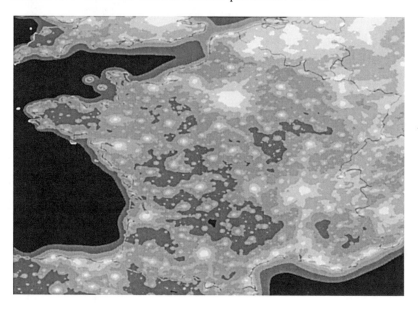

Figure 4.   Artificial sky brightness at sea level in France and Belgium in V band for aerosol content parameter $K = 1$.

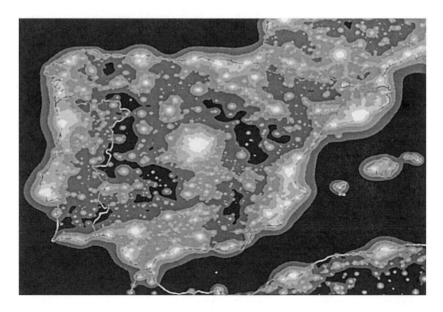

Figure 5.   Artificial sky brightness at sea level in Spain and Portugal in V band for aerosol content parameter $K=1$.

Figure 6.   Artificial sky brightness at sea level in Central Europe in V band for aerosol content parameter $K=1$.

Acknowledgments. We are indebted to Roy Garstang of JILA-University of Colorado for his friendly kindness in reading and refereeing this paper, for his helpful suggestions and for interesting discussions.

References

Berry, R. 1976, J. Royal Astron. Soc. Canada, 70, 97-115

Cinzano, P. 2000a, in Measuring and Modelling Light Pollution, ed. P. Cinzano, Mem. Soc. Astron. Ital., 71, 1, 113-130

Cinzano, P. 2000b, in Measuring and Modelling Light Pollution, ed. P. Cinzano, Mem. Soc. Astron. Ital., 71, 1, 93-112

Cinzano, P. 2000c, in Measuring and Modelling Light Pollution, ed. P. Cinzano, Mem. Soc. Astron. Ital., 71, 1, 239-250

Cinzano, P., Falchi, F., Elvidge, C.D., Baugh, K.E. 2000, MNRAS, 318, 641-657

Elvidge, C.D., Baugh, K.E., Kihn, E.A., Kroehl, H.W., Davis, E.R. 1997a, Photogrammetric Engineering and Remote Sensing, 63, 727-734

Elvidge, C.D., Baugh, K.E., Kihn, E.A., Kroehl, H.W., Davis, E.R., Davis, C. 1997b, Int. J. of Remote Sensing, 18, 1373-1379

Elvidge, C.D., Baugh, K.E., Hobson, V.H., Kihn, E.A., Kroehl, H.W., Davis, E.R., Cocero, D. 1997c, Global Change Biology, 3, 387-395

Elvidge, C.D., Baugh, K.E., Dietz, J.B., Bland, T., Sutton, P.C., Kroehl, H.W. 1999, Remote Sensing of Environment, 68, 77-88

Falchi, F. 1998, Thesis, University of Milan

Falchi, F., Cinzano, P. 2000, in Measuring and Modelling Light Pollution, ed. P. Cinzano, Mem. Soc. Astron. Ita., 71, 1, 139-152

Garstang, R.H. 1986, Publ. Astron. Soc. Pacific, 98, 364-375

Garstang, R.H. 1987, in Identification, optimization and protection of optical observatory sites, eds. R.L. Millis, O.G. Franz, H.D. Ables & C.C. Dahn (Flagstaff: Lowell Observatory), 199-202

Garstang, R.H. 1988, The Observatory, 108, 159-161

Garstang, R.H. 1989a, Publ. Astron. Soc. Pacific, 101, 306-329

Garstang, R.H. 1989b, Ann. Rev. Astron. Astrophys., 27, 19-40

Garstang, R.H. 1991a, Publ. Astron. Soc. Pacific, 103, 1109-1116

Garstang, R.H. 1991b, in Light Pollution, Radio Interference and Space Debris, IAU Coll. 112, ed. D.L. Crawford, Astron. Soc. of Pacific Conference Series 17, 56-69

Garstang, R.H. 1991c, The Observatory, 111, 239-243

Garstang, R.H. 2000, in Measuring and Modelling Light Pollution, ed. P. Cinzano, Mem. Soc. Astron. Ital., 71, 1, 71-82

Isobe S. & Hamamura, S. 1998, in Preserving the Astronomical Windows, IAU JD5, ed. S. Isobe, Astron. Soc. of Pacific Conference Series 139, 191-199

Sullivan, W.T., 1989, Int. J. of Remote Sensing, 10, 1-5

Treanor, P.J.S.J. 1973, The Observatory, 93, 117-120

Walker, M.F. 1973, Publ. Astron. Soc. Pacific, 85, 508-519

Preserving the Astronomical Sky
IAU Symposium, Vol. 196, 2001
R. J. Cohen and W. T. Sullivan, III, eds.

Using DMSP Night-Time Imagery to Evaluate Lighting Practice in the American Southwest

Christian B. Luginbuhl

United States Naval Observatory, Flagstaff Station, PO Box 1149, Flagstaff AZ 86002 USA

Abstract. The U.S. Defense Meteorological Satellites provide an opportunity to measure the uplight produced by artificial lighting on the ground. In this study DMSP data are used to measure the integrated at-detector radiance of a number of communities in the American Southwest in an attempt to evaluate the effectiveness of outdoor lighting codes. Use of DMSP data in this manner is complicated by many factors, and some of these are briefly discussed.

1. Introduction

The U.S. Air Force Defense Meteorological Satellite Program (DMSP) satellites orbit Earth in sun-synchronous, low-altitude polar orbits. One orbit is oriented such that the satellite circles the globe approximately over the sunrise/sunset terminator, the other passes over near noon and midnight. The principle purpose of the program is to monitor cloud conditions, but the night-time observations of city lights are what have the attention of the astronomical community.

Other workers have begun using DMSP data to measure uplight produced by cities, including Isobe & Hamamura (1998), Isobe (1998), Falchi & Cinzano (1999) and Cinzano *et al.* (2000). The present study evaluates the possibility of using DMSP data to measure the overall success of light pollution control efforts. Tucson and Flagstaff, Arizona, are two cities in the American Southwest that have a substantial history of light control efforts through outdoor lighting codes. Are these codes working?

2. This Study

The image used in this study is a cloud-free composite of the United States built from many DMSP midnight passes during the dark of the lunar cycle in March 1996 and January-February 1997 (Elvidge *et al.* 1999).

Brightnesses were measured from this image for a sample of towns and cities in Arizona, Utah, Nevada and New Mexico, covering a range in population from under 2000 to almost 2.5 million. The radiance values were summed within rectangular regions around each municipality.

Population figures were taken from the U.S. Census Bureau website, where estimated figures for July 1996 (released in June 1999) are listed. These figures are shown in Table 1 and Figure 1.

Table 1.   Radiance (10^{-10} watts/cm^2/sr/μm) and 1996 Population for Southwestern U.S. Cities

Fig. 1	City	State	Population	Radiance
CaV	Camp Verde	AZ	7552	3616
ChV	Chino Valley	AZ	6588	1714
Cot	Cottonwood	AZ	6937	9941
Dou	Douglas	AZ	15015	28288
Flg	Flagstaff	AZ	55094	31110
GiB	Gila Bend	AZ	1695	10447
Hol	Holbrook	AZ	5398	5951
Kin	Kingman	AZ	17270	29998
LHC	Lake Havasu City	AZ	39503	18912
Phx	Phoenix metro[a]	AZ	2427230	1666182
Pre	Prescott	AZ	49760	29693
Sed	Sedona	AZ	9109	5722
SiV	Sierra Vista	AZ	37434	28307
Tuc	Tucson [b]	AZ	472305	396799
TuM	Tucson metro [c]	AZ	729479	396799
Wic	Wickenburg	AZ	5312	4020
Wcx	Willcox	AZ	3533	4005
Wil	Williams	AZ	2706	3588
Win	Winslow	AZ	10420	10257
Bly	Blythe	CA	12982	14855
LVN	Las Vegas[d]	NV	577904	1086814
Mes	Mesquite	NV	6200	22970
Alb	Albuquerque[e]	NM	425526	438288
LaC	Las Cruces	NM	74779	55252
LVM	Las Vegas	NM	16437	18365
Ros	Roswell	NM	47559	45081
StF	Santa Fe	NM	66522	67701
StG	St George	UT	42763	43521

[a]Includes Apache Jct, Avondale, Chandler, El Mirage, Fountain Hills, Gilbert, Glendale, Goodyear, Guadalupe, Litchfield Park, Mesa, Paradise Valley, Peoria, Scottsdale, Surprise, Tempe, Tolleson, Youngtown.

[b]Includes Oro Valley, South Tucson.

[c]Includes 95% of Pima County population.

[d]Contains three saturated pixels. Includes Henderson, North Las Vegas.

[e]Includes Corrales.

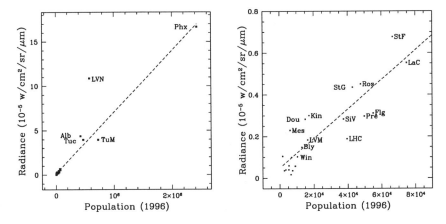

Figure 1.   Integrated radiance *vs.* population for all measured cities (left) and smaller cities (right). The dashed line is a first-order fit to the data below 80,000 population. Letter abbreviations are in Table 1.

3.   Discussion

Though there are many uncertainties in these data and their interpretation (see below), the measured integrated radiance of Flagstaff, AZ falls near the lower limit for cities of similar size. Flagstaff has had a long history of outdoor lighting controls, beginning in 1958 with the first lighting code addressing astronomical interests anywhere in the world, and followed by regular updates to the present time. The integrated radiance of Flagstaff is approximately that of an average town of population about 37,000, or about 67% as much as an average city of its size in these data.

But other communities without strict lighting codes appear fainter than the average as well (e.g. Prescott, AZ), and this leaves the question whether the moderate brightness of Flagstaff can be attributed to its lighting codes.

Another city with a similar long history of outdoor lighting codes is Tucson. There are few cities of similar size in the region against which to compare its brightness, and therefore the brightness of an "average" city of this size is hard to define. In Figure 1, the Census Bureau population figure and the integrated DMSP radiance place Tucson close to the line connecting the small cities to the single large metropolitan area of Phoenix; it also appears comparable to the similarly sized Albuquerque. But Tucson has a large polulation in adjacent areas of Pima County, outside the city limits on which the official population is based. Tucson planners estimate that presently the population of the Tucson metropolitan region is 95% of the Pima County population. Applying this fraction to the Census Bureau 1996 estimate for Pima County (767,873) would put Tucson at 729,479, and therefore considerably fainter *per capita*.

There is scant indication in these figures for a decreasing uplight *per capita* as population increases, as seen by Falchi & Cinzano (1999), nor is there any apparent tendency to decreased *per capita* output with increasing population, as reported by Garstang (1986). A first-order fit to the cities of less than 80,000 nearly exactly intercepts the largest and brightest city measured.

3.1. Sources of Error

Assuming there are no residual clouds in the composite image, and neglecting photometric calibration uncertainties in the DMSP photometric equipment and processing, the following are some of the effects that will influence the *per capita* apparent brightness of cities in the DMSP data:

Population Uncertainties. Population figures are uncertain for many communities, since what the Census Bureau defines as a given city does not always correspond with what appears to be the city extent in the DMSP image. The example of Tucson presented here, where the true figures may be more than 50% higher than Census Bureau figures, may be generally indicative of the magnitude of this source of error. A proper approach will require a detailed community-by-community population analysis.

Angular Dependence. The light emitted from cities is not likely to be independent of direction. Reflected uplight might be approximately lambertian (ignoring blocking effects) but direct uplight is likely to have strong angular dependencies. Garstang (1986) assumes an intensity dependence of direct uplight proportional to the fourth power of the zenith angle. Although the raw DMSP scans contain observations over a range of zenith angels on an east-west line, it is not clear how such a limited sample of altitude and azimuth measures can be applied to evaluating the entire upward hemisphere.

Ground-Level Obscuration. Variable amounts of obscuration from vegetation and structures will affect the apparent brightness of cities as viewed from different angles. This effect may increase toward the horizon but it has not been measured. The degree of this effect may also be correlated with city size as larger cities have taller buildings.

Albedo Variations. Varying reflectivity of the ground, the relation of such variation to lighting use, and the presence or absence of snow cause further uncertainities. These effects, particularly of snow, could be large.

Extinction. The zenith angle of DMSP observations as viewed from the ground varies from 0 deg to about 60 deg, leading to a 1 to 2 range in airmass between the source and detector. Atmospheric extinction coefficients vary from approximately 0.35 mag/airmass near the blue limit of the detector to 0.1 mag/airmass or less near the red limit. This effect has not been removed from these data, and sufficient information may not be available to do it.

References

Cinzano, P., Falchi, F., Elvidge, C.D., Baugh, K.E. 2000, MNRAS, 318, 641-657

Elvidge, C. *et al.* 1999, Remote Sensing of Environment, 68, 77

Falchi, F. and Cinzano, P. 2000, Mem. Soc. Astro. It., 71, 139-152

Garstang, R. H. 1986, PASP 98, 364

Isobe, S. 1998, paper presented at Division 4 meeting of the CIE, 24 October 1998

Isobe, S. and Hamamura, S. 1998, ASP Conf. Ser. 139, 191

Preserving the Astronomical Sky
IAU Symposium, Vol. 196, 2001
R. J. Cohen and W. T. Sullivan, III, eds.

Light Pollution and Energy Loss from Cairo

A. I. I. Osman

National Research Institute of Astronomy and Geophysics, Helwan,
Cairo, Egypt
E-mail : aibrosman@frcu.eun.eg

S. Isobe

National Astronomical Observatory, Mitaka, Tokyo, Japan
E-mail : isobesz@cc.nao.ac.jp

S. Nawar and A. B. Morcos

National Research Institute of Astronomy and Geophysics, Helwan,
Cairo, Egypt

Abstract. Light pollution is a serious problem, not only to astronomy but also to the environment and the economy. The amount of light energy loss from Cairo has been calculated using data obtained at Kottamia Observatory in 1980 and 1995. The amount of energy loss is much greater in 1995. Comparison with a number of other world cities is given. These results show that much money is being wasted. Recommendations to overcome the problem, and the corresponding savings, are given.

1. Introduction

The problem of light pollution and energy loss is one of the important problems to astronomy, the environment and the economy. Many investigators have studied light pollution (see e.g. Walker 1970, 1973, 1977, 1988; Treanor 1973; Berry 1976; Garstang 1984, 1986, 1988, 1991). However, only a limited number of investigators have studied the other face of light pollution, which is the electric energy loss. Hunter et al. (1989) have studied the electric energy loss due to light pollution in the USA. They have found that the total amount of electric energy loss is 17.4 billion kWh per year, costing US\$1.288 billion. Isobe and Hamamura (1998) have studied the city lights of Japan observed by the Defense Metreological Satellite Program (DMSP). They have deduced the relative indexes of the cities of Akita, Shizuoka, Hiroshima, Tokushima and Matsuyama. Their results show that the energy loss of city light has increased by 20 to 50 percent in 3 years.

In the present work the effect of light pollution at Kottamia Observatory from the city of Cairo has been studied. Light and electric energy loss for Cairo City estimated from ground observations are compared with space results obtained by DMSP satellites.

2. Light Pollution Studies at Kottamia Observatory

The light pollution at Kottamia Observatory from Cairo and nearby cities had been studied during two periods. The first study was carried out by Asaad et al. (1982) at different altitudes in the sky from 5° to 45°. The second set of observations was carried out during Jan-Feb. 1995 over the entire sky and the results have been published by Nawar et al. (1998). The average values are given in Table 1, accompanied by the value obtained in 1980 by Asaad et al. for comparison.

Table 1.   Light pollution at Kottamia Observatory in the visual band

Date	R
12-12-1980	0.05
Average of 7 nights in 1995	0.49

The expected values of light energy loss from Cairo can be estimated by using Walker's relation as follows:

$$\log R = -4.7 - 2.5 \log D + \log F, \tag{1}$$

where R is the ratio between the observed sky glow as measured in the direction of the source and natural background radiation at altitude 45°, D is the distance (in km) between the observatory and the source, and F is the total luminance (in lumen) of the outdoor lighting in the city.

From Walker's relation we can obtain:

$$F = 10^{(4.7+2.5 \log D + \log R)}. \tag{2}$$

The equivalent electric power loss from the source E in Watts can be calculated as follows:

$$E = F/L, \tag{3}$$

where L is the average lighting efficiency of the source in lumen/Watt. From relations (2) and (3) we can put E in the following form:

$$E = [10^{(4.7+2.5 \log D + \log R)}]/L. \tag{4}$$

If we know L and if we assume the average time for lighting is 10 hours per night during the year, then the average electric energy loss per year (T_y) can be calculated as follows:

$$T_y = E(\text{W}) \times 10 \text{ hours} \times 365 \quad (\text{Wh}), \tag{5}$$

or

$$T_y = C[10^{(4.7+2.5 \log D + \log R)}]/L \quad (\text{kWh}), \tag{6}$$

where C = 3.65.

From the National Energy Information Centre (EIC) the average lighting efficiency in Cairo is taken as 40 lumen/Watt in 1980 and about 70 lumen/Watt in 1995. The distance (D) from Kottamia Observatory to the centre of Cairo is about 60 km. Then by using relation (6) and the values of R given in Table 1, the equivalent electric energy losses per hour and per year from Cairo in kWh for 1980 and 1995 have been calculated and the results are given in Table 2. The amount of electric energy loss per year from Cairo increased by about 6 times from 1980 to 1995. If we assume that the annual rate of increase of electric energy loss from Cairo is constant, then the expected value of electric energy loss in the year 2000 will be 45 million kWh. Since the average price of 1 kWh of electric energy at Egypt is about US$0.06, then the annual loss per year increased from about US$0.375 million in 1980 to US$2.1 million in 1995.

Table 2.   The electric energy loss from Cairo per hour and year

Date	E(kWh)	T_y(MkWh)	Loss of Money (Million US dollars)
12-12-1980	1747	6.38	0.375
Average of 7 nights in 1995	9783	35.7	2.1

The light and electric energy loss for Cairo and other capitals in the world using DMSP images (Isobe and Hamamura 2001) have been tabulated in Table 3 for comparison with ground-based observations. It is noticed that the energy loss from Cairo is nearly equal to that of Paris and somewhat lower than that of London, but higher than all the other listed cities. The table shows that the value of energy loss from Cairo obtained from ground-based observations is about 30% higher than that obtained by DMSP satellite images. This difference may be due to atmosphere extinction, the fact that the DMSP measures only the direct light from the city and/or the fact that DMSP measures reflected light which makes a large angle with the horizon (60° to 90°, depending on the sub-satellite track). However, from the ground-based observations we have deduced the total light coming from the city at angles between 0° and 90° with the horizon.

Table 3.   Light energy loss from Cairo and other capital cities

City	Date	Observed Value 10^{-8} Watt/cm^2/st/ μm	Light Energy Loss 10^6 kWh
Cairo	1997.02.05	4.51×10^3	27.00
London	1997.01.13	4.84×10^3	29.00
Amsterdam	1997.01.13	1.07×10^3	6.43
Bruxelles	1997.01.13	9.64×10^2	5.78
Paris	1997.01.13	6.33×10^3	37.90
Tel Aviv-Yafo	1997.01.09	1.72×10^3	10.30
Amman	1997.01.09	8.77×10^2	5.25
Damascus	1997.01.09	4.98×10^2	2.98

3. Recommendations

So, the following recommendations should be done to overcome the light pollution problem:

1. Set restrictions and laws that require correct light systems.

2. Spread awareness about this kind of pollution among the people via environmental groups.

3. Replace all bad lighting systems with new systems that consume less energy, have a longer life and at the same time give adequate lighting.

References

Asaad A. S., Nawar S. and Morcos A. B. 1982, HIAG BULL. II A

Berry, R. L. 1976, J. Roy. Aston. Soc. Canada, 70, 97-115

Garstang, R. H. 1984, Observatory, 104, 169

Garstang, R. H. 1986, PASP, 98, 364

Garstang, R. H. 1988, PASP, 108, 159

Garstang, R. H. 1991, PASP, 103, 1109-1116

Isobe, S. and Hamamura, S. 1998, ASP Conference Series, Vol. 139, 191-199

Isobe, S. and Hamamura, S. 2001, paper presented in this symposium *Preserving the Astronomical Sky.*

Nawar S., Morcos A. B., Metwally Z. and Osman A. I. I. 1998, ASP Conference Series, Vol. 139, 151-158

Treanor, P. J. 1973, Observatory, 93, 117

Walker, M. F. 1970, PASP, 82, 672

Walker, M. F. 1973, PASP, 85, 508

Walker, M. F. 1977, PASP, 89, 405

Walker, M. F. 1988, PASP, 100, 496

Preserving the Astronomical Sky
IAU Symposium, Vol. 196, 2001
R. J. Cohen and W. T. Sullivan, III, eds.

Local and National Regulations on Light Pollution in Italy

Valentina Zitelli

Osservatorio Astronomico di Bologna, Via Ranzani 1, 40127 Bologna, Italy

Mario Di Sora

Osservatorio di Campo Catino Guarcino, Italy

Federico Ferrini

Dip. Fisica, Univ. di Pisa- sez. Astronomia e Astrofisica, P. Torricelli 2-56126 PISA, Italy

Abstract. The situation in terms of regulation of light pollution in Italy is here presented. The activity, started with the effort of amateur astronomers, lighting engineers and professional astronomers, is split into several actions oriented both to the local and national levels. One of the first goals was an ordinance of the city of Frosinone providing specific guidelines for lighting. The provinces of Valle d'Aosta and Veneto have followed up with local laws. The regional law of Valle d'Aosta includes a safeguard for nocturnal fauna from light pollution. The regional law of Veneto includes protection of the environment, observatories and astronomical sites. To avoid a proliferation of local laws, a national Italian bill was submitted and it is still under discussion. Bill No. 751 has been approved by the general assembly of the Italian Astronomical Society and provides more specific guidelines and standards concerning lighting in order to avoid unnecessary upward illumination and to reduce glare. At the same time a new technical standard, UNI 10819, has been settled after long debate among UNI (Italian National Standards Institute), the Italian Astronomical Society's Light Pollution Committee, lighting engineers and representatives of lighting manufacturers.

1. Introduction

It is unquestionable that the shift of our old economy, mostly based on agriculture, to one based primarily on products of industry has improved our standard of living, but it is also unquestionable that, as consequence, we lost our sensitivity to the ambient environment. We are not yet able to know, and to control, the environmental impact arising in each industrial production cycle.

Following the definition of *positive development* expressed by the World Committee on the Environment (W.C.E.D.) in 1987 as *a development able to satisfy the requirements of today without compromising the possibility for future generations to satisfy their requirements,* we need to insert the effects of urban lighting on night sky brightness as a man-made product that we must reduce before the future generation completely loses their view of the sky, and with it an important contact with our nature.

Several studies have been conducted (Walker 1970, 1973; Bertiau de Graeve and Treanor 1973) to verify the assumption that the total light emitted by a city is proportional to its population. Models have been constructed to allow calculation of the night sky brightness produced by cities at their centres and outskirts, obtaining the correlation between artificial lighting and sky brightness and population (Berry 1976), number of lumens used for street lighting and distance from the city centre (Garstang 1986). The most important observatories are carrying on long term campaigns to determine the total natural sky background, as well as monitor the level of sky pollution from artificial lights (Walker 1988, Mattila 1996, Cinzano 2000).

There are solutions to reduce this pollution of the night sky and in the following we describe actions in Italy to reduce it or at least bring the situation under control.

2. The Italian Economy

Nowadays 56 million people live in Italy, prevalently distributed near the most important cities, near the Po and Arno Valleys and along the 10000 km of coasts. Economic development and a steadily improving standard of living has produced an increase in the consumption of electrical energy and power for both public and private uses. The Italian electric company has measured over the past years 5% annual increase of the per capita amount of energy consumption used for urban lighting (Cinzano 1997). Another consequence of growth has been the expansion of urban areas. In most of the major Italian cities the darkness of the night sky is jeopardized. Because of the narrow shape of Italy, the dazzle shining up from urban areas also reaches small towns far from the emitting sources. Sky vision is thus diminished even for people living in small towns and in rural areas. Activities of amateur astronomers, often observing close to urban areas, are seriously jeopardized, but the situation is also at risk for professional astronomers, whose telescopes in effect have reduced equivalent collecting areas.

SAIt (Italian Astronomical Society), understanding the danger not only for astronomy but also for the loss of this natural heritage, promoted a committee to study and control the situation of the light pollution in Italy. At the same time SAIt members started several actions, suggesting ordinances and laws to both local and national authorities. A number of educational activities have also been organized.

3. Actions

The first Italian activity, in 1990 at the SAIt meeting held in Padova, was the appointment of a specific Committee to study the phenomenon of "light pollution" and to find some solutions to reduce it. In the same year a large opinion movement grew up, with the participation of professional astronomers, amateur astronomers, ecologists, politicians, lighting technicians and also some ordinary citizens. Local ordinances were submitted to several local governements. All the ordinances were aimed at the limitation of upward light to reduce one of the main contributions to the sky background and gave guidelines for the design, construction, operation and maintenance of lighting installations.

3.1. National Bill

On a national scale, to avoid proliferation of local laws that might introduce some difficulties for people and governments, in the past legistlature the Italian Parliament was presented with a national law still at the centre of political and technical-scientific debate. Bill No. 751, written by the SAIt Committee for the "Study of light pollution" with the help of lighting engineers, had been first approved in an SAIt general assembly.

The proposed law provides specific guidelines and standards concerning lighting in order to avoid unnecessary upward illumination and to reduce glare. In particular:

- it intends to limit the phenomenon *sky pollution* over the entire nation.

- it provides for reduction of the luminous flux when traffic on the roads is reduced (i.e. night time)

The Bill requires the following technical parameters for lighting installations:

1. Road lighting: cut-off fixtures with emission not exceeding 0 cd/klm above 90°

2. Ornamental and open lighting: fixtures with emission less than 15 cd/klm above 90°

3. Lighting of large areas: system fixtures with emission less than 10 cd/klm above 90°

4. Monumental and building lighting: it is forbidden to illuminate in an upward direction. When it is not possible to observe this prescription, it is necessary to limit the light flux to within the outline of the concerned building

5. Luminous signs and banners:

   (a) signs necessary for nocturnal activities (i.e. hotels, hospitals or police): can be lit during the night

   (b) other signs: must be turned off at 11 pm during winter time and at 00 pm during summer time

6. It is forbidden to use rotating or fixed searchlight-type beams for advertising purposes.

3.2. Regional Laws

In Italy there are several Provinces, mostly located at the borders of our country, that have a special status with local law. Among them, the Veneto Region and Valle d'Aosta have approved a law on light pollution. Veneto regional law No. 22 of June 1997 declared among its purposes the protection of the environment, observatories and astronomical sites. Valle d'Aosta law No. 17 of April 1998 expands the benefit of reduced sky pollution by including among its goals the safeguard of nocturnal fauna. Valle d'Aosta also introduced the concept of sky view as a common inheritance of all the people and not only limited to astronomers. In addition, Piemonte and Toscana have draft Bills.

3.3. Local Activities

Presently in Italy there are some tens of ordinances among Communal Regulations, Provincial Regulations and Prefectorial Circulars. In general in all texts is indicated the permitted upward flux emitted by lighting installations. Most of these Ordinances were approved after presentation of the results of long monitoring of the local sky by the writer of the proposed law. The Frosinone Ordinance was the first regulation applied to a city and has served as an important template that opened the way to other cities.

3.4. Technical Standards

The SAIt made a strong effort to match requirements of astronomers, lighting engineers and lighting manufacturers. A working group composed of the SAIt Committe and UNI (Italian National Standard Institute) representatives, after two years of long debate, settled in the spring of 1999 on the technical standard UNI 10819. This standard regards light and lighting, is available only for outdoor new lighting installations and for renewing the existent. UNI 10819 serves as a support to each national or regional law, defines the technical requirements of lighting installations, defines restrictions regarding the maximum upward flux emitted by the total lighting devices of the community to limit the upward scattered luminous flux.

The UNI standard classifies lighting installations in 5 classes:

- A - installation in a highly protected zone

- B - recreational activities lighting

- C - environmental and architectural lighting

- D - display lighting

- E - temporary lighting, such as luminaires for Christmas displays

It divides the national territory into three zones, characterized by the degree of sky protection:

- Zone 1 - Highly protected

- Zone 2 - Less protected, either by itself or as an annulus around Zone 1

- Zone 3 - Other

The standard specifies

(a) the requirements to reduce the luminance during the night, for instance introducing the idea of road declassification when traffic is reduced,

(b) technical conditions on the construction, installation and maintenance of lighting equipment, and

(c) parameters for town lighting plans to avoid uncontrolled road lighting.

Table 1 shows the values of maximum permitted flux, R_n, published in the standard UNI 10819 for each defined class of lighting installation and for each defined Zone. R_n is defined as the ratio between the nominal upward flux emitted from the lighting equipment and the total luminous flux. Although this standard does not undo light pollution, it permits control of lighting equipment and improves the quality of the installations.

Table 1. Maximum permitted values of upward flux (%)

Installation	Zone 1	Zone 2	Zone 3
A			
B			
C	1	5	10
D			
E	not allowed	not allowed	allowed

Moreover, by not defining specific components but only specific requirements, it does not limit technical developments in this field.

3.5. Education

Major efforts are being addressed to educational activities oriented to the people, government and light engineers. Events such as "The day of the off light" (a day without lighting) or "The star parks" (permanently protected sites for viewing the night sky) are promoted. Seminars and meetings are also devoted to the public and to specialists on lighting. At the end of 1998 a national meeting on "Technology and environment in the next century" was held in Italy. In this meeting, for the first time, light pollution was put at the same level as the other pollutions, such as noise pollution or atmospheric pollution, thanks to several actions that showed this phenomenon as another unwanted byproduct of modern technology.

All the educational activities have the purpose to explain that there is no conflict between astronomers and the activities of the public. The relationship "more light = more security" is wrong and must be substituted by the equation "better light = better security".

4. Conclusions

The actions in Italy to reduce sky pollution are the following:

- Bill No. 751 presented to Italian Parliament
- The law of Regione Veneto No. 22 of 27 June 1997
- The law of Regione Valle d'Aosta No. 17 of 28 April 1998
- Draft Bills of Regione Lombardia, Piemonte and Toscana
- Several Communal ordinances
- Standard guidelines, UNI 10819
- Educational activities, conferences, books.

References

Berry R. 1976, J. Royal Astronomical Society Canada, 70, 97-115

Bertiau F. C. S. J., de Graeve E. S. J., Treanor P. J. S. J. 1973, Vatican Observatory Publ. 1, 4, 159-179

Cinzano P. 1997, Inquinamento luminoso e protezione del cielo notturno (Venezia: Istituto Veneto di Scienze, Lettere ed Arti), 224 pp. (in Italian)

Cinzano P. 2000, Mem. Soc. Astron. Italiana, 71, 159-165

Garstang R.H. 1986, PASP, 98, 364-375

Mattila K., Vaisanen P., Appen-Schnur G.F.O. 1996, Astron Astrophys. Suppl. Ser. 119, 153-170

Walker M. F. 1970, PASP, 82, 672

Walker M.F. 1973, PASP, 85, 508

Walker M. F. 1988, PASP, 100, 496

Preserving the Astronomical Sky
IAU Symposium, Vol. 196, 2001
R. J. Cohen and W. T. Sullivan, III, eds.

Japanese Government Official Guideline for Reduction of Light Pollution

Syuzo Isobe

National Astronomical Observatory, Mitaka, Tokyo, Japan

Abstract. After 3-year discussions within a working group of the Japanese CIE and then 2-year discussion under a committee of the Environmental Agency in Japan, "Guideline for Light Pollution - Aiming for Good Lighting Environments" was published by the Environmental Agency. This is the first governmental guideline in the world and therefore a good example to be discussed in the other countries.

1. Introduction

The governmental environmental agency of Japan published "Guidelines for Light Pollution - Aiming for Good Lighting Environments" in March 1998. This is the first governmental guideline in the world which deals with reduction of light pollution. Here, I will describe the processes by which this guideline was set up and some important points considered within the guideline.

The IAU (International Astronomical Union) has been discussing light pollution and has been in contact with the CIE (Commission International de l'Eclairage, International Commission on Illumination) from just after the creation of IAU Commission 50, "Protection and Identification of Existing and Potential Observatory Sites". A CIE document "Guidelines For Minimizing Urban Sky Glow Near Astronomical Observatories" was published in 1984. After IAU Colloquium No.112 "Light Pollution, Radio Interference and Space Debris" held in Washington D. C. in August 1988, CIE set-up a technical committee within Division 4, TC4-21 "Interference of Light with Astronomical Observations", with Chairman Dr. Duco A. Schreuder, in 1989. I was appointed to be one of the members of TC4-21 because I was the Vice President of IAU Commission 50. I then contacted the CIE in Japan and the Lighting Society of Japan and succeeded in establishing a Working Group within the Society.

Following a request by some astronomers, the Bureau of Atmospheric Preservation of Japan, under the Environmental Agency, have adopted a star-watching programme from 1986, as one measure of *air* pollution (Isobe & Kosai 1998). This is a simple method to determine night-time sky brightness by taking a photograph of a known bright star within the frame pointing near the zenith. Over all of Japan there were 100 groups who participated in this programme at the beginning and 300 groups in the latest year. Through this connection, officers at the Bureau learned of the CIE activities and then joined the Working Group in the Lighting Society of Japan. They were very keen on the possibility that the

CIE resolution would become an ISO standard which might not be applicable to the current Japanese systems and regulations.

After 3 years' study within CIE Japan, they understood what problems existed and set up a Committee within the Environmental Association, which is totally controlled by the Environmental Agency. In contrast with the Working Group, the Committee had members from different fields, officers of local government, physicists dealing with atmospheric phenomena, and so on. Therefore, the Committee covered most fields relating to light pollution.

To have an idea of this guideline, it is good for the reader to ask for a copy of the Guideline either from the Environmental Agency or from the present author. An essential point of this Guideline is to give a check list to use when outdoor lighting fixtures are prepared. This is a perfect list: therefore if one follows this list, we may minimize light pollution. However, it is usually the case that when one builds a house or a building, one does not seriously take care over the lighting design. Therefore, the Guideline recommends that higher status be given to lighting engineers, who can manage good outdoor lighting to minimize light pollution.

The guideline reviews the problems relating to bad lighting and defines good lighting systems. It also requests good monitoring systems. However, it gives only three lines of the text for monitoring of night-time brightness by satellite measurement. Certainly, ground-based observation of sky glow is able to give a direct measure of light pollution, but the star watching programme needs many collaborating people all over the area to be measured. This is good to educate people for light pollution, but it cannot cover all the land area and also its accuracy is rather worse (0.2 magnitude for faint sky brightness to 1.0 magnitude for bright sky brightness).

Since there are so many parameters to be considered in order to have a good measure, the other ground-based observation does not give an accurate sky brightness as expected. Satellite measurements (Isobe & Hamamura 1998, 2001) detect light escaping to space, which is energy loss. This cannot be a direct measure of light pollution, since the density of reflecting materials varies from time to time, but it can provide a better measure than a star-watching programme. Satellites are obtaining digital data on the night-time brightness all over the world every day and therefore we can obtain its daily variation. The Guideline could not unfortunately stress this point.

It is hoped that readers of this article will study the Guideline and will set up similar Guidelines or regulations in each country; then those well-studied Guidelines will push the Japanese Guideline to be better.

2.   Table of Contents of the Guideline for Light Pollution

Table 1.   Contents of the Guideline for Light Pollution

Preface

I. Light pollution countermeasures guidelines
    1. Preparation of the guidelines
    2. Structure of the guidelines
    3. Roles and obligations of concerned parties, based on the guidelines
    4. Establishment of status of designers of lighting environments

II. "Light Pollution Countermeasures Guidelines"
    1. Definition of light pollution
    2. The "Sky Glow" problem
    3. Lighting environment which conforms to the characteristics of the area or region
    4. Those involved in the lighting environment guidelines
    5. How to use these Guidelines
    6. "Guidelines for Outdoor Lighting, etc."

Appendix
    Technical terms, abbreviations and symbols used in the guidelines

References

Isobe, S. and Kosai, H. 1998, Star Watching Observations to measure night sky brightness, ASP Conference Series, Vol. 139, pp 175-184

Isobe, S. and Hamamura, S. 1998, Ejected city light of Japan observed by a defense meteorological satellite program, in ASP Conference Series, Vol.139, pp 191-199

Isobe, S., Hamamura, S. and Elvidge, C. 2001, Educating the public about light pollution, in this volume

Preserving the Astronomical Sky
IAU Symposium, Vol. 196, 2001
R. J. Cohen and W. T. Sullivan, III, eds.

Outdoor Lighting Ordinances:
Tools to Preserve the Night Sky[1]

Donald R. Davis

*International Dark-Sky Association, 3225 N. First Avenue, Tucson,
Arizona 85719 USA*

Abstract. The dark night sky is rapidly being lost throughout the
world due to wasted outdoor night lighting. Following warnings by astronomers[1] and environmentalists, many communities are acting to preserve
the night sky by enacting outdoor lighting codes. Here we briefly recount
the history of codes and outline the essential elements of an effective code.
Additional information, including a handbook on how to write a good
outdoor lighting ordinance, is available from the International Dark-Sky
Association web site, www.ida.org .

1. Introduction

For much of the 19th century the major astronomical observatories of the world,
principally in Europe, were located close to major population centres – Paris,
Berlin, London – where most astronomers lived and worked. The increasing
quality of astronomical instrumentation coupled with the declining observational
environment near major population centres led to the search for prime observing
sites removed from cities. About a century ago sites such as Williams Bay, Wis-
consin (Yerkes Observatory), Mt. Wilson and Mt. Hamilton (Lick Observatory)
in California were identified and became major astronomical facilities.

Growth near observatories was relentless, however, particularly in the US
West. Mt. Wilson was the first to experience significant adverse effects on its
night sky quality due to explosive growth in the Los Angeles area. As shown in
Figure 1, the Mt. Wilson sky brightness is calculated to have increased at the
rate of nearly 0.3 mag per decade from 1910 to 1970. Owing to the loss of dark
skies (although the Mt. Wilson seeing is still exceptional), deep sky observing
terminated there in the early 1980s[2].

In the 1950s and early 1960s, other major astronomical sites in the US
– Mt. Palomar, Kitt Peak, McDonald Observatory – were thought to be too
remote from major cities to be significantly affected by light pollution. For
example, Meinel (1960) wrote, "For the foreseeable future there appears to be

---

[1]This paper is dedicated to Dr. Arthur Hoag (1921-1999), whose pioneering efforts led to the
first comprehensive outdoor lighting code.

[2]Technological advances combined with the exceptional seeing at Mt. Wilson enable outstanding
astronomical research to still be done from this site on moderately bright (by astronomical
standards) objects.

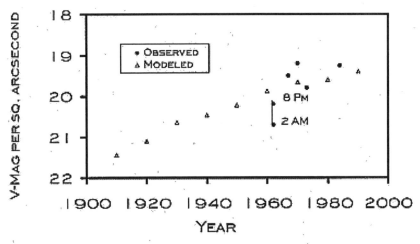

Figure 1.   Sky brightness at Mt. Wilson observatory. Comparison of model predictions based on Garstang (1989) with measurements (data provided by S. Baliunas, private communication).

no serious threat to astronomical conditions on Kitt Peak ...". Within a decade, though, the light pollution threat to many major observatories was apparent to all. Riegel (1973) summarized the problem that outdoor lighting posed to astronomy. Hoag and Peterson (1974) expanded on the light pollution issue and described methods for ameliorating it. In an amazingly short time, explosive population growth in the southwestern US and its concomitant light pollution threatened the observatories' world-class dark skies.

Responding to this threat, astronomers from Kitt Peak National Observatory (KPNO) and the University of Arizona led an effort to establish a comprehensive outdoor lighting code (OLC) in Tucson and Pima County, the first such ordinance in the world. Following the adoption of this code in 1973, a permanent Outdoor Lighting Code Committee composed of representatives from the astronomical community, lighting engineers, general business people, electrical company representatives, licensed contractors and building engineers was created to address ongoing light pollution issues. Other municipalities near astronomical facilities, such as Flagstaff, Arizona, rapidly followed suit; Table 1 summarizes the status of OLCs currently in effect to protect major observatories.

While preserving dark skies was initially an issue for astronomers, in recent years it has become a concern to a much larger number of people. The awareness of light pollution as an environmental issue increased as more and more people were deprived of cherished views of the night sky – the Milky Way is disappearing into the milky haze of light pollution. Interestingly, though, in the US the solution to the light pollution environmental problem has been to continue adoption of local ordinances instead of addressing the issue at the national level through the Environmental Protection Agency. Local codes have proliferated and there are now more than 150 in effect in the United States, including four state-wide codes. Numerous others exist worldwide and the number is increasing rapidly as people are beginning to protect their night sky.

In some countries, a comprehensive approach to the light pollution problems may be feasible on the national level. Australia has a national standard for "The Control of Obtrusive Outdoor Lighting." This standard is the first comprehensive approach to containing detrimental effects of poor lighting practices. Countries with a strong environmental awareness on the national level may find that national legislation is the most useful tool for effectively combating light pollution.

Table 1. Status of Outdoor Lighting Codes (OLC) for Protection of Selected Major Astronomical Observatories

Observatory	Date of First OLC	Date of Current OLC	Notes
KPNO, Kitt Peak, Arizona, FLWO, Mt. Hopkins, Arizona	1972	1999	a
USNO, Lowell Observatory, Flagstaff, Arizona	1973	1989	b
Mt. Wilson Observatory, Los Angeles, California	- - -	- - -	c
Palomar Observatory, San Diego, California	1984	1986	
Lick Observatory, San Jose, California	1980	1983	
Mauna Kea, Hawaii	1974	1988	
Siding Spring Observatory, Australia	1988	1988	d
McDonald Observatory, Austin, Texas	1999	1999	
Cerro Tololo Observatory, La Serena, Chile	1999	1999	e
Roque de los Muchachos Observatory, Canary Islands, Spain	1988	1988	f

---

[a]Significant changes in 1999 code.

[b]First use of lumens cap; LPS near observatories.

[c]No comprehensive OLC; the city of Los Angeles is switching its street lighting to full cutoff.

[d]This code is currently undergoing revisions.

[e]First "norm" or regulation passed this year, previously only recommendations were passed.

[f]Code applies to the islands of La Palma and Tenerife.

2. How to Write an Outdoor Lighting Code

To provide guidance for anyone wishing to create an OLC, the International Dark-Sky Association has developed an OLC Model Code Handbook, which is available on the IDA web site (www.darksky.org). This handbook provides information such as how to write a good code, protection of astronomical sites and discussion of the different types of illumination sources, with pros and cons for different applications. General guidelines for protecting the night sky through good lighting practices are also given by the Commission Internationale de l'Eclairage (CIE) 1997 technical report entitled "Guidelines for Minimizing Sky Glow."

A good outdoor lighting code includes the following elements:

- Minimize direct uplight. Light emitted above the horizontal plane directly contributes to light pollution, so outdoor luminaires should be full cutoff fixtures that emit no light above the horizontal plane.

- Use only the amount of light that is needed. Limits on the amount of light generated appropriate to the activities of the area are essential to controlling the overlighting situation that develops when outdoor lighting is used to attract people's attention (the "moth to the brightest candle" syndrome). Examples of lighting limits are given in the IDA Model Code Handbook.

- Curfews for unneeded light. Turning off lights when they are not needed significantly reduces light pollution (as well as saving energy) for much of the night. For example, turn off parking lot lights (except for needed security lighting) after business hours.

- Use low pressure sodium (LPS) lighting near astronomical observatories. LPS is the astronomer's choice for outdoor lighting since the illumination is only at one color, a golden yellow, and can readily be filtered out for many astronomical applications (see the article by Luginbuhl in this volume). A bonus is that LPS is one of the most efficient lighting sources available: it produces about eight times as much light for each watt of electrical energy consumed as does an incandescent lamp.

Preserving dark skies through outdoor lighting ordinances does not end with the passing of an ordinance – effective enforcement is crucial in order to reap benefits. Enforcement includes: (a) ensuring that plans for development projects comply with the code, and (b) field inspection for compliance once the lighting is installed. Obviously it is much better to recognize and fix non-compliant cases in the first step rather than the second one. The first and most critical step in enforcement comes in the design/plans review phase of development that requires outdoor lighting. Code compliance should be achieved through a knowledgeable lighting engineer and verified by an appropriate jurisdictional authority. Next, the actual installation must be inspected to ensure that what has been installed complies with the approved plan. Finally, a mechanism should be established for surveying the community to identify non-compliant lighting.

It should be noted that most non-compliant lighting is installed out of ignorance of the outdoor lighting code.

There can be various procedural difficulties in implementing an OLC. First, there may be multiple jurisdictions in a geographic area that need to adopt similar codes in order to protect dark skies. For example, many urban areas in the US are a patchwork of small communities: there are six "jurisdictional authorities" within Pima County, Arizona, where KPNO and the Whipple Observatory are located. Without effective enforcement and public education programmes, however, an outdoor lighting code is just another arcane example of bureaucratic rubbish.

The bedrock of code compliance is education: all segments of the population must be aware of the code, its purpose and how to comply with it. This education process includes training sessions for lighting professionals, public officials and developers, newspaper articles and public talks. Brochures summarizing key points of good lighting for distribution to the general public are also helpful. Finally, all optical observatories should have a knowledgeable staff member charged with responsibility for light pollution issues. Most such institutions will not be as fortunate as La Palma, which has a full-time staff for the "Oficina Tecnica por la Proteccion de la Calidad del Cielo," but the function of dark sky preservation should be a structured staff activity at all observatories. Resources sufficient to accomplish the task must be allocated, ideally as a line item in the observatory budget. Cerro Tololo, for example, is now committing around 1% of its resources to light pollution issues (M. Smith, private communication).

3. Do Lighting Codes Work?

How effective are outdoor lighting codes in protecting observatories? The answer to this question has not yet been quantified. Despite the call over a quarter of a century ago for systematic monitoring programmes at major astronomical observatories (Riegel 1973), no such programme has, to my knowledge, been implemented. The best datum is from Pilachowski *et al.* (1989), who measured the KPNO sky brightness near solar minimum (Figure 2). They found a zenith V-brightness of ~ 21.9 mag arcsec^{-2}, which was 0.04 magnitudes fainter than sky brightness predictions (Garstang 1989). They attributed the difference to the efficiency of the Tucson/Pima County Outdoor Lighting Code. Unfortunately, these measurements were not repeated at the 1996-97 solar minimum. Also shown in Figure 2 are sky brightness measurements from Mt. Hopkins (FLWO) and La Palma indicating that skies are still quite dark at those sites, despite significant population growth nearby. Lighting ordinances do seem to be protecting dark skies.

4. Summary

All major observatories should establish a continuing programme to measure sky brightness at their site. Such data are crucial, not only for knowing their effective limiting magnitude for scientific programmes, but also for establishing long-term sky brightness trends and for assessing how well outdoor lighting codes are doing their job of protecting dark skies.

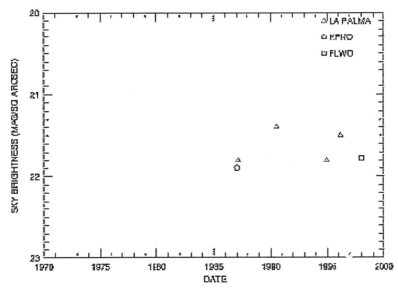

Figure 2. Zenith sky brightness in the *V*-band for La Palma, Kitt Peak National Observatory (KPNO) and the F. L. Whipple Observatory (FLWO) on Mt. Hopkins, Arizona.

The awareness of light pollution has just begun – it is comparable to the awakening to the problems of air and water pollution in the early 1960s. While light pollution does not pose the clear and present danger of air and water pollution, it does represent a more subtle threat to humanity. By cutting people off from their direct views of the stunning night sky, light pollution robs us of an awareness of the vast universe and the sense of awe and humility that it inspires. We have a heritage of connection to the heavens, a vital component of the natural world. As our dark skies disappear, we lose our appreciation of the historical significance that stars and planets have played in the development of our culture as navigation aids, as challenges to our sciences and as inspiration to our religions. Without night skies humanity is losing a critical factor in making us what we are today. This cannot be allowed to happen.

References

Garstang, R. H. 1989, *Ann. Rev. Astr. Ap.*, **27**, 19

Hoag, A. A., and Peterson, A. A. 1974, Kitt Peak National Observatory, April 1974

Meinel, A. B. 1960, In *Stars and Stellar Systems*, Vol.I, 154-175 [*Editor: p.175 is quoted*]

Pilachowski, C. A., Africano, J. L., Goodrich, B. D., and Binkert,W. S. 1989, *PASP*, **101**, 707

Riegel, K. W. 1973, *Science*, **179**, 1285

Preserving the Astronomical Sky
IAU Symposium, Vol. 196, 2001
R. J. Cohen and W. T. Sullivan, III, eds.

Plan of the Modification of Public Lighting in Frosinone in Accordance with the Rule for the Limitation of Light-Pollution and Power Consumption

M. Di Sora

International Dark-Sky Association and
Astronomical Observatory of Campo Catino, Guarcino (FR), Italy.
mario.disora@rtmol.stt.it

Abstract. The plan for modification of public lighting in Frosinone, in accordance with the rule for the limitation of light-pollution and power consumption, is summarized. The energy saving obtained is 40% per year, equivalent to over US$222,000.

1. Street Lighting Adaptation

In the city of Frosinone (50,000 inhabitants) there were about 7,100 street-lamps in 1997, subdivided as follows:

1. 6,400 fixtures for road lighting (2,000 cut-off);

2. 300 fixtures for artistic lighting with open optics lanterns;

3. 300 fixtures for ornamental lighting with spheres;

4. 100 beams for squares and monumental lighting with beams (prevalently symmetric).

The municipal technical office, in collaboration with the Astronomical Observatory of Campo Catino, has realized a plan to modify all street-lamps of the city within 5 years of 1997, in conformity with the rule against light-pollution. The originality of this plan lies in the fact that it is not necessary to substitute every fixture, but only to modify them, with reduced costs. Only when modification is not possible (about 600 street lamps) is there a total change of the fixtures (equivalent cost US$55,000).

The costs of this plan, divided by category, are as follows:

1. 4,000 fixtures modified with flat tempered glass, as shown in Figure 1 (US$5 each, total cost US$20,000);

2. 400 fixtures must be changed integrally (US$90 each, total cost US$36,000);

3. 300 fixtures (old lanterns) modified with conical metallic screens as shown in Figure 2 (each US$5, total cost US$1,500);

4. 150 spheres must be changed completely with electrical components (each US$75, total cost US$11,250);

Figure 1.   Modification of street lighting fixtures with a flat tempered glass to give more transmission of light onto the road, with drastic reduction of glare and light-pollution. Cost: US$5 each.

5. 150 spheres modified with only the adoption of shielded spheres as shown in Figure 3 (each US$30, total cost US$4,500);

6. 100 beams modified with rectangular metal screens as shown in Figure 4 (each US$2, total cost US$200).

The final cost of the plan is US$73,450, that is, 132,210,000 Italian lira.

2.   Savings in Power Consumption

Saving of power consumption is another very important aim of the rule passed by the town council of Frosinone. Every year this city spends about one billion lira (equivalent to over US$555,000) on power.

The rule approved in 1996 provides the following strategies and prescriptions for reducing power consumption:

1. Utilization of flux-reducers for installations with power lamps $\geq$ 150 W high-pressure sodium, or of timers for smaller installations (70 W and 100 W high-pressure sodium);

2. Substitution of all old fixtures of 250 W or 125 W mercury lamps with new fixtures using cut-off 150 and 70 W high-pressure sodium;

3. Reduction of power (from 150 to 100 W, or from 100 to 70 W) for installations which adopt the new spheres shielded in the upper half (made by

Figure 2.   Modification of an open latern "old style" with a cheap and functional conical metal screen. Dispersion flux is reduced by 50%. Cost: US$5 each.

Figure 3.   Modification of the most polluting light source in the world, the opaline sphere! We have utilized a new cheap shielded sphere made in Italy by MARECO LUC. Dispersion flux is reduced by 60%. Cost: US$20-40 each.

Figure 4.   Modification of a symmetric beam with a simple metal screen. No more light in the sky! Cost: US$2 each.

MARECO) or the new lanterns cut-off (made by NERI) or modified by the technical office;

4. Monumental lighting must be turned off after 11 p.m. in solar time and after midnight in summer time.

These measures will allow an annual saving 40% (400 million lira, equivalent to more than US$220,000). In this way it is possible to carry out an ulterior reduction of light-pollution (about 35-40%) thanks to limitation of the power engaged. All that, clearly, without introducing problems for road safety.

Preserving the Astronomical Sky
IAU Symposium, Vol. 196, 2001
R. J. Cohen and W. T. Sullivan, III, eds.

Sky Glow Measurements in the Netherlands

Duco A. Schreuder

Duco Schreuder Consultancies, Spechtlaan 303, 2261 BH Leidschendam, The Netherlands

Abstract. Sky glow was measured in the Netherlands in 1992 and in 1997. The 1992 measurements were photographic surveys, assessed in the Netherlands and in Japan, whereas the 1997 measurements were limit-star assessments. In all cases, the number of measurements was very small. The three sets of data are compared. The sky glow differs considerably over the country and during the night; there are no systematic differences between the results for different epochs and assessment methods. For measurements that allow conclusions that can be used in national policy making, the samples must be very much larger.

1. Introduction

In the Netherlands, just as elsewhere, light pollution is getting more attention, particularly from amateur astronomers and nature lovers. Big cities, industrial areas, sports stadia, airport facilities and - in the Netherlands the main source of light pollution - greenhouses create a 'sky glow' that stretches over most of the night sky. Furthermore, plant and animal life suffers from it. A survey of the problems of sky glow and their solutions is given in another paper to this conference (Schreuder 2001)

It is desirable to know how sky glow is distributed over the country in order to fall in line with more general environmental measures and policies, as well as to find the most favorable locations for amateur astronomers. As sky glow changes with weather, water content of the air, seasons and the solar cycle, it is essential to do measurements at all sites at the same moment in time and with similar weather.

For different purposes, different systems and methods to measure sky glow are in use. At present, none of these systems is standardized. It is proposed to standardize methods for different applications. For details see CIE (1997), Kosai et al. (1993). For simple area surveys, the most common methods are limit-star assessment, star counting and photographic surveys.

Usually, in astronomy photometric data regarding intensities are expressed in magnitudes (mag), those regarding luminances in magnitudes per square arc second (mag/arcsec^2). In lighting engineering, ISO units are used, such as candela (cd) and candela per square meter (cd/m^2) respectively. For conversion, the factor proposed by Crawford is used: "a luminance of 3.2×10^{-6} cd/m^2 corresponds to 26.33 magnitude per arcsec^2" (after Crawford 1997, Table III). This implies a natural background luminance of 3.52×10^{-4} cd/m^2.

2. Photographic Survey, 1992

Photographic surveys are made by inviting a large number of amateur observers (not necessarily amateur astronomers) to make pictures with a normal (fixed position) camera and with normal slide film of the zenith-area in their own neighborhood. The camera must have a focal length of 50 - 55 mm and an f-ratio faster than 2.0. The exposures are made on 400 ASA colour reverse slide films at a stop of 4 with an exposure time of 80 sec. From the relation in film density between the background and the tracks of stars from which the "magnitude" is known, the sky luminance can be assessed. The method has been developed in Japan (Isobe 1995; Kosai & Isobe 1991). The same method was used in the Netherlands on 7 February 1992, between 20.00 and 22.00 hours local time at about 100 locations. The results are reported in Schreuder (1994). They are summarized here.

The individual values ranged from 20.97 to 17.2 mag/arcsec^2. From the average over the nine postal zones, it seems that the rural areas in the North and the East of the country score about one magnitude better than the highly industrialized Western parts. The number of measurements is, however, far too small to draw any firm conclusions.

Some of the photographs were measured by Dr S. Isobe at the National Astronomical Observatory of Japan in Osawa, Mitaka, Tokyo, using a micro-densitometer and measuring one particular star in all exposures (Capella). The method is the same as used in the Japanese surveys.

3. The Dark Night of 5 April 1997

On 5 April 1997, the 'Stichting International Dark Sky Association Nederland' undertook a survey to estimate the extent of the problem. A large number of volunteers made an estimation of the weakest star they could see. The 'dark night' was accompanied by a series of activities aimed at the media (publications in daily newspapers and weeklies, radio and TV spots, a travelling exhibition etc.). In this report, the results of the observations of threshold star visibility are summarized. More details are given in Schreuder (1999).

In total, 136 observations divided over 89 locations within the Netherlands were used. A toilet paper roll was used to define the area of observation, which was the sky area was centred around the 'rectangle' of the Big Dipper (Ursa Major). The best of the two eyes was used. The area measures about 20° in angular diameter. Within that area, there are about 10 stars brighter than magnitude 5, the usual value given for the threshold for normal good eyes of untrained observers, a clear sky and no light pollution.

The data are combined within the nine post code areas. Again, the number per area was too small to do any statistical analysis. The threshold star observations were converted into sky glow luminances. In the 'worst' area, the sky was about 2.5 times as bright as in the 'best' area. The individual observations are, of course, much wider apart. The lowest value in the measurements is 0.564 mcd/m^2, whereas the highest is 6.12 mcd/m^2 - more than a factor of ten!

The data permitted an investigation of the way the light pollution decreased during the night. Is was found that during the night, averaged over all areas, the luminance dropped to about 33% of the evening value.

4. Discussion

It is interesting to compare the three sets of data. The Schreuder and Isobe data are from the same date, but are different selections of the material; the dark sky data are five years later. There is a certain similarity between the three sets of data. The average values are of the same order of magnitude, suggesting that all three methods are likely to give useful information. There is no clear sign of the fact that the light pollution increased considerably - at least according to subjective experiences - between 1992 and 1997. Clumping the data even further into 'industrial', 'mixed' and 'agricultural' areas does not seem to reduce the differences between the three sets of measurements.

In conclusion, light pollution as a whole is a serious problem in the Netherlands. The values found in these tests are systematically considerably higher than the natural background luminance. In most areas the sky luminance is often ten times as high! One does not need, of course, measurements for this statement: even the most cursory glances outdoors confirms this. However, in order to find a base for regulatory measures or for legislation, a more detailed, quantified, picture is required. For this, the measuring grid must be finer and the number of observations needs to be much, much larger - we estimate about a factor of one hundred larger. Such a project cannot be handled by a small number of volunteers: the project must be managed professionally and there are considerable sums of money involved. It seems, however, that the expenditure is worthwhile. It might be profitable to look at large, international bodies for financial support.

Acknowledgments. The author wishes to thank Dr S. Isobe for making measurements at his observatory of a large number of slides. The author wishes also to acknowledge the essential work of Hittie Dales, who was the main 'engine' behind the 'dark night'. Unfortunately, she did not live to see the results of the analysis.

References

Anon. 1997, *Control of light pollution - measurements, standards and practice* (Conference organized by Commission 50 of the International Astronomical Union and Technical Committee TC 4-21 of la Commission Internationale de l'Eclairage CIE at The Hague, Netherlands, on August 20, 1994). 1997, The Observatory, 117, 10-36

CIE 1993, Urban sky glow, A worry for astronomy. Publication No. X008. 1993. CIE, Vienna

CIE 1997, Guidelines for minimizing sky glow. Publication No. 126-1997. Vienna, Commission Internationale de l'Eclairage CIE

Crawford, D.L. 1997, Photometry: Terminology and units in the lighting and astronomical sciences; pp. 14-18 in: Anon. 1997

Crawford, D.L., ed. 1991, *Light pollution, radio interference and space debris* (Proceedings of the International Astronomical Union colloquium 112, held 13 to 16 August, 1989, Washington DC). Astronomical Society of the Pacific Conference Series Volume 17, San Francisco

Isobe, S. 1995, Energy loss of light ejected into space. Paper, 3rd European Conference on Energy-Efficient Lighting, 18th-21st June 1995, Newcastle upon Tyne, England

Kosai, H. & Isobe, S. 1991, Organized observations of night-sky brightness in Japan. National Astronomical Observatory, Mikata, Tokyo, Japan. In: Crawford, ed. 1991

Kosai, H.; Isobe, S. & Nakayama, Y. 1993, A global network observation of night sky brightness in Japan - Method and some result. In: CIE 1993

Schreuder, D.A. 1994, Comments on CIE work on sky pollution. Paper presented at 1994 SANCI Congress, South African National Committee on Illumination, 7 - 9 November 1994, Capetown, South Africa

Schreuder, D.A. 2001, Recent CIE activities on minimizing interference to optical observations, IAU Symposium No. 196, "Preserving the Astronomical Sky", 12-16 July 1999, Vienna, Austria

Schreuder, D.A. 1999. De donkere nacht van 5 April 1997. (The 'dark night' of 5 April 1997. Zenit (To be published). Leidschendam, the Netherlands, Duco Schreuder Consultancies

Preserving the Astronomical Sky
IAU Symposium, Vol. 196, 2001
R. J. Cohen and W. T. Sullivan, III, eds.

Light Pollution in Quebec

Yvan Dutil

Observatoire du Mont Mégantic

Abstract.
    As in any developed country, light pollution is a serious problem in Quebec. However, a specific cultural and economical situation forces us to use new tactics when it is time to discuss this problem. As an example, using energy loss as an argument against light pollution is very inefficient since the electricity is cheap (US$0.04/kWh) and clean (hydroelectricity). On the other hand, using the negative ecological impacts of light pollution appears to be the most effective way to create awareness in the population

1. Introduction

Interest in the fight against light pollution has grown in the last five years in Quebec. It began when professional astronomers realized the threat of light pollution that would be soon faced by the only research observatory in Quebec. Rapidly the amateur astronomers joined the battle. Now, the astronomical community at large is well organized and is starting to spread the word to the general public.

With a price of Can$0.06/kWh (Can$1 = US$0.68 at 08/99), the electricity in Quebec is cheap. This simple fact delayed the introduction of efficient lighting methods. Nevertheless, for some time, Hydro-Québec, the government-owned electricity producer has subsidized the change of streetlights to more energy efficient models. Unfortunately, this programme is now finished. Meanwhile, some cities have changed their street lighting system from the old "cobra" mercury-vapour type to a much more efficient "Helios" high-pressure sodium (HPS) full cut-off type, locally produce by Lumec-Schréder. However, most cities chose to limit the cost of the transformation and simply replaced the mercury-vapour lamp with a HPS lamp. Nevertheless, this street lighting looks good when compared to most systems used for domestic or industrial applications, such as the standard unshielded 175-W mercury-vapour "security light" used by farmers. Conservative calculations based on DMSP data (Isobe 1999) estimate the cost of energy loss for the whole of Quebec to be Can$45 millions per year (760 GWh/yr).

However, in a country where the sky is clouded two nights out of three, where there is 14 hours of darkness in winter and where the electricity is one of the cheapest in the world, we are fighting an uphill battle. Fortunately, we are bringing into the battle the ecological groups that are starting to consider light as another menace to the environment.

2. Educational Activities

In 1995 I established a web page (http://astro.phy.ulaval.ca/astro/pol_lum.html) about light pollution in Quebec. At that time, it was one of the first web pages written in French on this subject. Today, this site receives a few thousand visitors per year and is referenced throughout the world. In the last few years, many astronomical events (comets Hyakutake and Hale-Bopp, Perseid and Leonid meteor showers) have been used to awake the public to the problem. Ironically, the Znamya satellite reflector project was one of the most useful events to raise this issue, thanks to the tremendous media coverage it received.

Also in 1995, the Fédération des Astronomes Amateurs du Quebec (FAAQ) gave birth to their "Comité Ciel-Noir". In collaboration with the lighting industry and professional astronomers, this committee produced a document about the protection of the night sky. It is an intervention guide for those who want to establish a lighting regulation. This document is now broadly available in its printed or electronic version (http://www.quebectel.com/faaq/cielnoir.htm). In addition, this committee also produced a nice report that was published in *Québec-Science*, the largest popular science magazine in Quebec.

Two other astronomical institutions are also strongly involved in public education. The Astrolab du Mont Mégantic, the visitor centre of the Mont Mégantic Observatory, is probably the most committed to this task. In addition to its own education programme explaining the importance of the dark sky for astronomers, the Astrolab's future is preserved by protecting its own dark sky around Mont Mégantic through informing local authorities about actions to be taken. The Planétarium de Montréal has its own way to explain the problem to its public. During the presentation, they simulate the effect of light pollution on the visibility of the stars. Therefore spectators can directly compare between a light-polluted environment and a dark one. Employees of the planetarium have also written to local newspapers about the ecological problems caused by artificial lighting.

3. The Mont Mégantic Observatory

Jointly operated by Université de Montréal and Université Laval, the Mont Mégantic Observatory is the only major telescope (1.6 m) in eastern Canada. It is built in one of the last relatively dark regions in southern Quebec. Site selection analysis made in the seventies estimated the light pollution at about 25% of the natural background level. Now it is certainly higher, some observers estimating it as much as 50% of the natural background. In order to establish the main sources of pollution, we have estimated the relative contributions of nearby towns by using the Walker law (Walker 1977). This coarse study allows us to pinpoint the largest polluters at the observatory site. Small towns near the observatory are the strongest contributors; about half of the overall pollution comes from less than 25 km. An additional 30% comes from the city of Sherbrooke (population 120 000) situated at 55 km. Future analysis will be based on satellite data and a model of the atmosphere in order to establish more accurately the contributions of different sources (see Cinzazo 2001 in these proceedings).

Since about half of the light comes from the nearby but relatively unpopulated region, significant improvement could be expected at little expense, In addition, Lumec-Schréder is designing a cheap version of its "Helios" system, a full cutoff-light fixture, in order to facilitate its utilisation for domestic use. If farmers adopt this system instead of their monstrous mercury-vapour lamps, this alone may be enough to restore most of the original darkness of this rural sky.

To finish on a positive note, I shall note the interest of local communities around Mont Mégantic Observatory for protection of the night sky. At this moment, discussions between the Observatory, the Astrolab, the Fédération des Astronomes Amateurs du Québec, local and county authorities and the lighting industry (Lumec-Schréder) is taking place about future installation of appropriate lighting fixtures in the surrounding region.

4. Political Action

The first major political action against abusive lighting was carried out in July 1997. At that time, a new casino was inaugurated in the city of Hull, near Ottawa. The administration of the casino had the bright idea of using high intensity projectors to attract gamblers like moths! Overall, it took one month of protests from amateur astronomers and ordinary citizens upset by the horrendous beam of light, a petition of a few hundred names, motions by nearby city councils and many articles and editorials in newspapers to force them to shut down their system indefinitely.

Unfortunately, things are not always so easy. The next fight was organized against the proposal of the Commission de la Capitale Nationale, which wants to transform Quebec City into a "Ville Lumière". To do so, they plan to illuminate 63 sites around Quebec City: buildings, bridges, falls, cliffs, etc. An international protest campaign (support by IDA members) and a petition did not induce any positive reaction from the authorities. In addition, a group of ornithologists from Toronto (FLAP) alerted the Commission de la Capitale Nationale to the dangers to birds of lighting buildings. To the inquiry of a journalist, their only comment was: "the lamps used are too faint to create any problem". However, they naively admitted that the electricity bill for one of their sites was "only" Can$15 per day. If we extrapolate from this statement, the overall "Plan Lumière" will increase the upgoing light for the region of Quebec City by at least 10%; locally, the effect will be even larger. We have not given up yet, but in the short term it appears they will continue with their project with the blessing of the local and provincial authorities.

5. Enlisting More Troops

The astronomical community is very small in Quebec. Overall, there are about 15 professional astronomers and a few hundred amateur astronomers. Obviously, it will be extremely difficult to change the mind of the governments without some external support. Attempts have been made to get the attention of ecological groups. So far results have been limited but significant.

Actions such as those above should only be attempted if you have access to a complete and detailed description of direct effects of light pollution in ecological terms. We had access to such a document, thanks to the French astronomical community. The document "Impacts écologiques de l'éclairage nocturne" was written by François Lamiot following the "Premier Congrès européen sur la protection du ciel nocturne" in May 1998. This is a compilation of effects of artificial lighting observed on birds, insects and even snails! We have distributed it to a large number of ecological groups in Quebec. It can also be downloaded (http://ecoroute.uqcn.qc.ca/gen/bull/index.html) from the web site ÈcoRoute de l'Information supported by l'Union Québécoise pour la Conservation de la Nature. Following this intervention, some "green" associations have demonstrated their interest.

Finally, I want to emphasize the fact that media tend to be more receptive to the environmental impact of artificial lighting than to the astronomical aspect of the problem. Therefore, in order to use this particular sensitivity, volunteers should have access to a good database of *direct effects* of artificial light on wildlife and on human life. Such a database would be an invaluable weapon in the battle against light pollution.

6. Conclusion

The particular cultural and economical situation produces an especially high level of light pollution in Quebec. Fortunately, in the last few years we have seen an increase of awareness in the astronomical community. Now, this awareness is growing beyond the ranks of astronomers and is reaching the ecological community plus a segment of the general population. We expect to harvest soon the fruits of this education process.

References

Cinzazo, P. 2001, in these proceedings

Isobe, S. 1999, private communication

Walker, M. F. 1977, PASP, 89, 405-409

Preserving the Astronomical Sky
IAU Symposium, Vol. 196, 2001
R. J. Cohen and W. T. Sullivan, III, eds.

Observing Conditions from 1988 to 1999 at Huairou Solar Observing Station

Yuanyong Deng[1,2] and Yihua Yan[1,2,3]

[1]*Beijing Astronomical Observatory, CAS, Beijing 100012, China*

[2]*National Astronomical Observatories, CAS, Beijing 100012, China*

[3]*CAS-Peking University joint Beijing Astrophysics Center, Beijing 100871, China*

Abstract. This paper summarizes the change of astronomical observing conditions since 1986 at Huairou Solar Observing Station. Our results from 1988 to 1999 are as follows: quantitatively, stray light has increased and visibility has decreased; qualitatively, the seeing condition has become worse.

1. Introduction

Huairou Solar Observing Station (HSOS) is located on a small island near the north bank of Huairou Reservoir, 60 km north of Beijing. The water surrounding the station extends out about 2 km from the observing tower. The observing dome is on the top of a 31-meter tower. The elevation of HSOS is about 62 m, and the location is 40.°4 N and 116.°6 E. The mean temperature is 26°C in summer and -4°C in winter. The annual precipitation is about 500 mm.

The vector magnetic fields and line-of-sight velocity at the photosphere and chromosphere of the Sun are observed at HSOS. The main optical telescope is a Solar Multi-Channel Telescope (SMCT), which consists of a 60 cm Solar Nine-Channel Telescope, a 35 cm Solar Magnetic Field Telescope, a 10 cm Full-disc Magnetograph, a 14 cm Partial and Full-disc Hα Telescope and a 8 cm Full-disc Calcium Monochromator. HSOS plays an important role in the solar community.

Solar observations at HSOS began in 1986 and routine observations began in 1988. In addition to the scientific data, we have also recorded some parameters of observational conditions, such as seeing, stray light and visibility of the terrestrial atmosphere, etc. In this paper, we give a summary of these parameters.

2. Observations and Results

2.1. Amplification Factor

In our solar observation, the image collecting and processing system consists of a CCD Camera and corresponding frame grabber. Before digitizing a solar image,

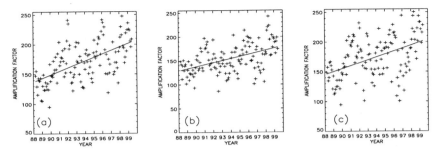

Figure 1. Change of amplification factor with time for HSOS solar observations. Each point is a monthly average over a daily time interval (a) times before 0200 UT, (b) 0200 - 0600 UT and (c) after 0600 UT. (local noon = 0400 UT)

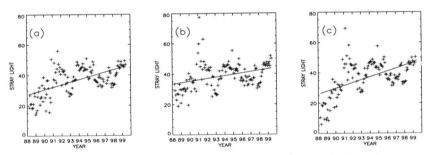

Figure 2. Change of stray light with time. Plots as in Fig. 1.

the amplification factor of the system must be determined by the intensity of received sunlight, in order to get highest accuracy. The weaker the intensity of sunlight, the larger the amplification factor and thus the larger the noise. Of course the annual change of solar declination causes a change of sunlight with the seasons, but it does not affect long-term statistics, such as we use in this paper. On the other hand, the terrestrial atmosphere can reduce sunlight by absorption, scattering, etc. Figure 1 shows the amplification factor from March 1988 to May 1999. Each point represents the average value within one month. It is obvious that the condition of the atmosphere has worsened in the past ten years. The solid lines are linear fits and have slopes indicating increases of 49%, 36% and 41% for (a), (b) and (c) respectively.

2.2. Stray Light

Generally, the stray light in solar optical observation comes from the Sun itself, the terrestrial atmosphere and the telescope. For our long-term statistics, we might think that most of the change of stray light would be caused by the atmosphere. Figure 2 shows the stray light recorded in our system from March 1988 to May 1999. The increase of stray light is obvious. The slopes for the fits indicate increases of 14%, 8% and 17% for (a), (b) and (c) respectively.

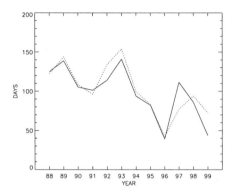

Figure 3.   The annual number of days with "good" seeing conditions. Solid line: morning (before 0400 UT). Dotted line: afternoon (after 0400 UT).

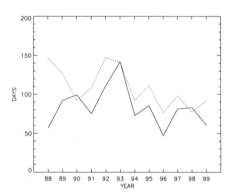

Figure 4.   The annual number of days with "good" visibility. Lines as in Fig. 3.

2.3.   Seeing

Every observational day we also record the seeing condition, as estimated by visual judgement. We divide seeing condition into six grades: best, better, good, bad, worse, worst. Figure 3 shows that the annual number of days with "good" grade or better seeing has declined significantly over the eleven years of records.

2.4.   Visibility

The visibility of terrestrial atmosphere is also recorded by visual judgement. Compared with the amplification factor discussed earlier, this parameter is less strict but more direct. We also divide the visibility into six grades: best, better, good, bad, worse, worst. Figure 4 shows that the annual number of days with "good" grade or better visibility has tended to decline.

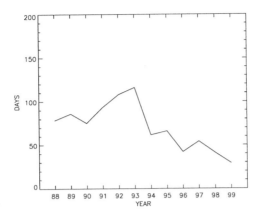

Figure 5.   The annual number of days with "good" image quality.

2.5.   Image Quality

Image quality is affected by all of the factors mentioned above. Visual judgement of this is also carried out every day. We again divide the image quality into the same six categories. Figure 5 shows the annual number of days with "good" grade or better image quality. One can see that the number of days with good-quality data has significantly declined.

3.   Discussion

The observational conditions at HSOS are becoming worse. We think this serious change comes from either a global change of weather or from local air pollution. Here we give a simple discussion of local factors that affect the atmosphere. The first factor is the expansion of the city dimensions. By the official plan of Beijing, the construction volume will increase 100% from 1990 to 2000 and 200% to 2010. By the most conservative estimates, the expansion rate of Beijing in the past ten years has been even larger than this plan. The second factor is the increasing number of automobiles. Experts estimated that the number of automobiles in Beijing is now 1,400,000, several times more than ten years ago. They warn that this number is growing by 200,000 every year. In addition, there are some others factors that seriously pollute the local atmosphere, such as the exploitation of natural attractions for tours around Beijing, etc.

Are there ways to deal with this bad condition? The permanent way is to reduce pollution, but this is a long-term sociological problem. For our current observations, modern technology provides some opportunities, such as more sensitive detectors, adaptive optics, correlation trackers, etc. In fact, as has been done at HSOS, these tools have been widely applied in astronomy. This is successful, but at the cost of a lot of bills and technical complexity.

Acknowledgments.   We thank Ms. Wang Guoping and Mr. Qi Hongwei for their assistance.

Preserving the Astronomical Sky
IAU Symposium, Vol. 196, 2001
R. J. Cohen and W. T. Sullivan, III, eds.

The Situation of Light Pollution in Germany

Andreas Hänel

*Museum am Schölerberg, Am Schölerberg 8, D-49082 Osnabrück and
DARK SKY, working group of the Vereinigung der Sternfreunde e.V.,
Germany.*
email: ahaenel@rz.uni-osnabrueck.de

Abstract. The central European climate in Germany is not ideal for serious astronomical observations. Increasing light pollution hampers thousands of amateur astronomers and millions of people from seeing starry skies and the Milky Way.

To study the sources of light pollution, we estimate the increase of light pollution in Osnabrück, a town of about 160,000 inhabitants located in northwestern Germany. We try to extrapolate these statistical data to Germany and discuss possible reasons for increasing light pollution though the energy consumption stagnates. Some enthusiasts of the German amateur association "Vereinigung der Sternfreunde" have formed a working group DARK SKY to exchange information on this matter. Finally, we present some activities to make the public aware of the problem.

1. Introduction

Weather conditions in Germany are not favorable for astronomical observations. Therefore astronomers at traditional observatories such as Hamburg, Potsdam, Göttingen, Bonn, Jena, Heidelberg or München concentrate most of their research at observatories in better climates. Yet thousands of amateurs must stay home and struggle against light pollution. Therefore they build small observatories outside of the cities, take their transportable telescopes to one of the rare dark places in Germany or travel to remote foreign places like Gornergrat, Haute Provence, Pico Veleta, or Namibia.

Millions of people have a latent interest in astronomy, but they have only rare chances to see the wonders of the starry sky because the sky is heavily light polluted. Coming to a planetarium they are fascinated by the projection technique *and* by the artificial dark sky. Many visitors see the Milky Way for the first time in the planetarium.

2. Estimating Light Pollution in Germany

Shocked by the predictions for the increase of sky brightness published by Sullivan (1984) and Riegel (1973) for the United States, we wanted to estimate this increase for Germany. No continuous measurements of sky brightness at German observatories are known. Instead, limiting magnitudes determined by an amateur observer of variable stars in Stuttgart can be used to estimate the increase of light pollution: in 1958 the limiting magnitude was 5.6 mag, in 1971 it increased to 5.2 mag and since then it has increased by about 0.1 - 0.2 mag (Marx 1972, 1997, priv. comm.).

The main contributors to light pollution are:

- Lights for advertising. These seem to play a minor role in Germany, because they are restricted for traffic safety reasons. Increasingly disturbing become the powerful moving spotlights used by entertainment establishments such as cinemas or discotheques. In rural areas these skybeams are visible over dozens of kilometers and they increase the brightness of the night sky considerably. Due to different judisdiction in the cities, many still shine, but some have been switched off, for traffic safety or environmental reasons.

- Direct and indirect lights from industry and private households, such as illuminated windows, or unshielded garden lighting or security lights. Their contribution is difficult to estimate.

- Street lighting is the most important contributor, though estimates in the literature vary between 14 % (cited by Schreuder 1991) to 50 % (Shaflik 1997), a value that seems more realistic.

As no data on increasing light pollution exist, we will estimate this from the increase of electric power consumption. Detailed information on energy consumption for street lighting in Germany exists for 1989, when $2.8 \; 10^9$ kWh were used in the (former) western Länder in Germany, while the total consumption of electricity was $3.9 \; 10^{11}$ kWh. In 1996 the energy for street lighting had risen to $3.5 \; 10^9$ kWh for the reunified Germany, while the total power consumption was $4.8 \; 10^{11}$ kWh (Bundesministerium für Wirtschaft 1996). From these data for Osnabrück and Germany it can be estimated that less than 1 % of the total electric energy is used for street lighting.

Detailed figures about street lighting in Osnabrück, a city with 160,000 inhabitants in northwestern Germany, are available since 1981, as provided by the Stadtwerke Osnabrück, the municipal energy supplier (private communication). Figure 1 shows the percentage increase for the number of luminaires, the power consumption and the power supply. While power supply and consumption remained more or less constant, the increase of the number of luminaires is remarkable. The main reason for this is the increasing number of residential areas where lighting has to been installed by the Stadtwerke. Power supply and consumption have remained constant for different reasons:

- Financial cuts in the public administration force the reduction of energy costs. Therefore street lights are reduced or even switched off at certain times (after 11 pm or midnight).

- More efficient lamps are used. In residential areas there is an increasing trend to compact fluorescent lamps or high-pressure sodium lamps. On main streets mercury vapour lamps are replaced mainly by high-pressure sodium lamps, sometimes by low-pressure sodium lamps. However, due to

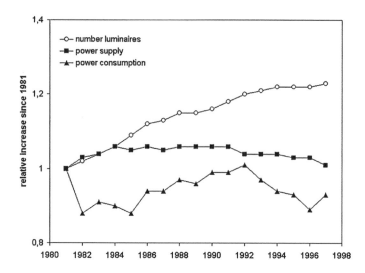

Figure 1.   Percentage increase of number of luminaires, power consumption and power for public street lighting in Osnabrück.

their monochromatic colour the low-pressure sodium lamps are not very much favoured.

Figure 2 shows the relative increase of electric power consumption in Osnabrück, Germany and the United States since 1950. We assume that the power consumption for light amounts to about 1 % of the total power consumption in Germany during the whole time interval. In addition, the light emitted towards the sky depends on the efficiency of the lamps, their shielding and the reflectivity of the streets. The increasing light emission postulated by Riegel (1973) is due to the fact that the number ratio of vapour/incandescent lamps increased from 0.1 in 1960 to 1.2 in 1970. In Germany incandescent lamps have rarely been used. Besides mercury vapour lamps, fluorescent lamps have often been used. We assume that the light efficiency of the lamps has increased from a mean of 40 lm/W (corresponding to 50 % fluorescent, 40 % mercury vapour, 10 % incandescent lamps) to 100 lm/W (50 % fluorescent, 25 % mercury vapour, 25 % sodium vapour) during the time interval considered. We then find the relative light increase for Germany given in Fig. 2. Therefore we suspect that the increase of light emission in Germany and perhaps in all Europe is not as high as estimated by Riegel (1973) for the USA.

In Germany street lighting is regulated by the Deutsche Industrie-Norm DIN 5044 which recommends on principal roads a minimal luminance between 0.3 and 2 cd/m^2, depending mainly on traffic density. Later at night, when the traffic density is less, illumination should be reduced. On the other hand there exist guidelines (Messung und Beurteilung von Lichtimmissionen, Länderausschuss für Immissionsschutz) for light emissions that indicate that vertical illuminance values of more than 1 cd/m^2 may already be considered as annoying.

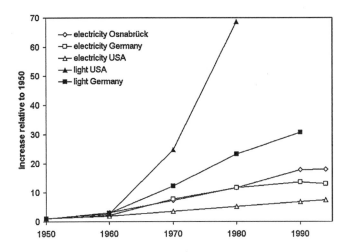

Figure 2.   Percentage increase of power consumption and increase of light pollution in Osnabrück, Germany and the United States.

3.   Aims for Reducing Light Pollution

Altogether, the following energetic and social factors influence the future increase of light:

- Light emission will increase due to:

  - more lamps with the extension of new residential or industrial areas
  - more efficient lamps
  - increasing demand for more security
  - decreasing costs for electricity (due to the liberation of the European market for electricity)
  - new lighting norms planned by the European Community, which seem to be less stringent than the national norms.

- The following reasons might produce a stagnation of lighting increase:

  - financial cuts at the public sectors
  - less stray light from more efficient fixtures
  - aims to reduce CO_2 production in the framework of a sustainable development (agenda item 21 from the UN conference in Rio de Janeiro)
  - introduction of ecological taxes, as planned in Germany
  - increasing public awareness of light pollution as energy waste, annoying light emission, and having negative influences, for example on nocturnal animals.

We think that one important step to reduce light pollution is to inform the public. Therefore in 1993 we started in Osnabrück with a poster presentation at a local fair on ecology, giving some information on the theme.

In 1995 some members of the association of amateur astronomers *Vereinigung der Sternfreunde (VdS)* formed a group called "DARK SKY" to exchange information on problems and possible solutions of light pollution. This group meets once per year on the occasion of the international telescope meeting at the Vogelsberg in the centre of Germany, which is one of the rare dark places. At the last meeting some biologists participated, because nocturnal insects and birds are disturbed by overwhelming night lighting. In a study by the Naturschutzbund NABU (Schanowski and Späth 1994), insects were collected at different sorts of lamps. It was shown that many fewer insects are attracted by sodium vapour lamps than by compact fluorescent or mercury lamps due to less ultraviolet emission. The group DARK SKY also issues press releases and information sheets. Main communication is via email and the general public can get more detailed information from our web site :

http://www.physik.uni-osnabrueck.de/astro/

In 1996 we proposed an experiment for the Astronomy On-line project of the European Southern Observatory (ESO) in collaboration with the European Association for Astronomy Education (EAAE). The aim of this project was to demonstrate the use and the advantages of the internet to young people (West and Madsen 1997). We proposed to measure light pollution in Europe by using the simple method of determining the limiting magnitude in the constellation Ursa Minor. However, during the peak phase of this project in November 1996, weather conditions were extremely bad, and only 25 reports were received, mainly from the southern parts of the continent.

Finally, a planetarium is another medium where light pollution can impressively be demonstrated to the public, just by raising the dome illumination a little bit. The worsening visibility of faint stars, the Milky Way, or a comet's tail is then dramatically demonstrated.

Acknowledgments. I would like to thank the Stadtwerke Osnabrück for providing the data for Osnabrück.

References

Bundesministerium für Wirtschaft 1996, Energie Daten '96 (and earlier years)

Marx, H. 1972, in *50 Jahre Volkssternwarte Stuttgart*, Verein Schwäbische Sternwarte ed., p. 41.

Riegel, K. 1973, Science, 179, 1285

Schanowski, A., Späth, V., 1994, Überbelichtet, Naturschutzbund Deutschland (NABU), Kornwestheim.

Schreuder, D. 1991, in *Light Pollution, Radio Interference, and Space Debris*, D.L. Crawford ed., Astr. Soc. Pac. Conf. Ser. 17, p. 25

Shaflik, C. 1997, Inform. Sheet 125, International Dark-Sky Association, Tucson

Sullivan, W. 1984, S&T, 67, 412

West, R., Madsen, C. 1997, The Messenger, 87, 51

Preserving the Astronomical Sky
IAU Symposium, Vol. 196, 2001
R. J. Cohen and W. T. Sullivan, III, eds.

Economic Imperative versus Efforts for Preserving an Astronomical Site

Hakim L. Malasan[1], Moch. Arief Senja, Bambang Hidayat and Moedji Raharto

Bosscha Observatory at Lembang and Dept. Astronomy, ITB, Bandung-Indonesia

Abstract. In this paper we report the gradual change of the physical environment of the Bosscha Observatory in Lembang, Indonesia. From photometric observations carried out in 1982 through 1993 we found a slight increase of the aerosol content in the atmosphere. However, light pollution from the expanding city of Bandung (pop. 2×10^6; 10 km away at 600 m altitude below the Observatory) has produced significant light scattering at zenith distances greater than $55°$ toward the south.

1. Introduction

The Bosscha Observatory at Lembang, Java, Indonesia has enjoyed a relatively quiet and stable environment from its founding in the 1920s until about 1980. Its environmental change in the 1980s came about as a result of relatively fast economic development in the 1970s. This stimulated demands for land, housing and accordingly lights, primarily due to the increasing numbers of a new emerging middle class society. Their view of land and housing as economic assets seemed to be more prevalent than the real need for accommodation. Efforts to protect important scientific and cultural sites have been undertaken by the central and by local governments (Govt. Act. No. 5, 1992). Government Regulation No. 10 (1993) contains the outlines for protection of scientific and cultural institutions. In these decrees and regulations, views of environmental-friendly, sustainable development have been incorporated.

However, perceptions pertaining to this aspect, coupled with weak control from the executive agencies, have in many cases created problems. Physical development driven by economic imperative dominated the attitudes of many decision makers in the 1980s in such a way that the protective spirit of the law was occasionally put aside. Nevertheless, there was a concerted effort to protect the Observatory in the early 1970s when Parliament and the central government took the stern decision to forbid the development of a hotel near the Observatory.

Because of the inherent topographical nature of the Observatory grounds, lying on the top of a rather steep hill that serves as a water-catchment area, the local government has established a development-free zone within a radius of 2 km around the Observatory. This regulation has unfortunately not been

---

[1]Momentarily at the Gunma Observatory, Japan

strictly observed due to increasing demands and unrealistic assessment about land use, as well as a narrow view with regard to physical constraints required by an astronomical observatory. The environmental engineers and scientists of the Institute of Technology of Bandung, as well as other environment-conscious groups, have made the case public by defending the views that preservation of lands immediately around the Observatory will not only do good for science, but it will also serve as a water reservoir for the people of the city of Bandung (population of 2 million, at an airline-distance of 10 km, 600 m below to the Observatory site). The view can be comprehended because of the nature of hills which surround the observatory complex.

Unfortunately, narrow interpretations of personal and human rights have occasionally brought conflict. The State Law No. 5 (1992) states that cultural and scientific sites, subject to a strict realistic definition, have to be protected. Protected areas around a scientific establishment include the supporting areas adjacent to it which make the institution function well. But here we are witnessing the confrontation of law versus economic imperative such as the increasing value of land and the pressure arising from population increase.

Stimulated by the volcanic eruption of Mt. Galunggung in 1982 (100 km away from the Observatory), which spread fine dust particles for several weeks, and by the increasing stray light, we decided to monitor atmospheric transparency.

2.   Efforts and Observations

It was not an easy task to convince the population that what they need is appropriate lighting directed groundward. We have therefore since the 1970s carried out efforts to preserve the optical window for our work by persuading the local community to reduce outdoor light consumption. Public cooperation has resulted in keeping the light level at the zenith relatively constant over the last 20 years, at about $20^{m}/\square''$ in the blue spectral region. Unfortunately, the advent of highways requiring brighter lighting has forced road engineers to introduce mercury lights. In addition, neon lights for advertisement and billboards have become more pronounced in the township of Lembang, 1 km away at an elevation of about 60 m below the Observatory. At the same time the previous use of incandescent lightbulbs has, unfortunately, been replaced by use of gas pressure lamps.

The expanding city of Bandung has produced more light output. Although it is partially screened by hills of altitudes higher than 700 m, we lost our view towards the southern and eastern horizons at elevation angles below 30° and 40° respectively. Observation below this level produces noisy results due to the increasing stray light.

The topography around the Observatory has formed a natural barrier which has made it impossible for expansion of residential areas toward the Observatory. Accordingly, West Java Province Law forbade housing development projects on any hills of slope more than 40° and, to ensure a development-free zone, areas within a radius of 2.5 km from the Observatory. The choice of the radius was again dictated by the topography of the region. The progress of technology, however, might change this as housing engineers claim that they technically can overcome the problem of building on steep slopes. But the conservationists maintain that preserving the hills creates an important natural asset for successful protection of water resources.

Disturbance by natural phenomena, such as volcanic eruptions, has occasionally disrupted the cleanliness of the atmosphere and changed seeing conditions considerably. The change of the atmospheric coefficient of extinction due to a nearby Galunggung volcano eruption has been reported by Hidayat and Malasan in 1985 and by Malasan et. al. (1984). The fine dust that prevailed in the atmosphere for weeks disrupted the atmosphere so much that observation of any kind was impossible. Assessment of the sky brightness has been given by Raharto (1995) and by Malasan and Raharto (1993), in which they described the observations of atmospheric coefficients of extinction from 1982 to 1993. Some changes were found and attributed to atmospheric pollution, in particular that caused by the increase of aerosol content in the atmosphere. There is also an indication that some change may be seasonal.

A thorough study of the sky brightness has been carried out by Senja (1999). The study, initiated by Malasan, was based on systematic photometric observations in 1994 on stars in IC 4665. The results indicate that in the U-band the sky brightness at the zenith is $18^{\mathrm{m}}/\square''$. The value deteriorates rather fast, reaching $14^{\mathrm{m}}/\square''$ at $z = 55°$ to the south. The increase of incandescent lights in the cities may have been the cause of the sky glow, in particular toward the south and east. The increase of the blue absorption coefficient from 0.5 in 1982 to 0.7 in 1993 has been attributed to the increase of aerosol content, if we apply the Hayes and Latham (1975) and Guitierez-Moneno et. al. (1982) procedures to our data. The value in the U-band has been observed to increase from 0.6 in 1989 to about 0.9 in 1993.

The seeing at Lembang is generally good. Within 30 minutes after sunset star images become steady with an average annual seeing of $\sim 1''$. Continous observations of close visual double stars have not shown any appreciable change, except for short occasions when the seeing becomes more than $1''.5 - 2''.0$, the cause of which remains unknown. We plan to continue monitoring it.

3. Conclusion

Governmental good intentions have sometimes been opposed by ignorance and narrow interpretations of personal and human rights. In the last two years, due to economic decline, the formerly protected landscape immediately around the Observatory has been opened for non-agricultural purposes. This might cause serious problems in the future. Up until now the telescopes at Bosscha Observatory can be employed to observe objects down from the zenith to about $z \sim 55°$ without serious problem. Beyond that point, observations at shorter wavelength regions will be troublesome. The double $24''-(61\text{-cm})$ refractor $(f/17)$ has fortunately not suffered too much.

To make the public aware of the beauty of astronomy and its needs, we continue to organize public meetings in order to disperse popular sciences. This effort is intended to implant a cultural identity that doing science is a noble endeavour. Scientifically, we are continuing to monitor the quality of the atmosphere and to increase our cooperative work through the Internet.

Acknowledgments. One of us (B.H.) would like to express his gratitude to the Dean, Faculty of Mathematics and Science, ITB, for the contribution toward his expenses to attend this meeting. The financial support and understanding from the Leids Kerkhoven-Bosscha Foundation in the Netherlands for accommodation and travel is also gratefully acknowledged, without which he could not have attended the meeting.

References

Guiterez-Moneno, A., Moreno, H. & Cortes, G. 1982, PASP, 94, 722

Hayes, A. S. & Latham, D. W. 1975, ApJ, 197, 593, 722.

Hidayat, B. & Malasan, H. L. 1985, Trans. IAU., vol. XIX B, p. 335

Malasan, H. L., Nurwendaya, C. & Hidayat, B. 1984, Proceedings, Institut Teknologi Bandung., 17, 11 (in Indonesian)

Malasan, H. L. & Raharto, M. 1997, Proceedings of the UN Meeting on Atmospheric Science, Held in Jakarta, Indonesia

Raharto, M. 1995, in *Star Watching*, National Astronomical Observatory, Tokyo, Japan, p. 61

Senja, M. A. 1999, Thesis, Dept. of Astronomy (in Indonesian)

Preserving the Astronomical Sky
IAU Symposium, Vol. 196, 2001
R. J. Cohen and W. T. Sullivan, III, eds.

Work for the Reduction of Light Pollution in Turkey

Z. Aslan

TÜBİTAK National Observatory and Akdeniz Üniversitesi, Antalya, Turkey

Abstract. In recent years Turkey has been losing the beauty of its night sky at a fast rate. The amount of outdoor lighting is increasing every day. University observatories at Ankara and İzmir are already irreversibly affected by light pollution. In this paper, work against light pollution in Turkey is outlined.

1. Introduction

The amount of outdoor lighting and, as a result, light pollution in Turkey is increasing at an incredibly fast rate in recent years. The university observatories at Ankara and İzmir are already irreversibly affected. The rate of light pollution is highest along the Mediterranean coast due to increased investments in tourism. Holiday villages and chains of new hotels are using decorative outdoor lighting. The sky brightness over Antalya increased from almost the level of natural background in 1986 to 23% higher in 1999 as seen from a distance of 30 km in a direction of 25 degrees altitude over the city. We have reason to believe that if the vertical light from the Antalya Riviera is controlled, the artificial sky glow could be cut down considerably. There are good and still dark potential observatory sites along the Mediterranean coast in the Taurus Mountains, which are now under threat. These include the recently established National Observatory, for which the site was selected in 1986 and construction began in 1992.

2. Work against Light Pollution

Through the initiative of TÜBİTAK (Scientific and Technical Research Council of Turkey) National Observatory, we have started a campaign against light pollution in Turkey. We got help from astronomers and from members of astronomy clubs in different cities in photographing examples of energy waste, poor lighting quality and poor lighting designs and installations. These are for use in our reports to local authorities, the press, and in our talks on television and radio, with the aim of getting light pollution to receive national attention. In reports and talks to local and central government officials responsible for outdoor lighting, due emphasis on energy waste has been rewarding.

A National Committee has been formed under the coordination of TÜBİTAK with members from the National Energy Conservation Centre (Ministry of Energy and Natural Resources), Highways Directorate, National Committee on Illumination and the Turkish Standards Institute. The committee has been working on a national awareness programme and on public education via the media, conferences, talks, information leaflets, etc. We emphasize energy waste, adverse effects on natural life, and the astronomical knowledge and culture we have inherited as a Mediterranean country, not to mention the value of the night sky as a "tool" for education. A sub-committee is studying local and national regulations on outdoor lighting, with a view to update these to include new standards. The Turkish Standards Institute is studying outdoor lighting and standards issued by other countries, in order to adopt national standards. The Committee is putting pressure on the central government to enact laws to curb light pollution, as well as on municipal governments to issue regulations to improve night-time visibility and to reduce urban sky glow. It is a slow and difficult process but we hope that the new standards and regulations will soon be available.

3.   Work for Potential Observatory Sites

In order to promote the protection of potential observatory sites, a plan is being made to measure the artificial skyglow from regions in the Taurus Mountains along the Mediterranean coast using a portable Celestron-8 telescope equipped with a photometer. We will then try to persuade the central and local municipal governments to adopt regulations to protect the sites.

Preserving the Astronomical Sky
IAU Symposium, Vol. 196, 2001
R. J. Cohen and W. T. Sullivan, III, eds.

Astronomical Sites in the Ukraine:
Current Status and Problems of Preservation

I. B. Vavilova[1], V. G. Karetnikov[2], A. A. Konovalenko[3],
O. O. Logvinenko[4], G. I. Pinigin[5], N. V. Steshenko[6],
V. K. Tarady[7] and Ya. S. Yatskiv[8]

[1] *Astronomical Observatory, Shevchenko National University,
   3 Observatorna St., Kyiv, 254053 Ukraine*
[2] *Astronomical Observatory, Odesa State University, Shevchenko Park,
   Odesa, 270014 Ukraine*
[3] *Institute of Radio Astronomy of the NASU, 4 Chervonopraporna St.,
   Kharkiv, 310002 Ukraine*
[4] *Astronomical Observatory, Lviv University, 8 Kyryla i Mephodija St.,
   Lviv, 290005 Ukraine*
[5] *Astronomical Observatory, 1 Observatorna St.,
   Mykolaiv, 327030 Ukraine*
[6] *Crimean Astrophysical Observatory, Naukove, Bakhchisarai,
   Crimea, 334413 Ukraine*
[7] *International Centre of Astronomical and Medical-Ecological Research,
   Golosiiv, Kyiv, 252022 Ukraine*
[8] *Main Astronomical Observatory of the NASU,
   Golosiiv, Kyiv, 252022 Ukraine*

Abstract.   The current status of optical and radio astronomical sites in the Ukraine and the problems of preservation are briefly reviewed. The problems of light pollution and the influence of technology can be solved using scientific and engineering methods. However the main problem of preservation is the economic one of maintaining infrastructure.

1. Introduction

There are two types of astronomical observatory in the Ukraine. Observatories of the first type were formed in the eighteenth and nineteenth centuries. Being now located in the big cities of Kyiv, Kharkiv, Lviv, Mykolaiv, Odesa and Poltava, these observatories are not involved in modern observational programmes but are used as administrative centres, and for publicity and training purposes. They have observatories located in suburbs and equipped with small optical telescopes.

The development of astrophysics in the twentieth century has resulted in the formation of a second class of observatory in the Ukraine, which includes the following: the Main Astronomical Observatory (MAO) in Golosiiv, Kyiv, the Crimean Astrophysical Observatory (CrAO) in Naukove, Crimea, the Observatory of the Institute of Radio Astronomy (IRA) in Hrakovo, Kharkiv, and the high-altitude observatory at the Peak Terskol (North Caucasus, Russia), which was founded by MAO in 1974 (now the "International Center of Astronomical and Medical-Ecological Research"). The astronomical sites in the Ukraine are listed in Table 1 and classified according to the above scheme.

Table 1.   Astronomical Sites in the Ukraine.

Title	Founded in	Type (1/2)	Situated in	Longitude Latitude	Altitude (m)	Field of Research
Astronomical Observatory, Lviv State University	1769	1	Lviv	24°01.8' 49°50.0'	330	solar physics astrophysics
Astronomical Observatory, Kharkiv State University	1808	1	Kharkiv	36°13.9' 50°00.2'	138	astrometry astrophysics
Mykolaiv Astronomical Observatory	1821	1	Mykolaiv	31°58.5' 46°58.3'	54	astrometry
Astronomical Observatory, National University	1845	1	Kyiv	30°29.9' 50°27.2'	184	astrometry solar physics astrophysics
Astronomical Observatory, Odesa State University	1865	1	Odesa	30°45.5' 46°28.6'	60	astrometry astrophysics
Gravimetrical Observatory	1926	1	Poltava	34°32.8' 49°36.3'	151	astrometry
Main Astronomical Observatory	1944	2	Golosiiv, Kyiv	30°30.4' 50°21.9'	188	solar physics astrophysics astrometry
Crimean Astrophysical Observatory	1945	2	Naukove, Peninsula Crimea	34°01.0' 44°43.6'	600	astrophysics solar physics radio astronomy
High-Altitude Observatory at the Peak Terskol	1974	2	Peak Terskol North Caucasus, Russia	42°30.0' 43°16.4'	3100	astrophysics solar physics
Observatory of Institute of Radio Astronomy	1985	2	Hrakovo, Kharkiv	36°56.0' 49°38.0'	150	radio astronomy

2.   Optical Astronomy and Light Pollution

2.1.   Positional Astronomy

University observatories in Kyiv, Kharkiv, Odesa (Table 1) were initially equipped with meridian instruments for the determination of star positions and were involved in the international programmes AGK3R, FK4, SRS etc. These observatories had practically finished their work by the 1990s. In 1995, the new automatic Axial Meridian Circle (AMC) equipped with CCDs was built in Mykolaiv Observatory (a former department of Pulkovo Observatory). The parameters of AMC are as follows: $D = 180$ mm and $F = 2480$ mm for the main telescope, and $D = 180$ mm and $F = 12360$ mm for the collimators. This unique meridian instrument of axial type has been recently used for observations of the HIPPAR-COS stars. At present, AMC is in the process of an upgrade that will allow the HIPPARCOS frame to be extended to objects down to 16^m. The main problem of preserving the Mykolaiv site is to reduce light pollution, so that the number of useful nights of observing (now 116 per year) does not decrease further.

2.2. Optical Astronomy with Telescopes of Small Size

Observational stations of university observatories (Lisnyky near Kyiv, Hrakovo near Kharkiv, Maiaky near Odesa and Briukhovychi near Lviv) are equipped with small reflectors ($D= 12,5$ to 100 cm), which are actively engaged in observations of comets, meteors, minor planets, satellites of planets and artificial satellites (Vovchyk, Blagodyr & Logvynenko 2001). In addition, photometric and polarimetric observations of stars are carried out with telescopes (AMT-3 with $D = 48$cm) at the Maiaky and Briukhovychi sites. The average number of clear nights for photometric observations is about 100 per year. Results of observations of Night Sky Brightness (NSB) and atmospheric transparency, which have been carried out at the Maiaky site during the years 1968-1997 are presented in Tables 2 and 3. The NSB ($erg/cm^2 \cdot sec \cdot A \cdot ster^2$) was measured at $\lambda = 0.36$ nm and $\lambda = 0.58$ nm after evening twilight (ev) or before morning twilight (mo), when the Sun was 17-20°below the horizon. Positive values of zenith distance are towards the Sun.

Table 2. Night Sky Brightness ($erg/cm^2 \cdot sec \cdot A \cdot ster^2$) as a function of zenith distance at the Maiaky site.

Date	$\lambda = 0.36$ nm					$\lambda = 0.58$ nm				
	60°	30°	0°	-30°	-60°	60°	30°	0°	-30°	-60°
27.06.68 ev	-6.34	-6.46	-6.50	-6.49	-6.42	-6.05	-6.07	-6.18	-6.16	-6.12
15.12.68 ev	-6.29	-6.32	-6.34	-6.38	-6.39	-6.05	-6.17	-6.19	-6.20	-6.09
13.10.69 mo	-6.52	-6.60	-6.58	-6.62	-6.65	-6.00	-6.19	-6.21	-6.22	-6.16
13.10.69 ev	-6.54	-6.62	-6.64	-6.63	-6.58	-6.06	-6.23	-6.18	-6.23	-6.11
Av. 1968-69	-6.42	-6.50	-6.52	-6.53	-6.51	-6.04	-6.16	-6.19	-6.20	-6.12
08.08.94 mo	-6.21	-6.45	-6.55	-6.58	-6.54	-5.85	-6.14	-6.23	-6.22	-6.10
08.08.94 ev	-6.36	-6.57	-6.62	-6.56	-6.48	-5.68	-6.13	-6.24	-6.17	-6.04
03.07.97 mo	-6.00	-6.35	-6.38	-6.38	-6.34	-5.80	-6.17	-6.26	-6.18	-5.93
03.07.97 ev	-6.06	-6.40	-6.48	-6.40	-6.28	-5.62	-6.05	-6.11	-6.08	-6.03
Av. 1994-97	-6.16	-6.44	-6.51	-6.48	-6.41	-5.74	-6.12	-6.21	-6.16	-6.02

Comparison of the NSB values measured during 1968-69 and 1994-97 demonstrates a significant variation during the 20 years at this site. The average NSB value has increased by about 0.1 mag at both wavelengths during this period. It can also be seen that the difference in NSB between zenith distances 0°and 60°was 0.1 mag in 1968-69 and 0.35 mag in 1994-97 at $\lambda = 0.36$ nm and 0.15 mag and 0.47 mag at $\lambda = 0.58$ nm, respectively. This means that these differences increase with increasing wavelength and there is a slight zenith distance effect.

Table 3.   Atmospheric transparency $P(\lambda)$ deduced at the Maiaky site during the years 1967-1997.

Date	λ 0.36 nm	λ 0.75 nm	Date	λ 0.36 nm	λ 0.75 nm
11/12.12.1967	0.44	0.71	29/30.06.1994	0.51	0.82
20/21.07.1968	0.42	0.66	09/10.08.1994	0.27	0.68
14/15.12.1968	0.51	0.72	22/23.08.1995	0.47	0.79
08/09.07.1969	0.48	0.75	19/20.09.1995	0.54	0.98
05/06.11.1969	0.44	0.69	02/03.07.1997	0.53	0.91
24/25.08.1993	0.28	0.66	07/08.09.1997	0.45	0.79

One can see from Table 3 that transparency values increased during 1968-97. The dispersion of these values also indicates degradation of the Maiaky site. The situation is similar for all the suburban observational sites of the Ukraine.

2.3.   Optical Astronomy with Telescopes of Medium Size

The Ukraine has optical telescopes of medium size, situated in CrAO (the 2.6-m Shajn telescope, two 1.25-m telescopes) and at the Peak Terskol (the 2-m telescope). These instruments are used for investigations of stars of various types, stellar systems, galaxies and Solar system bodies. Table 4 gives the results of measuring the extinction coefficients at these sites in comparison with MAO (Yatskiv 1999).

Table 4.   Coefficients of extinction in the UBV system for the Peak Terscol, CrAO and MAO sites.

System	Peak Terskol North Caucasus mean value	Peak Terskol North Caucasus best value	CrAO, Crimea best value	MAO, Golosiiv, Kyiv best value
U	$0^m.55$	$0^m.36$	$0^m.65$	$1^m.03$
B	$0^m.28$	$0^m.18$	$0^m.35$	$0^m.56$
V	$0^m.22$	$0^m.11$	$0^m.22$	$0^m.50$

The 2.6-m Shajn telescope (Cassegrain system), which began operations in 1960, is now the best equipped telescope in the Ukraine. There are about of 150 nights per year with moderate seeing. The photometric parameters of NSB in Crimea have been studied in detail by Lyutyi and Sharov (1982). They found that the B-V colour index of the night sky increases with the B-V colour extinction of the atmosphere, and there is no relation between NSB and atmospheric extinction.

The 2-m Ritchey-Chretien-Coude telescope has been in operation since 1995. The low atmospheric water vapour content and the high atmospheric transparency in the UV and IR wavebands allow us to infer that the Peak Terskol is one of the best observational sites in Europe (see Table 4). The mean seeing at the site is ~1 arcsec. The average number of clear night hours is ~1000 per year. Because of economic problems, the observational programme at the Peak Terskol is not regular. To increase the effectiveness of the 2-m telescope, a new CCD and fast telecommunication channel are needed. These will allow the instrument to take its part in international research projects.

3. Radio Astronomy and the Interference Struggle

The Institute of Radio Astronomy (IRA) of the NASU possesses the world's largest decametric radio telescope, the UTR-2 Array, with a frequency range of 8 - 32 MHz (Braude et al. 1978). Its huge effective area of 150,000 m^2 is more than the total effective area of all other existing radio telescopes. At a distance of 43 km from UTR-2, the decametric VLBI system URAN was built (Megn et al. 1997). It gives a highest angular resolution for decameter waves at the arcsecond level. Various investigations have been carried out with these instruments over almost 30 years, covering practically all objects in the Universe including the near-Earth vicinity, Solar system and Galaxy and more distant objects such as radio galaxies and quasars. Many priority results were obtained, confirming the great value of decametric radio astronomy.

The principal problem for decametric radio astronomy is the negative influence of transmissions from many broadcasting and other radio stations. It is not possible to regulate or restrict their operation. Because of the methods of decametric radio astronomy, the preservation of radio astronomical sites includes such matters as:

1. Creation of interference-immune high dynamic-range receiver systems, and

2. Use of special observational methods and data processing (Konovalenko et al. 1996).

Furthermore, it is important to carry out long-term interference monitoring. Fig. 1 illustrates some examples of such monitoring for intense interference in the UTR-2 frequency range. The results of deep interference monitoring in a relatively narrow band for very weak interference signals are shown in Fig. 2 (Konovalenko et al. 1997). According to these experiments, it can been concluded that interference pollution has not increased during the last decades, on the whole. More significant variations of interference level occur according to the Solar activity, the season, the day of the week and the time of day. Furthermore, even for this difficult decametric range there are rather clear frequency windows. This allows the reception of radio astronomy signals even in the presence of interference. Such experience of UTR-2 operations could be useful for the creation of a new generation of decametric radio telescope.

In recent years the IRA has been carrying out work to upgrade the centimeter-decameter waveband RT-70 antenna at Evpatoria, Ukraine. To preserve operations in the presence of rising levels of interference pollution, special filters were installed on the RT-70 for interference rejection near 325 MHz. The future direction of decametric and hectometeric radio astronomy is connected with radio telescopes outside the Earth ionosphere. This will allow us to exclude completely the strong influence of the ionosphere as well as that of terrestrial interference. There are some excellent spaceborne extremely low frequency projects (see, for example, Weiler et al. 1988), in which Ukrainian radio astronomers are actively involved.

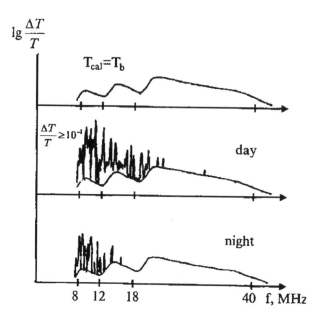

Figure 1.   Interference monitoring in the operating range of UTR-2 (calibration spectrum is at the top of Figure).

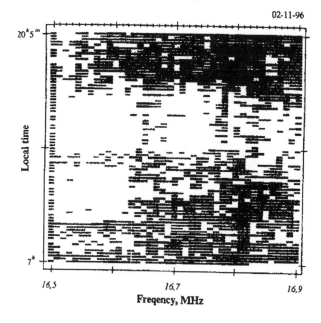

Figure 2.   Monitoring of weak hindering signals during night time near 16,7 MHz. White corresponds to interference with intensity 30 dB lower than the background level.

4. Conclusion

There are several tasks in preserving astronomical sites in the Ukraine. The problems of light pollution and the influence of technology can be solved by using appropriate scientific methods (some of them are mentioned in Paper). However, the main problem of preservation is connected with the economical situation. Most of the observating sites (CrAO, RT-22, observatory at the Peak Terskol, MAO etc.) are built as self-contained systems with their own water and energy supply, school and kindergarden. Thus, they are need comparatively large budgets. The preservation of these sites depends, in the first instance, on improving the level of their infrastructure.

As the first step in preserving these sites, the Ukrainian Astronomical Association has recommended the government of the Ukraine to give some observational sites the status of "National Property". This status is a guarantee of additional state funding. Such a status was given in 1999 to the following facilities:

- Radio telescope UTR-2 with the URAN interferometric system (IRA);

- 2.6-m telescope, RT-22, Solar tower telescope, satellite laser-ranging station Cimeiz-1873, gamma-telescope GT-48 (CrAO);

- Coherent-optical image processor of the Astronomical Observatory of Kharkiv University;

- Axial Meridian Circle (Mykolaiv AO).

Acknowledgments. I.B. Vavilova thanks the IAU for financial support to attend such an important symposium.

References

Braude, S. Ya., Megn, A. V. & Sodin, L. G. 1978, Antennas, 26, 3

Konovalenko, A.A. 1996, in *Large antennas in radio astronomy*, ESTEC, 1996, 139

Konovalenko, A.A., Sokolov, K. P. & Stepkin, S. V. 1997, Radio Physics & Radio Astronomy, 2, 188

Vovchyk, J., Blagodyr, J. & Logvynenko, O.O. 2001, in these proceedings

Lyutyi, V.M. & Sharov, A.S. 1982, Astern. Hz., 59, 174

Megn, A. V., Braude, S. Ya., Rashkovskij, S. L., et al. 1997, Radio Physics & Radio Astronomy, 2, 385

Yatskiv, Ya.S. & Kondratyuk, R.R. 1999, personal communication

Weiler, K.W., et al. 1988, A&A, 195, 372

Preserving the Astronomical Sky
IAU Symposium, Vol. 196, 2001
R. J. Cohen and W. T. Sullivan, III, eds.

Chelmos (Aroania): a New European Telescope Site for the 2.3-m Telescope of the National Observatory of Athens

D. Sinachopoulos, F. Maragoudaki[1], P. Hantzios, E. Kontizas

Astronomical Institute, National Observatory of Athens, PO Box 20048, GR-118 10 Athens, Greece

R. Korakitis

Dionysos Satellite Observatory, National Technical University of Athens, GR-157 80

Abstract. The National Observatory of Athens (N.O.A.) will install a new, advanced technology 2.3-m telescope in Greece. A favoured location for the installation of the new telescope is the top of Mount Chelmos. We present preliminary results concerning the conditions at this site. It is found that the site has a quite dark sky, a good percentage of clear nights (about 60%) and very good seeing conditions.

1. Introduction

The National Observatory of Athens (N.O.A.) will install a new advanced technology 2.3-m telescope in Greece. The telescope will have Ritchey-Chretien optics with a field of view of 1.04 degrees. Its construction was awarded to Carl Zeiss Jena GmbH.

The National Observatory intends to install it at the top of Mount Chelmos (Aroania), at approximate coordinates longitude 22°12′ East and latitude 37°59′ North. The site is near the town of Kalavryta in northern Peloponnese, at an altitude of 2340 m.

In satellite images (Fig.1, adapted from Cinzano et al. 2000), the site appears near the centre of one of the darkest spots of Southern Europe. We present below first results of our site tests (weather and seeing) and analysis of weather conditions based on satellite images.

2. Weather Conditions at the Chelmos Site

NOAA satellite weather maps (in visual and infrared) were studied in the period May 1998 to April 1999. The distribution of clear, partially clear and cloudy days is shown in Figure 2. The numbers inside the columns of the histogram indicate the actual number of days studied.

---

[1]Section of Astrophysics, Astronomy & Mechanics, Dept. of Physics, University of Athens, GR-157 84

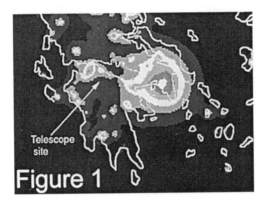

Figure 1.   Location of Chelmos site in satellite image from Cinzano et al. 2000.

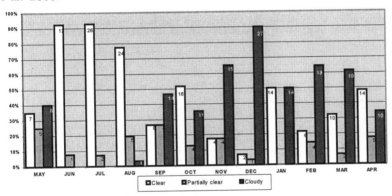

Figure 2.   Chelmos weather statistics based on observations with NOAA satellites.

It is worth noting that the top of Chelmos mountain lies often above the clouds. This means that several of the cloudy days indicated in Fig. 2 are in fact clear ones at the telescope site. This is also evident after a comparison with Fig. 3, which presents the distribution of clear and cloudy days at Chelmos based on visual observations for the period December 1998 to April 1999. These observations were made at the parking lot of the nearby Winter Sports Center at an altitude of 1650 m, which is practically always below the cloud layer. From these data we conclude that the total amount of clear nights at Chelmos is about the same as that at La Silla (Schwarz and Melnick 1993).

3.   Seeing Conditions at the Chelmos Site

Seeing measurements were taken using the stellar trace technique (Birkle et al., 1976). A ST8 CCD camera (1530×1020 pixels, pixel size 9μm) was used for the

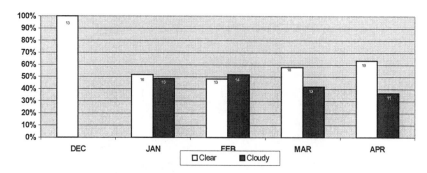

Figure 3.   Chelmos weather statistics based on visual observations.

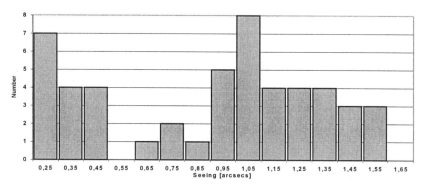

Figure 4.   First seeing measurements at the Chelmos site.

observations. The preliminary results are presented in Fig. 4, which shows the amount of seeing as it is estimated from several observations during ten winter nights. These data indicate very promising seeing conditions, with a mean value around 1 arcsec and, very often, as small as 0.3 arcsec.

Acknowledgments.   We express our thanks to the General Secretariat for Research and Technology for financing the new telescope project. We also thank Dr. P. Elias of the Institute for Ionospheric and Space Research of N.O.A., for his assistance in the reduction of NOAA images.

References

Birkle, K., Elsasser, H., Neckel, Th., Schnur, G. & Schwarze, B. 1976, A&A, 46, 397

Cinzano, P., Falchi, F., Elvidge, C.D., Baugh, K.E. 2000, MNRAS, 318, 641-657

Schwarz, H.E. & Melnick, J. 1993, "The ESO Users Manual", 20

Preserving the Astronomical Sky
IAU Symposium, Vol. 196, 2001
R. J. Cohen and W. T. Sullivan, III, eds.

The Impact of Light Pollution on a Proposed Automatic Telescope Network (ATN) and Vice Versa

John R. Mattox

Institute for Astrophysical Research, Boston University; Mattox@bu.edu

Stefan Wagner

Landessternwarte Heidelberg; swagner@lsw.uni-heidelberg.de

Gino Tosti

Perugia University; Gino.Tosti@pg.infn.it

Kent Honeycutt

Indiana University; HONEY@struve.astro.indiana.edu

Abstract. We are proposing to operate a Network of Automatic Telescopes (ATN) for CCD imaging to conduct diverse astrophysical investigations and as a resource for science education. Coordinated utilization of telescopes at diverse sites provides the possibility of obtaining continuous photometry without diurnal interruption. We describe this project and discuss how it is impacted by light pollution, and conversely, how it might mitigate the growth of light pollution.

1. Introduction

The automatic operation of optical telescopes began with a 50-inch telescope on Kitt Peak, constructed with NASA funding as the Remotely Controlled Telescope (RCT). The initial intention was to develop techniques for controlling telescopes in space. It was soon apparent that this was not a useful approach to learning how to control a space telescope — the dynamics were very different, as were the scales of the budgets (personal communication, Steve Maran 1999). The RCT telescope focus then shifted to an attempt to demonstrate the operation of an automated telescope — something which the Whitford Committee suggested in 1964 as a means to enhance the productivity of small telescopes (Maran 1967). A decade of effort resulted in one astronomical paper (Hudson et al. 1971) and the realization that a human telescope operator was much more cost effective than telescope automation with the technology available in 1969 (personal communication, S. Maran 1999).

During the 3 decades which have transpired since the RCT telescope experiment, remarkable advances in technology have occured and telescope automation is now straightforward. It is likely that the 1960s vision of the Whitford Committee of substantial gains in productivity through the automation of telescopes may soon be realized. The most important technological advances have been: (1) powerful, reliable, inexpensive, and compact computers; (2) intelligent controllers for mechanical motions; (3) charge-coupled devices (CCDs), and (4) the accumulation of experience in the most effective ways to control and use fully automated and unattended observatories.

A telescope equipped with servo motors and rotation encoders on both axes and driven by a computer with an accurate model of telescope flexure and pointing aberrations, can point anywhere on the sky with an open loop accuracy of $<10''$. Source acquisition is thus straightforward and easily automated. The CCD camera has liberated astronomers from the drudgery of the darkroom, and the anguish of the interpretation of non-linear photographic media. The CCD-based camera produces digital data with linear response, and with a quantum efficiency as high as $\sim90\%$, two orders of magnitude better than photographic film. Also, CCDs can provide simultaneous measures of sky brightness and comparison star brightness, which permit accurate differential photometry even with a partly cloudy sky.

A number of groups are operating automatic telescopes and some are developing plans for networks of automatic telescopes. Hypertext links to those with web pages are maintained at http://gamma.bu.edu/atn/auto tel.html. One of these groups is the Global Network of Astronomical Telescopes, GNAT; see http://www.gnat.org/$\sim$ida/gnat/. At least three manufacturers have designed telescopes of aperture 60 cm or larger which are capable of automated operation, Torus Technologies and DFM in the USA and TTL in England.

2. The ATN Project

The ATN project is developing a network of automated telescopes. To coordinate effort and disseminate information, a web site has been established at http://gamma.bu.edu/atn/. We anticipate that some telescopes in this network will be at sites compromised by light pollution. Therefore, we need to develop an understanding of how light pollution at each site will affect photometry in broad photometric bands, and the dependence of this on meteorological conditions. Conversely, this project can serve to publicize the impact of light pollution in the regions where participating telescopes are located, and thus help in its mitigation.

A coordinated network of automated telescopes at diverse sites will facilitate optical monitoring of the blazar class of AGN on sub-day timescales, a task which is not otherwise routinely feasible — although we have done this experimentally with a miniscule duty-cycle with the Whole Earth Blazar Telescope (WEBT) (see http://gamma.bu.edu/webt/).

Extensive blazar monitoring is expected to be extremely useful during the NASA GLAST mission (the next generation GeV gamma-ray telescope). It is currently scheduled to be launched in 2005 and to operate for a minimum of 5 years. NASA's URL for GLAST is http://glast.gsfc.nasa.gov. Multiwavelength monitoring can provide the opportunity to learn about blazars through correlating the variability of gamma-ray flux with flux at lower frequencies.

It is also expected that networks of automatic telescopes will be useful for studying other transient phenomena, e.g., binary stellar systems, gamma-ray bursts, quasar/galaxy lensing systems, microlensing events, and asteroseismology. A network which is sized to provide observing time for other areas of investigation will include more telescopes. Therefore, it will be more efficiently scheduled, and will provide better multi-longitude coverage for blazars and gamma-ray burst follow-up.

Automatic telescopes can also serve as a very valuable facility for science education. A network of automatic telescopes is being proposed by the Hands-On Universe Project (http://hou.lbl.gov/) to provide abundant, high-quality, CCD data for education.

3. Automatic Telescope Standards

Much remains to be done in the realm of software before automatic telescopes can execute a program such as the GLAST blazar monitoring. There are currently no standards in place to permit automated telescopes to be used coherently.

Therefore, an international working group is in formation to work with IAU Commission 9 on Instruments to develop standards for automatic telescopes. The web site is http://gamma.bu.edu/atn/standards/. These standards will expedite the creation and utilization of networks of telescopes for science and education.

The existence of a standard command set will form an interface between a specific Telescope Control System (TCS) and a higher level Observatory Control System (OCS). This will promote the development of telescope-independent OCS software, which will provide for instant robotization of additional new and refurbished telescopes that comply with the TCS standard.

The standards will also include appropriate protocol for Internet control of automatic telescopes. The existence of such a standard protocol will promote cooperative development and utilization of networks of robotic telescopes. This will provide for the utilization of more diverse facilities by all participants, increasing the range of projects possible, and the efficiency of telescope utilization.

References

Hudson, K.I., Chiu, H.Y., Maran, S.P., Stuart, F.E., Vokac, P.R. 1971, ApJ, 165, 573

Maran, S.P. 1967, Science, 158, 867

Preserving the Astronomical Sky
IAU Symposium, Vol. 196, 2001
R. J. Cohen and W. T. Sullivan, III, eds.

Bridges and Outdoor Lighting

Arthur Upgren

Wesleyan and Yale Universities

Abstract. Northeast Utilities, the electric power distributor for Connecticut, has devised a plan to light the Arrigoni Bridge across the Connecticut River at Middletown. This paper descibes the plan and its impact on the surrounding region. The floodlights will affect migratory birds, and teaching and research at the nearby Van Vleck Observatory on Wesleyan Campus. Examples of bridges at other sites are also discussed, including the suspension spans of New York and San Francisco, that bear only tracer lights, or no lights at all.

1. Introduction

Northeast Utilities, the electric power distributor for Connecticut, has devised a plan to illuminate the Arrigoni Bridge across the Connecticut River at Middletown, connecting that city with Portland just across the Connecticut River. The bridge, erected in 1938, features two steel arch spans, each 600 feet in length. It is shown in Figure 1.

The lighting plan calls for inductive discharge lamps, 172 of which are to be necklace or tracer lights of 55 watts, 3500 lumens each, that would outline the steel arches that form the superstructure of the bridge. In addition, 160 floodlights of 85 watts and 6000 lumens each, would be aimed upward, with an estimated 5% of the light falling on the arches and 95% or 912,000 lumens glaring directly up into the sky. An additional 28 lamps would illuminate the piers of the bridge below the roadway. These data are included in a brief summary of the plan, prepared in collaboration with the Lighting Research Center located at Troy, New York. The present paper describes the plan and its impact on the surrounding region and includes examples of bridges at a number of sites elsewhere (Figures 2 to 6).

2. Bridge Lighting

With few exceptions, prominent bridges such as the suspension spans of New York and San Francisco, bear only tracer lights, or no lights at all beyond a number that illuminate the roadway. The Hernando DeSoto Memorial Bridge, spanning the Mississippi River at Memphis, is a double-steel-arch twin of the Middletown bridge. It served as the prototype for the plan, but carries only tracer lights, some 200 bright HPS lamps of 6000 lumens each, as shown in Figure 2.

Figure 1.   The Arrigoni Bridge at Middletown, CT, showing the two 600-foot (183-m) steel arches. The lighting plans call for 172 tracer lights and 160 floodlights pointed upward, with an estimated 5% of the light illuminating the arches and 95% shining directly into the sky.

Of the bridges studied, only two incorporate upward-shining floods in their lighting design. The new Akashi Kaikyo Suspension Bridge in Japan, the world's largest, restricts upward-shining light to its two towers (Figure 5). The same appears to be the case for the George Washington Bridge in New York.

Northeast Utilities, through its associate, Connecticut Power & Light, has provided details, although the citizens of Middletown and Portland are still mostly unaware of the plan and its estimated cost, which is projected at $764,000 for the installation of the lights and an annual amount of $20,000 for lighting energy consumption and maintenance. A model of the bridge is on public display, but provides no lighting details. The floodlights will affect migratory birds, and teaching and research at the nearby Van Vleck Observatory on the Wesleyan campus may be impaired. The extent of the damage to the environment is still uncertain should this plan be approved and financed.

From the bridges depicted here and others, most, even famous ones, appear to have been limited to tracer lights of low light pollution levels. I know of only one other case of an intent to floodlight an entire span. This is the Humber Bridge, a suspension bridge across the Humber River at Hull, the longest bridge in England. I have been informed recently, that the plan to install 72 floodlights has been dropped. But it is vital for us to be aware of and forestall any emerging trend towards illuminating the superstructures of bridges.

Figure 2.   Hernando DeSoto Memorial Bridge at Memphis is a near twin of the Arrigoni Bridge and served as a model for the lighting plan. The 200 tracer lights are HPS 6300 lumens each. There are no floodlights.

Figure 3.   The Bronx-Whitestone Bridge, linking the boroughs of Bronx and Queens in New York, shows the necklace or tracer lights that are in common use throughout the city's many suspension bridges.

Figure 4.   The Golden Gate Bridge in San Francisco. This span incorporates only roadway lighting.

Figure 5.   The new Akashi Kaikyo Bridge near Kobe, Japan. Floodlights illuminate only the two towers, each 300 meters (984 feet) high, just equal in height to the Eiffel Tower.

Preserving the Astronomical Sky
IAU Symposium, Vol. 196, 2001
R. J. Cohen and W. T. Sullivan, III, eds.

Search for and Protection of Astronomical Sites in Developing Countries

François R. Querci and Monique Querci

Observatoire Midi-Pyrénées, 14 Av. E. Belin, 31400 Toulouse, France
e-mail: querci@obs-mip.fr

Abstract. The archives of meteorological satellites allow pre-selection of dry sites well adapted to astronomical observations (in the visible, infrared and millimetric ranges). The GSM (Grating Scale Monitor) technique then permits qualification of them as sites for future astronomical observatories. Such sites are found in new astronomical countries or in developing countries. At the same time, their protection from light pollution and/or radio interference has to be secured. In practice, once pre-selections are made, the governments of these countries ought to be alerted, for example by the IAU and/or the UN Office for Outer Space Affairs. The local site testing through GSM should be carried out in cooperation with scientists of these countries under the umbrella of the IAU. This could be an approach to help introduce astronomy and astrophysics into developing countries.

1. Introduction

Knowledge of short time-scale variations of stars has made some progress due to campaigns of non-stop observations with simple manual telescopes and photometers, conducted simultaneously around the world. During the last two decades, the automation of telescopes (remote-controlled or robotic telescopes) opened the way to study variability over the entire HR diagram (*e.g.* Henry 1999), especially at places where simultaneous variations of different characteristic times, from hours to years, are found (AGB, RGB stars, etc.). To follow these permanent or temporary variations over months or years and to understand their origin, networks of robotic telescopes nowadays seem to be the most appropriate technology.

The first discoveries of the visible counterparts of gamma-ray bursts and of many new Near Earth Objects demonstrate that networks of robotic telescopes are now and will be powerful in many scientific fields (*e.g.* ROBONET, GNAT, TORUS, NORT, etc., as described in Querci and Querci (2000)). Consequently, sites for networks have to be established at various longitudes in the two hemispheres. Excellents sites (Hawaii, Northern Chile, South Pole, etc.) and very good ones (Canary Islands, South Africa, India, Uzbekistan, etc.) are already at work. But are we sure that other excellent sites do not exist elsewhere, for example, in developing countries?

Astronomy and space science could contribute to the development of developing countries, as already seen some decades ago, in the Canary Islands, Chile and so on.

2. A Way to Select Sites in Developing Countries

A world-wide preliminary map of mean cloudiness (at 0.55 μm) was obtained from 12-year meteorological archives and with 250 km square meshes (Querci and Querci 1998a,b).

- A first step should be a cross correlation analysis between a worldwide map of high mountain summits (altitude 2400-3200 m or more) and a map of small-mesh meteorological archives (2 to 5 km) on cloudiness, humidity, dust storms and light pollution. It would permit the determination of 20 to 30 new meteorological excellent or very good sites adapted to optical, infrared, or millimetric observations.

- A second step should be a detailed analysis of the local atmospheric turbulence for these 20 to 30 pre-selected sites by a seeing-monitor or by a grating scale-monitor (G.S.M.) technique (Martin et al. 1994). The registration of the parameters L_0, the wavefront outer scale, r_0, the Fried parameter, τ, the speckle lifetime, and the isoplanetism angle for each of these sites should permit the selection of 8 to 10 sites besides those already classified as high quality observing sites.

- A third step should be the development of cooperation and analysis of the local facilities to implement and to maintain robotic telescopes and their equipment.

The two last steps could be a way to introduce astronomy and space science into developing countries through robotic equipment, especially the analysis of variable objects supported by hydrodynamical calculations.

Preliminary works on large meshes are in progress in many countries. Moreover, analysis with small meshes are in progress for the High Atlas Mountains (Morocco) and for Lebanese border mountains (Syria)(private communication).

3. Protection for the Selected Sites

In many developing countries, astronomy and space science are not developed at all and the search for sites is ignored. Consequently, potential sites might be polluted and lost for science in the future.

In a few developing countries, collaboration with astronomically-developed countries is in progress. So, site prospecting and subsequent site protection are now taken into account by national scientific authorities, contributing to the scientific and technical development of the country.

4. Conclusion

Prospecting for and the protection of potential future astronomical sites are very important for astronomy and space science in the next century. These sites, of which many excellent ones are in developing countries, have been suggested based on a worldwide mean annual cloudiness map.

We take the opportunity of this IAU/UN symposium to ask the questions: could such prospecting and protection be promoted

- by each developing country individually?

- by some developing countries or new astronomical countries grouped together inside regional astronomical organizations such as the Arab Union for Astronomy and Space Science (AUASS)?

- by international astronomical organizations such as the European Southern Observatory (ESO), Cerro Tololo Inter-American Observatory (CTIO), etc.?

- by the International Astronomical Union?

- by the UN Office for Outer Space Affairs?

Encouragement and help from international organizations would be certainly decisive for developing countries, for their own scientific benefit as well as that of the international community.

References

Henry, G.W. 1999, Techniques for automated high-precision photometry of Sun-like stars, PASP, 111, 845

Martin, P., Tokovinin, A., Agabi, A., Borgnino, J., Ziad, A. 1994, G.S.M.: a Grating Scale Monitor for atmospheric turbulence measurements. I. The instrument and first results of angle of arrival measurements, Astr. Astrophys. Suppl. Ser., 108, 173

Querci, F.R., Querci, M. 1998a, A method for searching potential observing sites, in *Preserving the Astronomical Windows*, eds. Syuzo Isobe and Tomohiro Hirayama, A.S.P. Conf. Series vol.139, p.135

Querci, F.R., Querci, M. 1998b, The network of oriental robotic telescopes (NORT), at the WWW address:
http://www.saao.ac.za/ wgssa/as2/nort.html

Querci, F.R., Querci, M. 2000, Robotic telescopes and networks: new tools for education and science, in *Eighth UN/ESA Workshop on basic Space Science: Scientific Exploration from Space, Mafraq, Jordan, 13-17 March 1999*, Astrophysics and Space Science, 273, 257-272

Preserving the Astronomical Sky
IAU Symposium, Vol. 196, 2001
R. J. Cohen and W. T. Sullivan, III, eds.

Aviation and Jet Contrails: Impact on Astronomy

H. Pedersen[1]

Copenhagen University Observatory, Juliane Maries Vej 30, DK 2100 Copenhagen, Denmark

Abstract. Attention is drawn to aspects of aviation that have a detrimental effect on ground-based astronomy. Depending on observing methods, science data can be influenced by an aircraft's emission of light, its thermal emission, exhaust products and condensation trail. Although these effects are mostly short-lasting for a given observing direction, they can be highly significant, and influence time-resolved astronomical observations. While the very young contrails can easily be recognized by ground-based or spaceborne observations, concern should also be given to older (hours, days) contrails, which have lost their characteristic linear shape. Contrails may grow to widths of tens of kilometers, and become almost indistinguishable from natural cirrus. As aviation increases, this may imply fewer photometric nights, in particular in the northern hemisphere, where by far the largest fuel consumption takes place.

1. Introduction

Already in 1977, the IAU recognized that airplane lights, heat, exhaust, and condensation trails could seriously degrade observational astronomy, and it was recommended (Graham Smith 1977) that aviation near observatories should be limited to below $10°$ above the horizon, and that low-flying aircraft should not come closer than 5 km.

It appears that these recommendations have been successfully adopted by Spain and Australia for some of their observatories, whereas many other countries have not adopted the rules or do not enforce them.

While in principle the IAU recommendations should provide substantial protection, the volume of aviation has risen greatly, since the late 1970s. According to the International Civil Aviation Organization (ICAO), as many as 17,000 commercial jets were in operation in 1997, carrying 1.4 billion passengers [1]. To this one must add small aircraft (<9 t) and military aviation, rescue services, etc. During the coming years aviation is expected to grow significantly, with only modest gains in fuel economy or other emission standards. The Intergovernmental Panel for Climate Change (Penner et al. 1999) has discussed scenarios with up to 9 times higher fuel consumption by the year 2050, compared to 1990.

---

[1] guest researcher

Some airliners already fly on routes close to astronomical observatories and in the future this may become more commonplace. This is to a large extent due to improved navigation aids, which no longer oblige a pilot to fly from one radio beacon to the next. And considering the enormous amounts of fuel combustion products, and their long residence times in the upper air, it can be expected that aviation will generate long-term effects that will slowly deteriorate general observing conditions. It is therefore becoming evident that the current IAU recommendations offer insufficient protection for astronomical observations.

2.   Aircraft Lights, Heat Emission

What are the major effects on observational astronomy? Stroboscopic and fixed aircraft lights may contaminate wide-angle imaging, and with the advent of wide-angle CCD imaging, the problem will be felt more. However, this effect, and the actual appearance of aircraft in solar imaging, for example [2], is a minor problem compared to the influence of engine exhaust and the resultant condensation trail (contrail). Heat emission can be traced far behind an airliner, where it may limit good 'seeing' conditions. Using the power law dependency given by Schumann et al. (1998) it can be estimated that the temperature difference with ambient air is 0.1 K or more at a distance of 25 km behind the aircraft, corresponding to 100 seconds of flight time. During this time an exhaust plume in an orthogonal jet stream field will have swept over $\sim$1000 square degrees, as seen from a mountain top observatory.

3.   Young Contrails

The condensation trail that forms in the wake of many flights is possibly the greatest threat to astronomy. As the trail sweeps over the sky, optical and infrared observations are affected by increased sky background, lesser transmission, and possibly by spectroscopic signatures.

The formation of a contrail depends on a range of parameters, which are fairly accurately described by the Schmidt-Appleman formula (see Schumann 1996). At a typical flight altitude of 10,000 m, contrails form when the air temperature is about -50 °C for 0% RH, while for 100% RH contrail formation sets in already at -40 °C. The exhaust's content of carbon soot plays an important role for contrail formation by furnishing cloud condensation nuclei. Also sulphuric aerosols may influence contrail formation. Most often the contrail evaporates after a few seconds, leaving soot in suspension. However, for supersaturated conditions the contrail may acquire water from ambient air. Under such circumstances it may persist for several hours, during which it will spread and most likely loose its linear shape. The optical thickness of the cloud may well exceed one.

Other constituents of aircraft exhaust, such as carbon oxides, nitric oxides, sulphur dioxide and unburned fuel slowly pile up in the atmosphere, or change it chemically (over the major flight corridors, anthropogenic NOx has already doubled the upper troposphere's concentration of those molecules). However, on short terms the effects are probably too small to be seen spectroscopically.

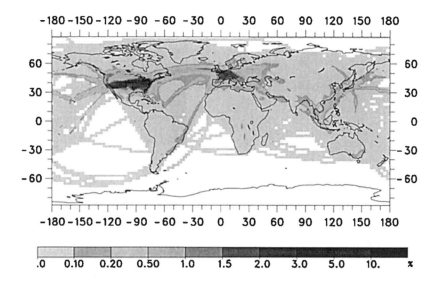

Figure 1.   Time averaged global coverage by young, persistent contrails, according to modeling by the German Aerospace Centre (DLR). Apart from Antarctica, almost no land areas are protected against young contrails. Courtesy DLR.

4.   Average Contrail Cover

For reason of their influence on global climate, contrails and their microphysics are being studied at several sites, in particular at the German Aerospace Centre (DLR) at Oberpfaffenhofen and the NASA centre at Langley, Virginia, USA.

To assess the amount of sky covered by contrails, systematic observations are required, and these are best carried out from space. The 1-km resolution AVHRR detector on board NOAA satellites has proven particularly useful. One image, from February 11 1999, shows as many as 70 contrails covering most of the sky over some New England states [3]. Another, from May 1995, shows a large set of contrails over southern Germany [4]. This latter image has been used to test DLR's automatic contrail recognition algorithms, which have then been applied to a much larger data set. The result is a map showing the average noon-time contrail coverage over Europe during 1996 (Figure 17 of Mannstein et al. 1998). In the most affected regions, more than 1 percent of the sky is covered by young contrails. Older, distorted contrails cannot yet be measured; this will require increased temporal and spectral resolution.

Apart from this observational approach, it is possible to model the mean contrail coverage making use of the Schmidt-Appleman criterion, knowledge of high-altitude fuel consumption and meteorological data (Sausen et al. 1998). For example, the programmes developed at DLR allow one to estimate what could be gained by changing typical flight altitudes as s function of geographic latitude. As to the time-averaged situation, Figure 1 shows that only the southern oceans are almost entirely free of young, persistent contrails, while Europe,

North America, and parts of East Asia are covered by several percent. Present day modelling is, however, unable to predict the lifetime of a contrail, once generated.

5.   Cirrus Formed by Contrails

Groundbased and spaceborne observations have revealed that some persistent contrails survive for many hours, and develop into extended cloud formations. An early observation was that of Georgi (1960), who described a contrail spreading into a 20° wide cirrus band. The NOAA satellites have followed one cluster of contrails for 17 hours, during which it spread to cover 35,000 square kilometers (Minnis et al. 1998). Another contrail, imaged by the ISIR instrument (on board Space Shuttle) grew to 30 km width [5]. These results were obtained in fairly limited studies, so there is no reason to believe that they represent extremes. Existing techniques seem unable to distinguish a contrail in its late stages of development from natural cirrus. Some of these clouds may be optically thin, almost invisible to visual detection techniques, but yet have strong infrared spectral absorption (Smith et al. 1998).

This anthropogenic cloud formation has long since given rise for concern. Changnon (1981) noted a loss of sunny days over the mid-western USA, and provisionally tied this to an increased abundance of cloud condensation nuclei caused by jet aviation. A similar rise of high-level clouds over Salt Lake City around 1960 coincided in time with a rapid rise in jet aviation (Liou, Ou, & Koenig 1990). Recently, Boucher (1999), working on data from 1982 to 1991, detected a significant rise in cirrus over extended regions of the northern hemisphere, and linked this to aviation. Many others have pointed out that increased cirrus cover will lead to higher night-time temperature, hence further evaporation and cloud formation.

6.   Degradation of Observing Conditions

Degradation of astronomical observing conditions has already been noticed at European observatories. Mt. Wendelstein Observatory, belonging to Munich University, is placed at 1845-m altitude in the Alps, some 75 km SE of that city. The observatory web site [6] states: *Mid 1970: Increasing air pollution caused by the increasing air traffic prevents further observations of the solar corona and worsen all other observations.* It has been shown (Schumann 1999) that the loss of coronal visibility at Mt. Wendelstein is not due to optically thick clouds, but is more likely caused by contrails. Also Pic du Midi remains affected by anthropogenic cirrus, as do many other European and North American observatories.

7.   Protecting the Best Sites

It is evident from inspection of Figure 1 that the west coast of South America is strongly protected against interference by old contrails. This absence of degraded and fuzzy contrails is a valuable asset for photometric and other critical observations.

The threat from young contrails is, however, on the increase. More than 70 commercial and military flights pass daily along the Chilean coast. Typically 10 of these are at night-time. A year-long study at Paranal has tabulated their closest approach to the ESO Very Large Telescope (VLT). The vast majority pass west of the mountain, 20° - 25° above the horizon (Pedersen 1998). Considering the typical upper troposphere wind speed, the engine exhaust and its eventual contrail will pass the observatory's zenith $\sim$10 min later, and remain in the field-of-view for $\sim$5 sec.

This interruption of high-quality conditions may not be noticed right away by the observer, but nevertheless degrades observations with non-statistical scatter, high sky background and/or increased absorption. Usually, such data will be rejected, but if the astronomical target itself is variable, the drop in intensity caused by a contrail may pass for true data.

This is obviously not desirable, so for large ground-based observatories evasive action has to be taken, if rerouting of the aircraft corridors is not an option. Visual detection of the instantaneous aircraft positions is possible, but experience shows that contrails are hard to see against a dark sky. Also, such observations are difficult to bring into a useful, digital form. Therefore, for the ESO VLT an instrumental (all-sky CCD) system is being considered (Sarazin, private communication). All night long at one-minute intervals this would log the celestial position of any aircraft above the observatory horizon. The system could draw upon upper troposphere meteorology to derive expected moments of contrail passage, and supply these warnings to human observers and automatic telescope scheduling processes. At the same time, quantitative information is generated on the general sky conditions, i.e. natural cloudiness, extinction coefficients and sky background.

Active systems, depending on lidar or radar, are also possible, but should in general be avoided as they in themselves emit unwanted radiation. On an even more general level, it might be asked if contrail formation can be suppressed. This would limit the most immediate effects on astronomy. Careful selection of flight altitudes or routes appears as one possibility, and is already used by military aircraft. Some of these, for example the B2 Stealth bomber, are equipped with a rear looking lidar [7] to warn the pilot against 'conning', (military jargon for contrail formation). To enforce similar systems on civil aviation will not be easy, since commercial incentives seem absent. Chemical additives (e.g. chlorosulphonic acid) have the potential of achieving the same goal. They work by lowering the surface tension of water, thereby limiting droplet sizes to below 1 micron. Special fuels like CS_2 also prevent contrail formation. Unfortunately, the properties of all known contrail inhibitors are unsuitable for application in large volume.

8.   Discussion

The absence of nearby air routes has long been considered a selection criterion for new observatories, and some countries have successfully adopted legislation to protect their sites accordingly. Nevertheless, the present huge volume of aviation inhibits full protection of some established sites. To some extent, observers worldwide can take preventive measures, but obviously professional observers, and for that matter amateur astronomers and the general public too, would be better served by a less contaminated sky.

It is desirable that the level of tropospheric emissions be moderated in some way, so as to limit accumulation of exhaust products. Currently, aircraft emissions are limited only for take-off and landing, and then only for NOx, CO, and unburned hydrocarbons. Emission of soot at flight level is just one of several quantities that could be limited by international agreement. Furthermore, not only commercial flights should abide to stricter standards, but also military flights.

In the mean time, astronomers have good reason to follow carefully the trends in night-time cloudiness, and to develop evasive observing strategies.

Acknowledgments.   I am grateful to M. Sarazin and his staff at ESO for support in the research on aviation near Paranal, and to H. Mannstein and DLR colleagues for discussions.

References

Boucher, O. 1999, Nature, 397, p.30.

Changnon, S.A. 1981, Journal of Applied Meteorology, 20, p.496-508.

Georgi, J. 1960, Zeitschrift Meteorologie, 14, p.102.

Graham Smith, F. 1977, IAU/UNESCO - DG/2.1/414/44

Liou, K.N., Ou, S.C., and Koenig, G. 1990, "Air Traffic and the Environment", U.Schumann, ed., Lecture Notes in Engineering, Springer, p.154-169.

Mannstein, H., et al. 1999, Int. J. Remote Sensing 20, p.1641-1660.

Minnis, P., et al. 1998, Geophys. Res. Letters, 25, p.1157-1160.

Pedersen, H. 1998, ESO Report, Reference No. VLT-17443-1679.

Penner, J.E., et al. 1999, "Aviation and The Global Atmosphere", Cambridge University Press.

Sausen, R., et al. 1998, Theor. Appl. Climatol. 61, p.127-141.

Schumann, U. 1999, "Contrail Cirrus", Report No. 114, DLR, p.1-22; to appear in *Cirrus*, ed. D. Lynch (Oxford University Press, in press).

Schumann, U., et al. 1998, Atmospheric Environment, 32, p.3097-3103.

Smith, W.L., et al. 1998, Geophys. Res. Letters, 25, p.1137-1140.

[1] http://www.icao.int/icao/en/jr/5306_r3.htm

[2] http://physics.usc.edu/solar/direct.html

[3] http://www.osei.noaa.gov/Events/Unique/Other/UNIcontrails042_N5.jpg

[4] http://www.dfd.dlr.de/app/iom/1999_03/

[5] http://isir.gsfc.nasa.gov/main.html

[6] http://bigbang.usm.uni-muenchen.de:8002/USM/WDST/history/

[7] http://www.ophir.com/airborne.htm

Part 3
Space Debris

Preserving the Astronomical Sky
IAU Symposium, Vol. 196, 2001
R. J. Cohen and W. T. Sullivan, III, eds.

The Space Debris Environment - Past and Present

W. Flury[1]

ESA/ESOC, Robert Bosch Str. 5, 64293 Darmstadt, Germany
wflury@esa.int

Abstract. The mass and number of Earth-orbiting human-generated space debris have increased steadily since the beginning of space flight. Recent voluntary measures for debris mitigation applied by space operators have not stemmed the increase. The debris hazard for manned and unmanned missions is still low, but rising. More effective but also more costly measures, such as selective deorbiting of used stages, will be necessary to avoid a run-away situation. Internationally agreed codes for debris management and control are needed to solve this global space environment problem.

1. Introduction

Since the launch of Sputnik 1 in 1957 there have been more than 4000 successful rocket launches. Regular tracking of Earth-orbiting objects has detected nearly 25,000 objects. This space surveillance activity is carried out by the USA and by Russia using global networks of radars and optical sensors. There are currently 9,000 catalogued objects in orbit, with a total estimated mass of 2,000 tons in LEO. There are currently estimated to be more than 100,000 objects larger than 1 cm in orbit. Figure 1 shows the growth in catalogued debris from 1957 to 1999.

> 'Orbital debris is defined as any man-made Earth-orbiting object which is non-functional with no reasonable expectation of assuming or resuming its intended function or any other function for which it is or can be expected to be authorized, including fragments and parts thereof. Space debris comprises orbital debris and reentering debris.'
> *IAA: International Academy of Astronautics*

Sources of debris include defunct spacecraft (21%), rocket upper stages (18%), operational debris (12%), fragments from breakups (43%), exhaust products of solid fuel rockets and materials produced by ageing effects (e.g. particles such as paint flakes, or sodium and potassium droplets from the leaking RORSAT cooling system (Kessler et al. 1997)). In 1998 there were 6 breakups, generating

---

[1]This summary of Dr. Flury's presentation is based on the tranparencies he used. For further information, see the review paper by Flury (1998), the Technical Report on Space Debris (A/AC/105/720) of the Scientific and Technical Subcommittee of COPUOS (which Dr. Flury played a leading role in finalizing), and the UNISPACE III Report (A/CONF.184/6).

more than 400 fragments, only 84 of which are catalogued. In 1999 there have already been 3 more breakups.

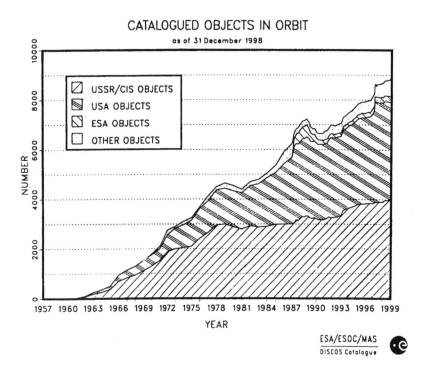

Figure 1.   Catalogued objects in space since 1957. To be detected, the objects must be at least ~10 cm in size in LEO and at least ~1 m in size in GEO.

This papers summarizes the current space debris environment and outlines measures which are needed to reduce the growth of the debris population.

2.   The Situation in Low Earth Orbit (LEO) and the Geostationary Ring (GSO)

About 85% of catalogued objects are in low Earth orbit (between 200 km and 2000 km altitude), with maxima of spatial density near 1,000 km and 1,500 km altitude. Some altitude bands may have reached critical density. More than 140 fragmentations are known to have occured in LEO. Other indicators of the degrading LEO environment include:

- Impacts on spacecraft returned from space (e.g. SolarMax, NASA's Long Duration Exposure Facility LDEF, ESA's EURECA, the solar arrays of the HST, more than 60 windows replaced on Space Shuttles, etc.)

- Evasive manoeuvres of the Space Shuttle and other spacecraft (ERS-1 and SPOT-2)

- Collision of Cerise with fragment of H10 of Ariane V16 (July 1996)

- More than 200 shields of International Space Station (ISS)

- Trails on astronomical observations

Atmospheric drag provides a natural removal mechanism from LEO, but is effective only at the lowest altitudes.

In the geostationary ring there are about 700 catalogued objects, including about 150 upper stages of rockets. Two fragmentations are known (Titan upper stage and EKRAN spacecraft). Relative velocities are much lower than in LEO. There is no natural removal mechanism from GEO, however, so objects brought into the geostationary ring will generally remain in its vicinity indefinitely. The growth of debris in GEO is shown in Figure 2. The annual launch rate of about 25-30 spacecraft is also increasing. There is a growing trend to place several satellites in the same longitude window (colocation), due to sustained demand for orbital positions in GEO.

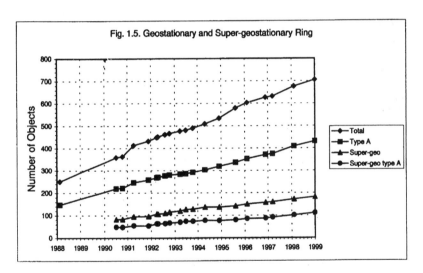

Figure 2.   Catalogued objects in the geostationary ring since 1988. The objects are at least 1 m in size.

3.   Long Term Evolution and Debris Mitigation

The current evolution of the debris population is determined by launch activities and breakups. If the spatial density of debris becomes large enough, population growth will be dominated by collisions and the avalanche effect will lead to uncontrolled growth. The future evolution will be strongly dependent on applied debris mitigation measures.

Current measures are directed to avoid creating new space debris by avoiding breakups, avoiding mission-related debris and use of graveyard orbits. These measures are in the long term not sufficient to avoid a run-away situation. Selective deorbiting of large objects (spacecraft and rocket upper stages) will become necessary. There is a high cost to deorbiting safely from high-energy orbits, without intersecting the zone of operational GSO spacecraft.

4.   Activities of International Organizations

The Inter-Agency Space Debris Coordination Committee (IADC) was founded in 1993, with the purpose of exchanging information on space debris research, to review progress of cooperative activities, to facilitate opportunities for cooperation in debris research and to identify and evaluate mitigation options. The activities of the IADC, the IAA (International Academy of Astronautics) and other international bodies such as COSPAR have made positive contributions to space debris research and education.

Space debris was included as an agenda item for the thirty-first session of the Scientific and Technical SubCommittee of UNCOPUOS, in February 1994. A multi-year work plan was agreed at the thirty-second session. The technical report on space debris will be submitted for approval during UNISPACE-III.

Conclusions

- The space debris population is steadily growing (satellite constellations)

- Current risk levels are still low but are increasing

- Orbital regions most at risk are in LEO (800 km - 1,600 km) and in GEO

- Most endangered regions are also the most useful regions

- Clean-up is economically not feasible: PREVENTION IS NEEDED

- Current debris reduction measures are not sufficient to avoid a run-away situation

- More efficient measures are needed, including selective deorbiting and lifetime reduction of spacecraft and rocket bodies

References

Flury, W. 1998, The Space Debris Environment of the Earth, ASP Conf. Ser. 139, 49-64

Kessler, D. J. et al. 1997, The Search for a Previously Unknown Source of Orbital Debris: the Possibility of a Coolant Leak in Radar Ocean Reconnaissance Satellites, JSC-27737, NASA JSC (Houston, USA)

Technical Report on Space Debris, Text of the Report adopted by the Scientific and Technical Subcommittee of the United Nations Committee on the Peaceful Uses of Outer Space, United Nations Publication A/AC.105/720, July 1999

Preserving the Astronomical Sky
IAU Symposium, Vol. 196, 2001
R. J. Cohen and W. T. Sullivan, III, eds.

UN Discussions of Space Debris Issues

Luboš Perek

Astronomical Institute, Bocni II 1401, 141 31 Praha 4, Czech Republic

Abstract. The role and activities of COPUOS, the UN Committee for the Peaceful Uses of Outer Space, are briefly described.

1. Introduction to COPUOS and its Activities

Soon after the launch of the first artificial satellite, the General Assembly of the UN established COPUOS, the Committee for the Peaceful Uses of Outer Space. The COPUOS was entrusted with the task to act as a focal point for space matters, to support international cooperation and to consider legal problems which might arise in space activities.

During the 1960s and 1970s, the COPUOS elaborated five international treaties and several sets of principles which have served the space community very well and with remarkable foresight. Nevertheless, the first phenomenon which had not *and could not* have been anticipated was the growing importance of space debris.

Some documents containing information on space debris have been put before the COPUOS and its Scientific and Technical Subcommittee or its Legal Subcommittee in the past. The first document dates back to December 1979 (Mutual Relations of Space Missions, 1979). A detailed treatment of space debris appeared in 1982 in a Background Paper for the conference UNISPACE 1982 (*The World in Space* 1982). The study dealt with collision probabilities, planned decay of inactive satellites in low Earth orbit (LEO) and with their disposal orbits in GEO (geostationary orbit). It also commented on concerns of astronomers over effects harmful to observations.

The COPUOS invited COSPAR (Committee on Space Research) and the IAF (International Astronautical Federation) in 1988 to prepare a study on environmental effects of space activities. The study dealt with the origin and impact of space debris and with possible preventive measures (Perek & Bauer 1988).

At the session of the COPUOS Scientific and Technical Subcommittee in 1991 several states proposed to put space debris on the agenda but consensus was reached only on a request addressed to Member States to present results of their national research on space debris. The situation changed in 1994 when consensus was finally reached and space debris became an agenda item. It was agreed that a Technical Report on Space Debris would be elaborated addressing all major technical aspects. Two organizations were asked for assistance: space agencies of launching countries, represented by the Inter-Agency Space Debris

Coordinating Committee IADC[1] and the scientific community represented by the IAA (International Academy of Astronautics). The Technical Report was finalized by the COPUOS Scientific and Technical Subcomittee at its session in 1999 (UN doc. A/AC.105/720). It contains details on the measurements, on modeling of the space debris population and on space debris mitigation measures. Possible further actions concerning space debris will be considered and decided upon by COPUOS at its forthcoming sessions.

2. Recommended Resolutions for UNISPACE III

It was proposed to address two recommendations to UNISPACE III. The first recommendation should express an appreciation of the work done by the UN and a support for future activities. Specifically it was proposed to join efforts with the American Institute of Aeronautics and Astronautics (AIAA), which adopted at a Workshop held in Bermuda, April 1999, the following wording:

> "The Workshop participants strongly support work being done by the United Nations, the Inter-Agency Space Debris Coordinating Committee (IADC), the International Academy of Astronautics (IAA), and others to develop guidelines designed to minimize the creation of new debris objects.
>
> The Workshop recommends that existing and future debris minimization guidelines be applied uniformly and consistently by the entire international spacefaring community. In addition, Government licensing agencies are encouraged to promote such compliance among the space community in their respective countries.
>
> In addition, minimizing the creation of new debris, the problem of on-orbit debris must be addressed. Mitigation of debris on orbit can be addressed in at least two ways. First, by moving large debris, such as satellites at the end of operational lifetime, out of the way of active satellite orbits and second, by the active removal of visible, but untracked smaller debris. Some aerospace companies are not including deorbit capabilities on their spacecraft, and hence these spacecraft will contribute to the problems of orbital congestion and debris well past their operational lifetimes."

The second recommendation deals with a problem which may become important in the future. At present, there are no methods for removing from orbit those objects which have no maneuvering capability. This includes almost the entire population of orbiting debris of a total mass of 2000 to 3000 tons.

The AIAA Workshop, quoted above, also adopted the following wording:

---

[1]The IADC was founded in 1993 by ESA, Japan, NASA, and the Russian Space Agency, RSA. China joined in 1995, the British National Space Center, the Centre National d'Études Spatiales of France, and the Indian Space Research Organization in 1996, the German Aerospace Center in 1997 and the Italian Space Agency in 1999.

"While the economic justification and consequences to the space environment of implementing debris removal technologies must be better understood, continued development of such technologies should be encouraged. Governments are strongly encouraged to invest in basic pre-competitive technology that could be further developed and applied by commercial operators."

By adopting these two recommendations, this meeting of UNISPACE III should express its point of view on preventing the generation of new debris, as well as on the need to develop and prove methods for removing existing small debris from orbit.

References

Mutual Relations of Space Missions, UN doc. A/AC.105/261 of 7 December 1979

The World in Space, Prentice Hall, Englewood Cliffs, NJ, 1982, p.311-360

Perek L. and Bauer S. 1988, *Environmental Effects of Space Activities*, UN doc. A/AC,105/420, of 15 December 1988

UN doc. A/AC.105/720

American Institute of Aeronautics and Astronautics: Fifth International Space Cooperation Workshop *"Solving Global Problems"*, held at Bermuda, 11-15 April 1999

Preserving the Astronomical Sky
IAU Symposium, Vol. 196, 2001
R. J. Cohen and W. T. Sullivan, III, eds.

Impact of Space Debris and Space Reflectors on Ground-Based Astronomy

D. McNally[1]

University of London Observatory,
Mill Hill Park, London NW7 2QS, UK

Abstract. Astronomical fields under observation are being increasingly crossed by satellites. Such crossings either leave a "trail" on a photographic or CCD image, or corrupt a photographic observation. Such trailing/corruption may render an observation useless for scientific analysis. There is also the much more serious problem posed by suggestions to put solar reflectors in space for Earth illumination, artistic, celebratory or advertising purposes and, by extension, to longer term suggestions of ways to utilise dark time rather than twilight time. It is only a matter of time until the solar reflector becomes proven technology. The time left for decisive action may be very short.

1. Introduction

Ground-based astronomy is concerned with the detection of faint cosmic signals. In such an activity one is concerned to maintain adequate signal/noise (S/N) ratio. Anything which degrades S/N is to be minimised. Unfortunately operational satellites, defunct satellites and large debris in orbit can reflect sunlight after it has become sufficiently dark on the ground to begin serious astronomical observation. Satellites still illuminated by the Sun will be detected if they cross a field under observation: if a photographic plate or CCD camera is being used, a bright satellite will leave a trail across the image; if a photometer is being used, the entire observation will be lost. An astronomical image carrying satellite trails may not be entirely useless: degradation will depend on the purpose for which the image was taken. However, it now has unfortunately become accepted that modern Sky survey plates will be trailed.

There have been recurring suggestions for placing reflectors in space to reflect sunlight. While such suggestions have not yet been implemented successfully, the development of solar reflector technology has reached the point of trial. Were such technology to be put into commercial use, e.g. for advertising purposes, the outlook for ground-based astronomy would indeed be bleak.

Space communications have entered the age of laser communication links. ESA uses this technology for the SPOT4 mission. This has brought about a new situation for optical astronomy, akin to that of radio astronomy, where ground-

---

[1]Current address: 17 Greenfield, Hatfield, Herts AL9 5HW, UK

based optical observatories will be swept by laser communication beams just as ground-based radio observatories are swept by radio communications. Thus far laser communication is taking place in the far red at a wavelength close to 800 nm and, in the case of SPOT4, for intersatellite communication.

Another "possible" hazard to optical astronomy may have been revealed (Ohishi 2001): the HAPS radio relay balloon project. This is clearly a project to be evaluated for its possible hazard to urban observatories located near large centres of population.

2. Measures of Satellite Trails

Satellite trails do not give good statistical data, since image-taking activity varies from year to year and the distribution of times at which images are taken varies from night to night. However, in any collection of plates the numbers will rapidly dwindle as the selection criteria are increased in number.

If one looks at an entire sample taken over a period of years, then one must accept inhomogeneity of data. Since satellite trailing is most likely within two hours after sunset or two hours before sunrise the sample will be biased by the number of plates taken within these periods. However, a large data set, while inhomogeneous, may be used to look at long-term trends. Tritton and Norton (private communication, 1995) have examined the UK Schmidt Plates for satellite trails for the period 1977/93. However, because of a change of filter, only the data for 1985/93 were selected. The results are set out in Table 1 (McNally and Rast 1999):

Table 1.   Trail and Satellite number data

Year	Number of trails[1] per 60 min exposure	Number of Spacecraft[2] on orbit	I_T	I_S
1985	2.5	1500	1.1	0.9
1987*	2.3	1624	1.0	1.0
1989	2.7	1749	1.2	1.1
1991	2.4	1916	1.0	1.2
1993	2.9	2084	1.3	1.3

* Year used to normalize data.
[1] Tritton and Norton (1995). [2] Orbital Debris 1995, p.20, Fig.1.2.

Trail rate is given as the mean annual rate per 60-min exposure. The number of spacecraft on orbit is also given. Indices I_T and I_S are normalised values to further illustrate the trend. It is clear from the table that the satellite on orbit population increases steadily at 4% per annum. The satellite trail rate, while varying erratically for reasons given above, shows an increasing trend at 2% per annum. McNally (1997) has unsuccessfully sought increases in trail rate associated with major satellite breakups. We may conclude that trailing is related to satellite numbers. In a visual inspection of the Mt. Palomar/ESO B-survey of the northern sky, we found that while there is an average of one trail per plate, 50% of plates were untrailed, 25% carried one trail and 25% carried multiple trails (in one case carrying 11 trails). Overall, one in two plates exhibit trails.

Murdin (1991) has pointed out that if those plates in the UK Schmidt collection taken within one to two hours of sunset or sunrise are considered, all plates are trailed, with an average of five trails per plate.

These figures suggest that doubling up satellite numbers will lead to significant increase in trail rates. Plates with 11 trails will soon not be a noteworthy rarity but a common occurrence. As trail rate increases, the chances increase that any image carrying trails will be unsuited for its scientific purpose.

One should also note that certain satellites, e.g. Mir and TiPS, are both bright and extended. TiPS subtends a maximum angle of 7′ on the sky, which is comparable with the field of many commonly used CCD cameras. While TiPS is just on the limit of naked eye visibility, Mir is readily visible to the naked eye. The International Space Station will certainly be both a readily visible and an extended object.

It is not easy to translate such statistics into the outlook for photometry. Clearly photometric fields of $15'' \times 15''$ are very much smaller than the $6° \times 6°$ deep sky survey fields. On the other hand photometric observations are susceptible to the presence of far fainter satellites, namely those several magnitudes fainter than the object being measured. In a photometric field, satellites of brightness down to the sky limit could be involved for the faintest observations. On the other hand, for a deep sky survey plate, a satellite moving at about $0.5° \ s^{-1}$, will cross a sky survey plate in less than 17 s. Clearly only the brighter satellites are of significance for such plates.

3.   Other Hazards from Space

There have been numerous proposals over the years to put highly illuminated objects in space: objects easily visible with the naked eye from the ground, e.g. the Ring of Light, the Star of Tolerance, Znamya 2 and 2.5. Thus far such objects are usually proposed in the form of a solar reflector and therefore confined to twilight hours. Only Znamyas 2 and 2.5 became reality. To date none has been successfully demonstrated, but all pose a severe threat to future astronomy. Were they to become a feature of "civilised" living, for celebratory, artistic or advertising purposes, then the limitations of the solar reflector would become evident and constraining. There has been a proposal already, calling for space holograms for advertising. Holograms would be a feature of the dark sky, not just the near twilight sky as with solar reflectors. Holograms could feature throughout the dark hours. Since the brightness of reflectors is now being assessed in terms of 10-100 times the brightness of the full Moon, one does not suppose the proposed holograms will be significantly fainter! The natural Moon is of sufficient brightness to rule out deep sky astronomy for a full half of every month. Were many man-made "super moons" to become a feature of the night sky then optical astronomy from the ground would be seriously emasculated. This is not thinking unrealistically. Significant optical astronomy would not be possible in the bright sky such reflectors and holograms would create.

Others are looking hard at other uses of solar reflectors, for example to collect solar radiation for conversion to microwaves and then transmission to

Earth for large-scale power generation. It is already admitted that such programmes are likely to ruinously damage ground-based radio astronomy and give optical and ground-based infrared astronomy very serious problems. The proposals are there now. The need for alternative sources of power generation is not in dispute, but the cost to sustained astronomical research would be the end of ground-based astronomy as it is now known.

4. Action

What action can be taken? In the case of operational satellites, defunct satellites and large debris, there is nothing that can be done on the basis of existing technology. Satellites will continue to increase in number on decadic timescales and deorbit only on century or millennial timescales. Trailing of fields under observation will continue to be a worsening problem. It is probably unreasonable to expect that satellites can be made less reflecting. Astronomy at both optical and infrared wavelengths will have to become used to the idea that, in the periods within two hours of sunset or sunrise, trailing will be the norm and that norm will only grow more unfavourable. We can develop techniques which could offset some of the consequences of trailing, but it is too early to predict how comprehensive such techniques should become. Sadly, the price to be paid for rapid accumulation of satellites on orbit will be paid by astronomy through loss of telescope usage.

In the case of bright solar reflectors and their putative progeny, it is not too late to take action, though time is undoubtedly very short before a solar reflector is successfully demonstrated. We need urgent discussion to ensure that such potentially disastrous creations do not emasculate optical and ground-based infrared astronomy. Astronomy, like any science, is critically dependent on high quality S/N data. Without a sustained flow of such data, astronomy will surely wither away. Wide agreement on this issue must be sought on an international basis. The need for agreement is very urgent. The next trial of Znamya, Znamya 3, is thought to be planned for 2000 or shortly thereafter.

Astronomy is now confronted with the prospect of space laser communication at optical (far red) wavelengths. Such a communications system has already been implemented by ESA with SPOT4. This is a situation which merits detailed discussion to assess how significant a threat this might be for the future and to examine how close a parallel such laser communication may create with the current situation for radio astronomy.

Finally HAPS (High Altitude Platform Systems), the placing of radio communication relay balloons over major centres of population at a height of 20 km, may have repercussions on optical astronomy, by reflecting city lights and by being an obstruction. Given a balloon dimension of 150 m, an observatory 20 km from the sub-balloon point would see it as an object with an angular extent of 15′, again comparable with the field of common CCD cameras. Further investigation is urgently needed to assess the impact for HAPS on urban observatories.

5.   Conclusion

Trailing of astronomical fields of view is an increasing hazard. While we may live with current levels of trailing, that may not be an option in the future. Trailing will continue to be a hazard for of the order of several hours per location per night, and the increasing loss of efficiency in operating telescopic facilities will become unacceptable. We have to recognise that that loss is imposed by the actions of others. Interference with the legitimate activities of one group by the legitimate actions of others is something the UN expects to resolve. Satellite trailing of astronomical fields is a classic example of such a situation.

There are those who argue that the ultimate solution for astronomy is to transfer all observational work to the far side of the Moon. The exciting possibilities that such a proposal raises should not blind us to the fact that it is not an option for astronomy now, nor in the foreseeable future, and to so argue verges on deception. Since it is only a matter of years until a successful solar reflector will be demonstrated, it is now urgent to examine the consequences for astronomy, particularly as an initial use for such reflectors may be for advertising. Should it be established that advertising from space is acceptable, then the consequences for astronomy at optical and infrared wavelengths could be severe. The limitations of solar reflectors would become all too apparent and advertisers would seek to exploit technology which could utilise the entire night sky. Should the sky become full of bright artificial Moons in perpetuity, the science of astronomy could not compete: even spectroscopic observations could not survive.

Astronomy also must urgently investigate the impact of laser communication. Radio communication has not been a happy precedent and the parallels do not engender confidence. The appearance of HAPS is an indication that new creative schemes will be continually proposed and that astronomical science will have to exercise continued vigilance.

I hope our fears prove groundless. Past experience does not encourage optimism. Yet the future for astronomical science holds such exciting promise. The fact that astronomy continues to survive in spite of growing adverse environmental impact surely means that we should not despair.

References

McNally, D. 1997, Adv.Space Res., 19, 399-402

McNally, D. and Rast, R.H. 1999, Adv.Space Res., 23, 255-258

Murdin, P. 1991, In IAU Colloquium 112, *Light Pollution, Radio Interference and Space Debris*, Ed. D. L. Crawford, Astr. Soc. Pacific Conf. Ser., Vol. 17, p.141

Ohishi, M. 2001, in these proceedings

Orbital Debris - a Technical Assessment, 1995, National Research Council, National Academy Press.

Tritton, S. and Norton, L. 1995, private communication.

Preserving the Astronomical Sky
IAU Symposium, Vol. 196, 2001
R. J. Cohen and W. T. Sullivan, III, eds.

Observations of Artificial Space Objects in Lviv Astronomical Observatory

Jeva Vovchyk, Jaroslav Blagodyr and Olexandr Logvinenko

Astronomical Observatory of Lviv State University, 8 Kyryla and Mefodij st., 290005, Lviv, Ukraine, e-mail eve@astro.franko.Lviv.ua

Abstract. Space debris is the price which humankind must pay for entering space. And as this debris is dangerous in different ways, there must be ways to identify, catalogue, and predict the positions of, these cosmic bodies. That is why it is necessary to observe all artificial cosmic objects that are in space. One way of observing debris is by photometry. Photometric observations give light-curves from which one may deduce information about the position in space, the form, size, and other parameters of the object. Since 1975 a research group at Lviv Astronomical Observatory has been working at the problem of recording light-curves of different artificial cosmic objects. Four electrophotometers were constructed and developed. Many light curves of different artificial objects were measured with these electrophotometers. All light curves are collected in the local computer data bank and are available for use by anybody who needs such information.

1. Introduction

Space debris includes all those artificial objects which are flying in space but never again will be useful for humankind (Flury 1991). These objects are dangerous to spacecraft, they adversely affect our atmosphere and they give misleading information to astronomical observatories. That is why it is so important to monitor space debris.

There are four different ways to observe any artificial satellite: radio, optical location, position, and photometric observations. The information received can be collected in a databank of artificial satellites (such as exists in the Institute of Astronomy Russian Science Academy or in NASA). As a rule such databanks are based on the orbital parameters of the objects (which are obtained from position measurements). However the orbital information is not always sufficient to identify the object, and then other types of observation are needed. For example, information about the shape and size of the object could be obtained from photometric observations.

2. Photometers for Observing Artificial Satellites

Since 1975 a research group at Lviv Astronomical Observatory has worked on the construction of electrophotometers for observing artificial satellites (Blagodyr at

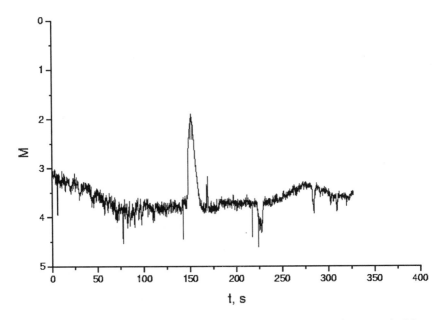

Figure 1.   The light-curve of the TOPEX measured on 10th May 1999. The zero of time is at 22h13m26,34s UT.

all, 1996). A family of four electrophotometers have been constructed. Each of the instruments has a different number of optical channels (one, two, four and/or eight) and different filters. Observations can be made in integrated light or in the BV system.

The multichannel instruments are mounted on four-axis astronomical mountings and have 10-cm objectives. That gives the posibility to observe objects in low Earth-orbit (distances from some hundreds to some thousands of kilometers). The single-channel photometer is mounted on the telescope with twin-axis equatorial mounting and is intended for observing artificial satellites in the geostationary orbit – the distance to these objects is 36000 km.

In all instruments the resulting signals are recorded by computer in "real-time" and all instruments have time-channels with 0.01-s time accuracy.

3.   Results of Observations

Many light curves of different artificial satellites were measured with these electrophotometers. All light curves are collected in the local computer databank and are available for use by anybody who needs such information.

Examples of the results are shown in Figures 1 and 2. Fig.1 shows the light-curve of the object TOPEX-POSEIDON received on the 10th May 1999. The begining of the light-curve (zero on the time axis) is at 22h13m26,34s UT.

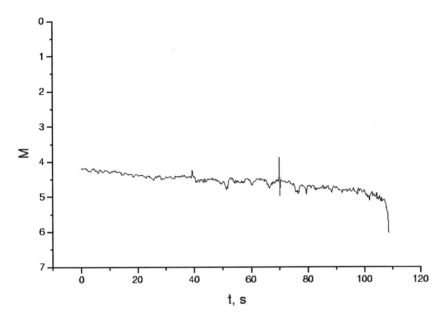

Figure 2. The light-curve of the satellite ERS2 measured on 19th May 1999 at 20h20m06,18s UT.

The TOPEX was launched in 1992 and is stabilized in space. Hence its light-curve should be quite smooth with no changes. Fig.2 shows the light-curve of the satellite ERS2, which is stabilized in space. This light-curve was received on the 19th May 1999 at 20h20m06,18s UT. From the internet (where one can find photographs of many artificial satellites) we know that TOPEX has one solar panel, and hence we conclude that the outburst of light which we see on the light-curve 150 s after the begining is just the reflection of this panel. There are other outbursts on the light-curve which we will not discuss at present.

Solving the geometrical problem Sun-satellite-observer under the condition that the reflection is from the mirror could give the position in space of the panel reflecting the light. Taking into account that on the light-curves one can see many different changes of light, and solving the different problems for each change, one can calculate the position of each detail and then the whole object. That is very simple example of how one can use the light-curve for solving the problem of defining the orientation of the satellite in space.

References

Flury, W. 1991, Adv.Space Res. v.II, N 12, 67-79

Blagodyr, Ja.T. et al. 1996, Turkish Journal of Physics 20, 879-882

Part 4
Threats to Radio Astronomy

Preserving the Astronomical Sky
IAU Symposium, Vol. 196, 2001
R. J. Cohen and W. T. Sullivan, III, eds.

The Future of Radio Astronomy: Options for Dealing with Human Generated Interference

R. D. Ekers and J. F. Bell

ATNF CSIRO, PO Box 76 Epping NSW 1710, Sydney Australia;
rekers@atnf.csiro.au   jbell@atnf.csiro.au

Abstract. Radio astronomy provides a unique window on the universe, allowing us to study non-thermal processes (e.g. galactic nuclei, quasars, pulsars) at the highest angular resolution using VLBI, with low opacity. It is the most interesting waveband for SETI searches. To date it has yielded three Nobel prizes (microwave background, pulsars, gravitational radiation). There are both exciting possibilities and substantial challenges for radio astronomy to remain at the cutting edge over the next three decades. New instruments like ALMA and the SKA will open up new science if the challenge of dealing with human generated interference can be met. We summarise some of the issues and technological developments that will be essential to the future success of radio astronomy.

1. Telescope Sensitivity

Moore's law for the growth of computing power with time (i.e. a doubling every 18 months) is often quoted as being vitally important for the success of the next generation of radio telescopes (working at cm wavelengths) such as the Square Kilometre Array (SKA). It is worth noting that radio astronomy has enjoyed a Moore's law of it own, having exponentially improved in sensitivity with time as shown in Figure 1. In fact the doubling time is approximately 3 years, and has been in progress since 1940, giving an overall improvement in sensitivity of 10^5 to the present time. New instruments and planned upgrades, listed in Table 1, will continue this improvement into the next century. However, we are approaching the fundamental limits of large mechanical dishes and the noise limits of broadband receivers systems. We need to look to other means of extending this growth further into the future.

Why do we want to be on this exponential growth curve? Fields of research continue to produce scientific advances while they maintain an exponential growth in some fundamentally limiting parameter. For radio astronomy, sensitivity is definitely fundamentally limiting. An interesting question is whether there are other parameters for which an exponential growth could be maintained for a period of time. The first and most obvious point about exponential growth is that it cannot be sustained indefinitely. Can we maintain it for sometime into the future? There are two basic ways to stay on the exponential curve:

1. Spend more money, and

2. Take advantage of technological advances in other areas.

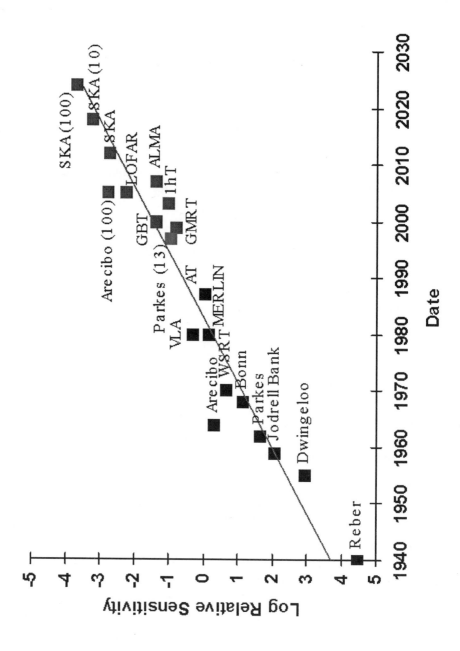

Figure 1.   Exponential growth of Radio Telescope sensitivity. Boxes indicate the sensitivity attained when the systems were first commissioned. This diagram does not indicate their present capabilities. Projected capabilities are shown for ALMA, LOFAR, SKA (with 1, 10 and 100 beams) and Arecibo (with 100 beams). The current 13-beam system on Parkes is also shown.

Table 1.   Summary of New Facilities

	D(m)	Area(m^2)	Freq(GHz)	Date
ALMA	64 x 12m	7.2 x 10^3	30.0 - 900	2007
GBT	100m	7.8 x 10^3	0.30 - 86	2000
1hT	512 x 5m	1.0 x 10^4	1.00 - 12	2003
VLA	27 x 25m	1.3 x 10^4	0.20 - 50	2002
GMRT	30 x 45m	4.8 x 10^4	0.03 - 1.5	1999
SKA	undecided	1.0 x 10^6	0.20 - 20	2015
LOFAR	10^6 x 1m	1.0 x 10^6	0.03 - 0.2	2003

1.1. International Mega Science Projects

International cooperation is now needed for dramatic improvements in sensitivity, because no one country can afford to do it alone. ALMA, the Atacama Large MM Array being developed by the USA, Europe and Japan is an example. It will cost around $US700M spread over 1999-2007 and will provide an unprecedented opportunity to study redshifted molecular lines. The Square Kilometre Array (SKA) which will work at centimetre wavelengths is likely to be a collaboration of ten or more nations, spending $US500M over 2008 to 2015.

There are two basic approaches to funding such large projects:

1. User pays, where member countries pay for slices of the time, or

2. The member countries build the facility and make it openly accessible to all.

Optical astronomy has moved very much down path 1, as the Keck, VLT and Gemini type facilities demonstrate. Radio astronomy has traditionally followed path 2, as have projects like CERN. For future facilities there may be pressure to move more into the user pays regime, possibly bringing about substantial change in the dynamics of the radio astronomy community.

In the past, considerable extensibility of instruments was attained because each country learned from the last one to build a telescope. That evolution is very clear from Figure 1. Since we are likely to be moving to internationally funded projects, that path to extensibility is much more restricted and designers must think very carefully about designing extensibility into the next generation telescopes.

1.2. Extensibility Through Improved Technologies

For a given telescope, past and present extensibility have been achieved by improvements in three main areas:

System Temperature: Reber started out with a 5000 K system temperature. Modern systems now run at around 20 K, meaning that if everything else were kept constant, Reber's telescope would now be 250 times more sensitive than when first built. There are possibilities of some improvements in future, but nothing like what was possible in the past.

Bandwidth: Telescopes like the GBT (Green Bank Telescope) having band-widths some 500 times greater than Reber's, will give factors of 20–25 improvement in sensitivity. Some future improvements will be possible, but again they will not be as large as in the past.

Multiple Beams: Whether in the focal or aperture plane, multiple beam systems provide an excellent extensibility path, allowing vastly deeper surveys than were possible in the past. Although multiple beam systems have been used for a number of experiments, the full potential of this approach is yet to be exploited. A notable example that has made a stride forward in this direction is the Parkes L-band system (Stavely-Smith et al. 1996). The fully sampled focal-plane phased array system being developed at NRAO by Fisher and Bradley highlights the likely path for the future.

Using these three methods, a small telescope like the Parkes 64-m has remained on the exponential curve and the forefront of scientific discovery for 35 years (shown in Figure 1 in 1962 and in 1997). Other telescopes have of course undergone similar evolution and we only highlight Parkes as an example. Scope for continuing this evolution looks good for the next decade, but beyond that more collecting area will be needed. A 64-beam system installed in 2010, would allow Parkes to stay on the curve for some time. The technology to do this is probably only 3-4 years away from providing a realisable system, making it possible to jump well ahead of the curve. Putting a 100-beam system on Arecibo by 2005 is possible and would allow Arecibo to jump way out in front of the curve as it did when first built in 1964.

The relevance of Moore's law in this context is that, if it continues to hold true for the next 1-2 decades, it will provide the necessary back-end computational power to realise the gains possible with multiple beams.

2.   Key Technologies, Driving Future Developments:

HEMT receivers which are wide-band, cheap, small and reliable, allowing us to build low-noise systems with many elements.

Focal plane arrays giving large fully-sampled fields of view will allow rapid sky coverage for survey applications and great flexibility for targetted observations, including novel possibilities for calibration and interference excision.

Interference rejection allowing passive use of spectrum outside the bands allocated to passive uses. High dynamic range linear systems, coupled with high temperature superconducting or photonic filters will allow use of the spectrum between communication signals. Adaptive techniques may allow some co-channel experiments, by removing the undesired signals, so that astronomy signals can be seen behind them.

More computing capacity may result in much more of the system being defined in software rather than hardware. This may lead to a very different expenditure structure, where software is a capital expense and computing hardware is a considered as a consumable or running cost.

Fibre/photonic based beamforming and transmission of recorded signals will revolutionise bandwidths and signal quality, especially for high resolution science.

Software radio and smart antenna techniques which will allow great flexibilty in signal processing and signal selection.

3. Interference Sources and Spectrum Management

It is important to be clear of what we mean when we talk about interference. Radio astronomers make passive use of many parts of the spectrum legally allocated to communication and other services. As a result, many of the unwanted signals are entirely legal and legitimate. We will adopt the working definition that interference is any unwanted signal, getting into the receiving system.

If future telescopes like the SKA are developed with sensitivities up to 100 times greater than present sensitivities, it is quite likely that current regulations will not provide the necessary protection. There is also a range of experiments (eg redshifted hydrogen or molecular lines) which require use of the whole spectrum, but only from a few locations, and at particular times, suggesting that a very flexible approach may be beneficial. Other experiments require very large bandwidths, in order to have enough sensitivity. Presently only 1-2% of the spectrum in the metre and centimetre bands is reserved for passive uses, such as radio astronomy. In the millimetre band, much larger pieces of the spectrum are available for passive use, but the existing allocations are not necessarily at the most useful frequencies.

3.1. Terrestrial Sources of Interference

Interference can arise from a wide variety of terrestrial sources, including communications signals and services, electric fences, car ignitions, computing equipment, domestic appliances and many others. All of these are regulated by national authorities and the ITU (International Telecommunication Union). In the case of Australia, there is a single communcations authority for whole country and therefore for the whole continent. As a result there is a single database containing information on the frequency, strength, location, etc. of every licensed transmitter. This makes negotiations over terrestrial spectrum use simpler in principle.

3.2. Space and Air Borne Sources of Interference

Radio astronomy could deal with most terrestrial interfering signals by moving to a remote location, where the density and strength of unwanted signals is greatly reduced. However with the increasing number of space borne telecom and other communications systems in low (and mid) Earth orbits, a new class of interference mitigation challenges is arising - radio astronomy can run, but it can't hide! The are several new aspects introduced to the interference mitigation problem by this and they include: rapid motion of the transmitter, more strong transmitters in dish sidelobes and possibly in primary beam, and different spectrum management challenges.

There is also an upside to the space borne communication systems in that they help to develop the technology that makes space VLBI possible, which leads to the greatest possible resolution.

A classic example of the problems that can arise is provided by Iridium mobile communications system, which has a constellation of satellites transmitting signals to every point on the surface of the Earth. Unfortunately in this case, there is some leakage into the passive band around 1612 MHz, with signals levels up to 10^{11} times as strong as signals from the early Universe.

3.3. Radio Quiet Reserves

Radio quiet reserves have been employed in a number of places, with Green Bank being a notable success. For future facilties such as the SKA and ALMA, the opportunity exists to set radio quiet reserve planning in process a decade before the instruments are actully built. Radio quiet reserves of the future may take advantage not only of spatial and frequency orthogonality to human generated signals, but also time, coding and other means of multiplexing. These later parameters may be particularly important for obtaining protection from space borne undesired signals, a number of which illuminate most of the Earth's surface.

4. Radio Wavelength Fundamentals

Undesired interfering signals and astronomy signals can differ (be orthogonal) in a range of parameters:

- Frequency

- Time

- Position

- Polarisation

- Distance

- Coding

It is extremely rare that interfering and astronomy signals do not possess some level of orthogonality in this 6-dimensional parameter space. We therefore need to develop sufficiently flexible back-end systems to take advantage of the orthogonality and separate the signals. This is of course very similar to the kinds of problems faced by mobile communication services, which are being addressed with smart antennas and software radio technologies.

5. Interference Excision Approaches

There is no silver bullet for detecting weak astronomical signals in the presence of undesired human generated signals. Spectral bands allocated for passive use provide a vital window, which cannot be achieved in any other way. There is a range of techniques that can make some passive use of other bands possible and in general these need to be used in combined or complementary ways.

Screening to prevent signals entering the primary elements of receivers.

Front-end filtering (possibly using high temperature superconductors) to remove strong signals as soon as they enter the signal path.

High dynamic range linear receivers to allow appropriate detection of both astronomy (weak signals below the noise) and strong interfering signals.

Notch filters (digital or analogue) to excise particularly bad spectral regions.

Decoding to remove multiplexed signals. Blanking of periodic or time dependent signals is a very succesful but simple case of this more general approach.

Calibration to provide the best possible characterisation of interfering and astronomical signals.

Cancellation of undesired signals before correlation, using adaptive filters and after taking advantage of phase-closure techniques (Sault et al. 1997).

Adaptive beam-forming to steer nulls onto interfering sources. Conceptually, this is equivalent to cancellation, but it provides a way of taking advantage of the spatial orthogonality of astronomical and interfering signals.

5.1. Adaptive Systems

Of all the approaches listed above, the nulling or cancellation systems (which may be adaptive) are the most likely to permit the observation of weak astronomical signals that coincide in frequency with undesired signals. These techniques have been used extensively in military, communications, sonar, radar, medicine and other fields (Widrow & Stearns 1985, Haykin 1995). Radio astronomers have not kept pace with these developments and in this case need to infuse rather than diffuse technology in this area. A prototype cancellation system developed at NRAO (shown in Figure 2) has demonstrated 70 dB of rejection on the lab bench and 30 dB of rejection on real signals when attached to the 140-foot telescope at Green Bank (Barnbaum & Bradley 1998). Adaptive Nulling systems are being prototyped by NFRA in the Netherlands. However their application in the presence of real radio astronomy signals is yet to be demonstrated and their toxicity effects on the weak astronomical signals need to be quantified. Arguably the best prospect for testing cancellation schemes in the near future lies in recording baseband data from exisiting telescopes, containing both interfering and astronomy signals (Bell et al. 1999). A number of algorithms can then be implemented in software and assessed relative to each other.

6. The Telecommunications Revolution

We cannot (and do not want to) impede the telecommunications revolution, but we can try to minimise its impact on passive users of the radio spectrum and

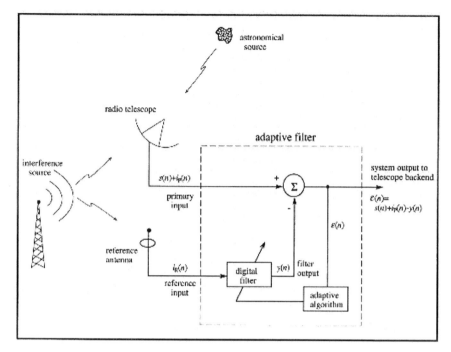

Figure 2.   Example of an adaptive cancellation system. From Barn-baum & Bradley (1998).

maximise the benefits of technological advances. The deregulation of this indus-try has had some impact on the politics. Major companies now play a prominent (dominant?) role in the ITU (International Telecommunication Union). Pro-tection of the bands for passive use must therefore addressed and promoted by government.

7.   Spectrum Pricing

There may be some novel ways in which spectrum pricing could evolve in order to provide incentives for careful use of a precious resource. Radio astronomy and other passive users cannot in general afford commercial rates and therefore need government support. One possibility would be to have a 'green tax' which could be used to fund interference management and research.

Such strategies do not come without cost. While the long term economic cost may be relatively small, upfront R&D costs to an individual company may compromise their competitiveness. This issue must therefore be addressed at national or international policy level.

Unlike many other environmental resource-use problems, spectrum over-use is both reversible and possible to curtail. This leads to certain political advantages because politicians like to have problems which they can solve and this is a more soluble problem than many other environmental problems.

8. Remedies

Siting Radio Telescopes Choosing remote sites with natural shielding helps, but doesn't protect against satellite interference. We can establish radio quiet zones, using National government regulations. This is easier for fixed than for mobile transmitters. The far side of the Moon and the L2 Lagrangian point are naturally occurring radio quiet zones but are very expensive to use.

OECD Megascience Forum: Task Force on Radio Astronomy The goals of the OECD Megascience Forum are complementary to IAU efforts, providing a path for top-down influence of governments which would otherwise not be possible. This task force aims to promote constructive dialogue between regulatory bodies, the international radio astronomy community, telecommunications companies and government science agencies. It will investigate three approaches favoured by the Megascience Forum: technological solutions, regulation and radio quiet reserves.

Environmental Impact In the meter and cm bands $<2\%$ of the spectrum is allocated to passive use! 98% is already used: resource use has been extravagant. Almost all the spectrum at wavelengths longer than 1cm is now polluted, and the situation is rapidly deteriorating at shorter wavelengths. However, the situation is reversible and shareable in more creative ways! in contrast to most other pollution problems.

9. Funding of Radio Astronomy

University-based radio astronomy research in the USA has suffered relative to astronomy at other wavelengths for two reasons:

1. The centralised development at NRAO has made it difficult for many universities to remain involved in technical developments.

2. Space-based programs in infrared, optical, UV, X-ray, and Gamma-ray bands have a rather different funding structure, where access to research funds is based on successful observing proposals. Radio astronomy has no access to such funds and therefore is a relatively uneconomical pursuit for astronomers. This method of funding is being taken up in other countries and radio astronomy needs to find a way to join the scheme.

More globally, radio astronomy has suffered relative to other wavelengths, because data acquisition, reduction and analysis is unnecessarily complex. Researchers have to spend a lot more effort in data processing than in other areas of astronomy. For example, in many other bands, fully calibrated data lands on the researcher's desk a few days after the observations were taken. Radio astronomy needs to finds ways to move into this regime (as Westerbork have done to some degree), but at the same time preserve the vast flexibility that can be derived from measuring the electric field at the aperture.

10. Conclusions

The possibilities for the future of radio astronomy are good, but there are some challenges issues for the community to consider and address:

- The whole radio spectrum is needed for redshifted lines

- About 2% of spectrum below 50 GHz is reserved for passive use by regulation, so we must develop other approaches

- We cannot (and don't want to) impede the telecommunications revolution

- Radio astronomy has low credibility until we use advanced techniques

- It is essential to influence government policy

- Astronomers should have a uniform position

- Threatening language doesn't help

- Is interference harmful?

References

Barnbaum, C. & Bradley, R. 1998, AJ, 116, 2598

Bell J. F., et al. 1999 "Software radio telescope: interference mitigation atlas and mitigation strategies", in *Perspectives in Radio Astronomy: Scientific Imperatives at cm and m Wavelengths*, Eds. M.P. van Haarlem & J.M. van der Hulst (Dwingeloo, NFRA)

Haykin, S. 1995, "Adaptive Filter Theory" Prentice Hall

Sault, B., Ekers, R., Kewley, L. 1997 "Cross-correlation approaches to interference mitigation" Sydney SKA workshop, proceedings on www: http://www.atnf.csiro.au/SKA/WS/

Smolders, A. B., 1999, "Phased-array system for the next generation of radio telescopes", in *Perspectives in Radio Astronomy: Scientific Imperatives at cm and m Wavelengths*, Eds. M.P. van Haarlem & J.M. van der Hulst (Dwingeloo, NFRA)

Staveley-Smith L. et al. 1996, PASA, 13, 243 "The Parkes 21cm Multibeam Receiver" (An overview of the science and overall system design of the multibeam receiver)

Widrow, B. & Stearns, S., 1985, "Adaptive Signal Processing" Prentice Hall

Interference Mitigation Web pages
    http://www.atnf.csiro.au/SKA/intmit/

ATNF SKA web site
    http://www.atnf.csiro.au/SKA/

Preserving the Astronomical Sky
IAU Symposium, Vol. 196, 2001
R. J. Cohen and W. T. Sullivan, III, eds.

Radio Astronomy and the International Telecommunications Regulations

Brian Robinson

Research Fellow Emeritus, Australia Telescope National Facility, P.O. Box 76, Epping, NSW 2121, Australia

Abstract. For forty years radio astronomers have had access to frequency bands allocated by the International Telecommunication Union (ITU) - initially a League of Nations body (from 1919) and then a United Nations body (since 1945). Hard work and skilful negotiation by a handful of radio astronomers since 1959 have ensured viable access to scarce spectral allocations. There have been many battles won, some key battles lost. The next treaty conference of the ITU is in the year 2000.

1. Treaty Conferences

The ITU holds treaty conferences every few years, with preparatory technical meetings in the years between. The conferences are called World Radio Conferences (WRCs) - originally called World Administrative Radio Conferences (WARCs).

Radio astronomy has argued its case strongly at many World Radio Conferences - in 1959, 1963, 1971, 1979, 1988, 1992 and 1997, for example. Some conferences were not attended - and a high price is still paid for those oversights.

At World Radio Conferences accredited bodies may speak and submit documents. But only governments can vote on matters to be incorporated in the resolutions, which become a treaty document - the Radio Regulations - for each country to sign.

Traditionally the delegates to World Radio Conferences were mainly government representatives, but in the last decade we have seen large numbers of industry representatives on the delegations. At WRC-1997 the U.S. had 99 delegates, Japan 81, France 74 and the U.K. 47; most developing nations could send only two delegates.

Third World countries are worried by the direction the ITU has been taking. They fear that, by the time they get onto their feet, the highly industrialised countries will have the whole radio spectrum exclusively sewn up. They see that the number of cellular phones in industrialised countries has grown from 144 million in 1996 to 274 million in 1998.

The ITU does not regard all services equally - some get PRIMARY allocations. Other services got SECONDARY allocations, meaning that they must not interfere with a Primary service. Lower down the scale of protection are the services who just get in, with FOOTNOTE allocations: they must not interfere with Secondary or Primary services. But "must not interfere" is loosely defined

- the ITU does not face up to the fact that some services (e.g. passive space research, remote sensing and radio astronomy) work at minuscule signal levels compared to transmissions for broadcasting, mobile communications, radar, etc.

An allocation in an ITU treaty that two services share a band does not guarantee protection for a service using very low signal levels. Furthermore the ITU has been blind to scores of footnotes in the Radio Regulations that state "Emissions from space and airborne stations can be particularly serious sources of interference to the radio astronomy service."

The ITU's engineering arm is ITU-R (long known as CCIR, the Comité Consultatif International des Radiocommunications), which produces "Recommendations" and standards carrying much weight at WRCs. Many ITU-R recommendations specify constraints on sharing between services. However, conclusions of ITU-R are reached by agreement between members of "study groups"; this usually works well, but there have been notorious cases of deliberate disagreement.

2.   Successes

The efforts of a few dedicated radio astronomers, supported by government delegates from their countries, have achieved many notable successes over the last forty years. A contributory factor was that radio astronomers were usually technically well ahead of the communications industry and working in short wavelength bands that no one else had begun to use.

2.1.   WARC-1959

Although falling considerably below the hopes of radio astronomers, the 1959 WARC laid a good basis for Radio Astronomy, notably the exclusive "passive" band at 1420 MHz for 21-cm hydrogen line observations. Prior to 1959 the ITU only recognised active radio "services ", such as broadcasting, fixed networks, radio navigation, radar, mobile communications, amateur use and meteorological aids. In 1959 there was little commercial or military demand for frequencies above 1000 MHz, and allocations (in footnote form) of reasonable bandwidth were possible.

For radio astronomy the main burden at WARC-1959 fell on Charles Seeger (Secretary of URSI Sub-Commission Ve). John Findlay (1988) describes the support of The Netherlands and the United Kingdom in 1959, and the opposition of the United States (see Walter Sullivan (1959), John Lear (1959) and John Finney (1959)).

2.2.   WARC-1963

During the preparations for WARC-1963, the ITU "formally recognised IUCAF as an active participant in the work of CCIR and the ITU" (Smith-Rose 1961). IUCAF (the Inter-Union Commission on Frequency Allocation for Radio Astronomy and Space Science) had been formed in 1960 to present the frequency allocation requirements of URSI, IAU and COSPAR, the three ICSU unions concerned with space research and radio astronomy.

The discovery of the main OH lines at 1665 and 1667 MHz was announced during WARC-1963 and an allocation was made on the spot. The 1612-MHz and 1720-MHz OH lines were discovered the following year, and footnote allocations were made at WARC-1971.

2.3. WARC-1971

At WARC-1971, which allocated frequencies up to 275 GHz, the hydrogen line band was extended down to 1400 MHz to cater for redshifts in emissions from more remote sources, and the main OH lines at 1665 and 1667 MHz received Primary allocation.

The advances in centimetre-wave and millimetre-wave radio astronomy had come along at just the right time - Townes and his collaborators had found ammonia at 23.7 GHz and water vapour at 22.235 GHz in 1968, while Zuckerman, Palmer, Snyder and Buhl found formaldehyde at 4.829 GHz in 1969. Then in 1970 Jeffert, Penzias and Wilson discovered the carbon monoxide line at 115.271 GHz.

The richness of the mm-wave spectrum in molecular lines also gave radio astronomy an entree into the top end of the radio spectrum, which other services had not begun to use (Horner 1971). Bands were allocated to ammonia at 23.7 GHz and hydrogen cyanide at 86.3 and 88.6 GHz. Footnote allocations were made to hydroxyl (1611.5-1612.5 and 1720-1721 MHz), formaldehyde (4.829 GHz), excited hydrogen (5.763 GHz), formaldehyde (14.489 MHz), water vapour (22.235 GHz), excited hydrogen (36.466 GHz) and carbon monoxide (115.271 GHz). The highest radio astronomy allocation was at 240 GHz.

On the other hand, WARC-1971 allocated broadcasting satellites the band 2500 - 2690 MHz, immediately adjacent to the 2690 - 2700 MHz radio astronomy allocation (this is discussed later). And an allocation to an aircraft microwave landing system at 5000 MHz was adjacent to the 4990-5000 MHz radio astronomy band (see later).

Recommendation Spa2-8 from WARC-1971 was entitled "Relating to the Protection of Radio Astronomy Observations on the Shielded Area of the Moon". Arthur C. Clarke (1961) had suggested the lunar shielded area as a location for Radio Astronomy. From 1974 CCIR studied the need to keep the radio spectrum quiet on the far side of the Moon.

2.4. WARC- 1979

An enormous international effort went into WARC-1979, the first "WARC General" after 1959, i.e. covering the whole spectrum. This WARC opened up the radio spectrum to 275 GHz.

Horner's report (1980) says of radio astronomy and space research:

> The overall outcome is widely considered to be satisfactory for these sciences; indeed the general opinion is that they have received very favourable treatment, but some provisos will be mentioned below.

> There was a strong and cohesive radio astronomy group ... five (radio astronomers) were involved for various periods as representatives of IUCAF.

... in the case of radio astronomy there was a significant increase in the number of (frequency) table allocations. ... In space research ... there are considerable gains in allocations for sensing of the Earth's surface and atmosphere by both radiometric and radar techniques.

WARC-1979 allocated Radio Astronomy sixteen bands, between 322-328.6 MHz and 105 -116 GHz. Allocations by footnote were made to 18 bands, between 140 GHz and 348 GHz. The WARC also approved a Dutch proposal for "Article 36" in the Radio Regulations, setting out the basic operation and needs of Radio Astronomy. There were also sections on "Radio Astronomy in the Shielded Zone of the Moon" and "The Search for Extraterrestrial Emissions" plus new definitions of interference and "unwanted emissions".

2.5. WARC-1992

At WARC-1992 the status of Space Research and Radio Astronomy was significantly enhanced between 137 and 3000 MHz, and above 13.5 GHz. Delegates from 125 countries clearly recognised the importance of scientific use of the radio spectrum in the face of increasing pressures from telecommunications, broadcasting and navigation interests - particularly where the proposed transmissions were from satellites.

IUCAF delegates became immersed in the projected characteristics of Digital Radio Broadcasting, High Definition TV, the "Big-LEOs" and the "Little LEOs" - the explosion of mobile communications via Low-Earth-Orbit satellites.

2.6. WRC-1997

At WRC-1997 a major issue was the allocation of a radar band at 94 GHz for the Earth Exploration Satellite Service (EESS). IUCAF was heavily involved in the placement of this allocation and setting limits on the number of satellites.

2.7. WRC-2000

A full report on IUCAF's preparations for WRC-2000 has recently appeared (Baan 1999). An IUCAF-CRAF-CORF position paper is being produced, addressing 12 WRC agenda items - such a High Altitude Platforms at 48 GHz, allocations to the Mobile Satellite Service (MSS) and MSS feeder links, ... There is to be an epic review of allocations from 71 - 275 GHz to EES (passive) and Radio Astronomy, with consideration of protection zones around mm-wave observatories.

2.8. IAU and CCIR Priorities for Radio Spectral Lines

During the 1970s there was a flood of discoveries of molecular spectral lines. By 1976, 220 molecular lines were known in the radio spectrum. IAU Commission 40 set up a Working Group (chaired by Brian Robinson) to determine the priorities to be given to the molecular lines during preparations for WARC-1979. By 1979 about 600 molecular lines were known between 0.8 and 346 GHz. An IAU list of 30 key spectral line frequencies was agreed to at the 1979 IAU General Assembly in Montreal (Robinson & Whiteoak 1979, Westerhout 1979). The IAU submitted the list of key spectral lines to CCIR in Geneva, who incorporated it in CCIR

Recommendation No. 314. CCIR Recommendations carry great weight in the deliberations at a WRC.

The IAU Working Group continued to evaluate the key spectral lines at every IAU General Assembly for the next 21 years. Each time the revisions were adopted by CCIR (now known as ITU-R). This Symposium needs to reactivate the IAU/ITU(R) listing of key spectral lines.

2.9. Filtering the GMS Meteorological Satellite

A bright spot in the 1970s was successful negotiations with Japan on the out-of-band transmissions from the Geostationary Meteorological Satellite (GMS), which would have extended sidebands to cause harmful interference to radio astronomy observations in the OH band 1660 to 1670 MHz. Effective filtering of widespread unwanted sidebands was shown to be feasible.

During design of the GMS satellite, Brian Robinson (an IUCAF Correspondent) contacted the Japanese Meteorological Agency and encouraged them to build a filter to protect the OH band. Hughes Aircraft Company engineers said that such a filter could not be built, so the Japanese decided to design and build it. The satellite was launched in 1977. Observations of the GMS satellite are reproduced in Robinson & Whiteoak (1979). In the radio astronomy band the filter gave 60 decibels attenuation of the GMS sidebands.

2.10. The Illegality of GLONASS Transmissions

Prior to WARC-1992, Radio Astronomy use of the most important 1612 MHz satellite line of OH had been authorised by Footnote 352K inserted in the Radio Regulations at WARC-1971. The Footnote warned that "Emissions from space and airborne stations can be particularly serious sources of interference to the radio astronomy service." But, later, WARC-1979 had inserted Footnote 352A reserving the whole band 1610-1626.5 MHz on a worldwide basis for "airborne electronic aids to air navigation and any directly associated ground-based or satellite-borne facilities."

In April 1983 the Soviet Government advised, using the regular ITU "Article 14" coordination procedures, its plan to operate a GLObal NAvigation Satellite System - shortened to GLONASS. Only one country responded to the announcement of this satellite system - but later than the 45-day cut-off for objections. Then radio astronomy observations of the OH line were closed donw by the horrific interference produced by these GLONASS satellites near 1612 MHz. Two "footnote" ITU allocations had equal legal status, and radio astronomers could only lament their oversight.

In May 1991 Russia approached the ITU seeking improved status for GLONASS in the band 1597-1617 MHz. This time IUCAF was immediately alert and organised opposition from 10 countries to the Russian request to ITU - which then lapsed. IUCAF argued that GLONASS should confine its emissions to the band 1559 to 1610 MHz that had been allocated by WARC-1979 to Radionavigation Satellites and Aeronautical Radionavigation.

IUCAF confronted the Russian Space Forces in Moscow in October 1991. After much negotiation, the Russians undertook to eliminate the out-of-band emissions of the GLONASS system "beginning from 1994 in process of replacement of the Space apparatuses with new ones equipped by improved filters."

At WARC-1992 the situation changed dramatically, when Radio Astronomy at 1612 MHz gained full Primary status in the band. GLONASS with Footnote status could not interfere with a Primary Service.

In June 1992 IUCAF confronted the Russian Space Forces again in Moscow and arranged that a delegation from the Space Forces and the Institute of Space Device Engineering would visit the Jodrell Bank telescope in November 1992 to witness the interference for themselves. The Russians were amazed at what they saw.

At a further meeting with IUCAF in Moscow (June 1993), the Space Forces recognised the illegality of their transmissions and agreed to move their operating frequencies away from the 1610.6 - 1613.8 MHz Radio Astronomy band. A final agreement with IUCAF was signed in November 1993. All GLONASS satellite transmissions will be out of the band by the year 2007.

3.   Other Successses

3.1.   The SSU Series Satellites

In April 1976 the U.S. launched three "SSU" satellites, which produced strong interference at 1420 MHz - signals a billion times stronger than the interstellar hydrogen emissions. After Canada protested that the signals were in contravention of the ITU allocations (Argyle et al 1977), the data dumps from the satellites ceased while the satellites were over the Pacific Northwest and Canada.

3.2.   Broadcasting Satellites at 22 GHz

In 1991 IUCAF was advised of Japanese plans to license a broadcasting satellite at 22.6 GHz with a bandwidth of 120 MHz. The 22.21-22.5 GHz band is a Primary allocation to space research, radio astronomy and earth exploration. Protests were made by eight countries in the Asia-Pacific region. The Japanese proposal failed to gain ITU registration (under "Article 14" procedures).

4.   Failures

In a number of cases Radio Astronomy has failed to gain ITU allocations or achieve protection from incompatible services.

4.1.   Continuum Bands

The discoveries of atomic and molecular lines were well timed in terms of the WARC timetable, which gave them plenty of impact. Since the frequencies were fixed and immovable, the allocation had to be just there. But from WARC-1959 on, requests were also made by radio astronomers for continuum bands. A bandwidth of 2.5% was sought in each octave. The requests for continuum bands have, over all the years, got nowhere.

4.2.   INMARSAT Satelllites

At WRC-2000 an old problem will be an extension band desired by INMARSAT around 1660 MHz. The IUCAF Report to ICSU in 1992 discussed "sharing problems in the band 1660 to 1660.5 MHz and out-of-band interference in the band 1660 to 1670 MHz from aircraft communicating with INMARSAT satellites." The 1992 Report said: "Discussions with INMARSAT on these problems have been unproductive." Nothing has changed!

4.3. Iridium Satellites

Did the emptying of the Radio Astronomy band by GLONASS satellites open the door to allow Motorola Iridium satellites to move in?

The struggle about Iridium satellites began in 1991 at CCIR meetings, where the spurious and unwanted emissions from the satellites were ill-defined but threatening. The name Iridium came from the original plan to have 77 LEO satellites; the Iridium nucleus is orbited by 77 electrons. Later Motorola reduced the proposal to 66 satellites, but have not yet renamed the system Dysprosium! The danger posed by Iridium is discussed by Abbott (1996) and Ponsonby (1996). Scientists negotiating with Motorola were surprised by the dictatorial style of the would-be satellite operators - even the Russian Space Forces were more pliable about GLONASS interference!

The U.S. National Radio Astronomy Observatory signed an agreement with Iridium in 1994. All other observatories adopted a strong defensive position behind the IUCAF banner. In 1998 Arecibo Observatory signed a Memorandum of Understanding with Iridium on much better terms than accepted by NRAO. Later in 1998 Australia signed an agreement with Motorola allowing use of Iridium mobile phones subject to "the licensee must not cause harmful interference to Australian radio astronomy services." At this Symposium we have heard of an improved agreement negotiated with Motorola by CRAF on behalf of European countries.

In 1999 the full constellation of Iridium satellites is in orbit, but very high costs to users have greatly inhibited demand. Also, acquiring the satellites has been difficult when users are inside buildings!

5. Partial Successes

An ongoing fight has been about ITU Recommendation 66 of WARC-79. Within ITU activities, "Recommendations" are the highest level of directive. Also World Radio Conferences are the highest level of authority. Organs of the ITU like ITU-R must take note of Recommendations, especially if they come from a WARC.

When WARCs allocate frequency bands to different users it is hard to avoid incompatible neighbours. The hyper-sensitivity of radio telescopes to low-level "unwanted", spurious and out-of-band emissions leaves them particularly exposed to these emissions. WARC-1979 issued "Recommendation 66" as a directive to CCIR to carry out studies of the Maximum Permitted Levels of Spurious Emissions "as a matter of urgency".

CCIR did absolutely nothing! The vested interests of the satellite operators made sure nothing happened! Thirteen years later the directive to CCIR to examine spurious and unwanted emissions was renewed by the 1992 WARC. For several years an ITU-R Task Group aimed to produce a set of protection levels that could be approved by WRC-1997. The Task Group could agree only on compromise "design objective" levels which were two or three orders of magnitude too high to protect Radio Astronomy.

At WRC-1997 Recommendation 66 was again on the agenda: WRC-1997 recognised the inability of the ITU-R Task Group to propose adequate protection for Radio Astronomy. A new Task Group has therefore been set up, to report to a future WRC. In summary there has been pitifully little progress in the 20 years since Recommendation 66 was first introduced at WARC-1979.

6.   "Narrow Squeaks"

6.1.   The Needles Project

In 1960 Project WEST FORD - the "Needles Project"- presented a great threat to radio astronomy, optical astronomy and space research, and called for a combined response by URSI, IAU and COSPAR. At that time IUCAF was set up as an ICSU Inter-Union Commission (see Smith-Rose 1960).

Many articles were published to point out the dangerous pollution of space around the Earth by the Needles Project. Anticipating new communications and navigation systems with powerful transmitters in earth satellites, Lilley (1961) concluded: "The pursuit of basic science and the progress of space radio technology represent needs of man which must be advanced. For the impending interference a simple solution exists: allocation of clear frequency bands for basic science. This action is imperative and must ultimately rest on national and international agreements."

The 1961 IAU General Assembly passed a resolution:

> ... maintaining that no group has the right to change the Earth's environment in any significant way without full international study and agreement;
>
> the International Astronomical Union gives clear warning of the grave moral and material consequences which could stem from a disregard of the future of astronomical progress,
>
> and appeals to all Governments concerned with launching space experiments which could possibly affect astronomical research to consult with the International Astronomical Union before undertaking such experiments and to refrain from launching until it is established beyond doubt that no damage will be done to astronomical research.

Fortunately, the Needles experiment was short-lived and other such experiments have not followed. It is appropriate that IAU Symposium 196 revisit these resolutions made 38 years ago. Many threats now come from commercial projects.

6.2.   ATS-6 at 2700 MHz

After WARC-1971 a broadcasting satellite ATS-6 was built to operate adjacent to the radio astronomy band at 2700 MHz. IUCAF alerted radio astronomers to the danger. A filter was added to the satellite at the last moment to protect the Radio Astronomy band.

6.3. Microwave Landing System at 5 GHz

A microwave landing system (MLS) for civil aircraft was proposed to replace the old system at 200 MHz. The new MLS system operates at 5 GHz, adjacent to the radio astronomy band. By 1999 only five of the MLS systems have been installed at airports. Fortunately, the MLS system has been supplanted by the use of GPS, particularly differential-GPS, as an aircraft landing aid.

7. Spread Spectrum Modulation

Many satellite communication or navigation systems use various forms of spread-spectrum modulation. Some basic spread-spectrum systems generate a wide spread of unwanted sidebands, whose intensity falls off very slowly [as the square of $(\sin x)/x$] to each side of the central frequency. Examples of satellites causing interference from these unwanted sidebands were the first series of GPS satellites and GLONASS satellites.

The widespread interference can be avoided in well-designed modulation systems. One approach has been suggested by Ponsonby (1991). Also, Delogne & van Himbeeck (1995) showed that out-of-band radiation can be carefully controlled in well-designed spread-spectrum systems. Direct Sequence Spread Spectrum Modulation can be perfectly confined in a nominal bandwidth of 1.5 times the "chip rate". The contribution of digital signal processing to efficient use of the radio spectrum is discussed further by Delogne and Bellanger (1999).

8. The Future

Radio Astronomy has great visions for the 21st Century. Kellerman (1997) has described many of the proposals, and elsewhere in this volume Ron Ekers describe other projects.

But the communications and broadcasting industries also have extensive and ambitious plans that will often compete for the same spectral bands. Over many decades, the ITU's attempts to accommodate incompatible spectral demands has been a succession of poor compromises. Can the ITU do a more competent job in the 21st Century? That seems unlikely now that commercial interests have much more influence on ITU decisions than national governments. The ITU needs to have the foresight to maintain in the radio spectrum the clarity of particular windows on the Universe that mankind needs to enhance and extend our understanding of the evolving Universe about us.

8.1. Electromagnetic Environmental Impact Statements

At Lille in 1996, URSI considered interference to radio astronomy from satellites and:

> 'called on the ITU and affiliated national and regional administrations to encourage the use of modulation schemes that minimise harmful interference, to require pre-flight testing of satellite transmission systems, to devise rulemaking that prevents new users from disrupting existing users, and to require electromagnetic environmental impact statements before operation is authorized.

This URSI proposal for electromagnetic EIS needs to be discussed at IAU Symposium 196, and considered by IAU at the General Assembly in the year 2000. We trust that UNESCO will support this proposal.

8.2.   Radio Quiet Reserves

Butcher (this volume) has discussed the OECD Megascience Forum suggestion of radio-quiet reserves for proposed radio astronomy arrays of very great sensitivity.

Many people at this IAU Symposium were at the 1992 Exposition in Paris, where UNESCO was exploring the possibility of designating a few selected observatories as World Heritage Sites. Has anything yet come of this?

A variation on this approach is suggested by the Antarctic Treaty of 1959, which followed the extensive scientific work carried out in Antarctica during the International Geophysical Year 1957-58. The Treaty has been supplemented by the 1991 Madrid Protocol, with a commitment that Antarctica and its environment is a natural reserve, devoted to peace and science, and stressing cooperation. Mineral resource activity is prohibited. The Antarctic Treaty states that in the interests of mankind, Antarctica should continue forever to be used exclusively for peaceful purposes. Military uses are prohibited. No future actions can be used to support prior national claims, and no new national claims can be made.

Radio-quiet reserves could be protected by a similar treaty, with a guarantee of low levels of interference and exclusion of satellite and airborne emissions.

8.3.   Dedicated Passive Spectrum

There is another approach: the relentless exploitation of land can be limited by the declaration/consecration of national parks. Could the exclusive "passive" bands required by Radio Astronomy, Space Research and Earth Exploration be kept in a pristine state as "international parks" in the radio spectrum, under the protection of UNESCO? This issue needs to be followed up with IAU, URSI, COSPAR, ICSU and UNESCO.

Acknowledgments. I have received much help in this review from Jim Cohen, Sir Francis Graham-Smith, Fred Horner, Willem Baan, John Whiteoak and Emile Blum. I have also received strong support from the IAU Secretariat, the URSI Secretariat and ICSU.

References

Abbott A. 1996, Nature, 380, 569

Argyle E., Costain C. H., Dewdney P. E., Galt J. A., Landecker T. and Roger R. S. 1977, Science, 195, 932-933

Baan W. A. 1999, Radio Sci. Bull., 289, 13-19

Clarke A. C. 1961, Harper's Mag., Dec. pp 56-62

Crane P. C. 1985, NRAO News, 24, 13

Dellinger J. H. 1961, URSI Info. Bull., 128, 78-79

Delogne P and Bellanger M. 1999, Radio Sci. Bull., 289, 23-28

Delogne P and van Himbeeck C. 1995, Radio. Sci. Bull., 275, 23-29

Findlay J. W. 1988, URSI Info. Bull., 246, 14-19

Finney J. 1959, New York Times, CIX (Oct. 17), 47

Horner F. 1971, URSI Info. Bull., 181, 9-22

Horner F. 1980, URSI Info. Bull., 212, 14-16

Kellerman K. I. 1997, Sky Telesc. 93(2), 26-33

Lear J. 1959, Sat. Rev., 42 (Oct 3), 47-49

Lilley A. E. 1961, Astron. J., 66, 116-118

Ponsonby J. 1991, J. Navig., 44, 392-398

Ponsonby J. 1996, Nature, 381, 550

Robinson B. J. and Whiteoak J. 1979, Proc. Astron. Soc. Aust., 3, 396-400

Smith-Rose R. L. 1960, URSI Info. Bull., 123, 130-131

Smith-Rose RL. 1961, URSI Info. Bull., 128, 76-80

Sullivan W. 1959, New York Times, CIX (Sept. 20), 27

Westerhout G. 1979, Trans. IAU XVII(B), 245-247

Isn't a window overlooking the universe worth $100,000,000 to the people? (That's what the requested radio astronomy channels might be worth in a TV auction.)

... If science doesn't get the right to use these radio channels for the future benefit of all the people, the channels will be grabbed up in time if not immeditately by the military and by whichever of the fiercely competing commercial interest prove able to pay the highest price.

J. Lear

Saturday Review, October 3, 1959, vol. 42, pp.47-49.

Preserving the Astronomical Sky
IAU Symposium, Vol. 196, 2001
R. J. Cohen and W. T. Sullivan, III, eds.

Radio Astronomy and the Radio Regulations

R. J. Cohen

University of Manchester, Jodrell Bank Observatory, Macclesfield, Cheshire SK11 9DL, UK

Abstract. This article gives a brief introduction to the status of radio astronomy within the International Telecommunication Union (ITU), the body which coordinates global telecommunications. Radio astronomy entered the ITU arena in 1959 as a relative latecomer. By its nature, radio astronomy does not fit into the ITU system very well: regulators are hoping to facilitate commercial development of the radio spectrum, whereas astronomers are hoping to retain quiet frequency bands through which to study the Universe at ever higher sensitivity. Nevertheless there are major long-term goals which radio astronomers can realistically hope to achieve via the ITU in the years ahead, including more favourable frequency allocations and better regulatory protection. The prospects for radio astronomy at the forthcoming World Radio Conference WRC-2000 are reviewed. It is vital that radio astronomers participate in force at this WRC.

1. Introduction

I am proud to be a radio astronomer. During my scientific career the typical measurement sensitivity has increased nearly a million-fold, from a tenth of a Janksy to microJanskies, and the angular resolution by a similar amount, from arcseconds to tens of microarcseconds. Radio astronomy now achieves the highest sensitivity and the highest angular resolution of any branch of astronomy, and it does so relatively cheaply using ground-based facilities (Kellerman 1997). But we face growing problems of radio interference which threaten to halt or even reverse the steady advances we have enjoyed (Cohen 1999).

Once radio astronomers had the radio spectrum more or less to themselves and any interference problems were local. The electrification of the railway line which passes my observatory, Jodrell Bank, was big news forty years ago, but it did not affect other radio observatories. Today the radio spectrum is crowded and we face global threats from satellites. Our experiences with the GLONASS satellite system showed just how effectively a branch of radio astronomical research could be halted worldwide by just a few small satellites (Galt 1990, Ponsonby 1991, Combrink, West & Gaylard 1994). Nowadays more than 100 satellites are launched each year and there are plans for thousands more, covering ever more bandwidth. The radio spectrum has become big business.

Radio astronomy does not really fit in alongside the telecommunication industry. The cosmic signals we study are extraordinarily weak and we usually have to observe them in the presence of much stronger manmade signals. It is exactly like trying to observe sources as faint as stars while there are many other sources as bright as the Sun in the sky. Cosmic sources are usually noise-like and their power levels and frequencies are fixed by nature. Because we are conducting research, the outcome of the measurements cannot be predicted in advance, so we are particularly vulnerable to interference. We can only survive the telecommunication revolution if commercial developments are well regulated.

The rules governing the use of radio are made by the International Telecommunication Union (ITU), a body which began as an intergovernmental organization, but which is increasingly influenced by multinational corporations. In this article I explain something about the nature of the ITU, and the status of radio astronomy within the ITU framework. I also highlight some of the problems which are developing, and raise the question of whether we need to be looking outside the telecommunication community to get protection for our work.

2. The International Telecommunication Union

The ITU was established long before radio astronomy existed. Its origins can be traced back to the first International Telegraph Convention of 1865, which set up the International Telegraph Union, with 20 member countries. As communications evolved so did the acronym. Since 1934 "ITU" has meant the International Telecommunication Union and since 1948 the ITU has had its headquarters in Geneva. Today the ITU is an intergovernmental organization within which governments and the private sector coordinate global telecommunication networks and services.

The modern ITU has a complex structure with three sectors (Radiocommunication, Telecommunication Standardization and Telecommunication Development), all governed by the Plenepotentiary Conference. The Radiocommunication Sector, ITU-R, holds World Radio Conferences (WRCs) at which the Radio Regulations are agreed and revised. The Radio Regulations set out rules for use of the radio spectrum. The regulations are a kind of treaty between nations, to allow everyone to use the radio spectrum without getting in each others' way. The basic principles underlying the regulations are to respect your neighbours' use of the spectrum and to respect existing radio services when introducing new services or other changes to the Radio Regulations.

The Radio Regulations contain legalistic definitions of telecommunication, radiocommunication, radionavigation, harmful interference, and so on. The ITU divides the world into three regions for administative purposes: Europe and Africa are in Region 1, the Americas in Region 2, and the Asian-Pacific countries in Region 3. The radio spectrum up to 275 GHz is divided into frequency bands which are allocated to various radio services. Allocations can have primary or permitted status, or secondary status, and both the allocations and the allocation status can vary between regions. The regulations are further qualified by footnotes which can spell out different categories of service within specific countries, or special conditions attached to the use of certain frequency bands by certain services. It is a legal maze.

3. Radio Astronomy in the Radio Regulations

Radio astronomy entered the Radio Regulations in 1959, when it was defined in Article 1 as "astronomy based on the reception of radio waves of cosmic origin". The radio astronomy service also received its first frequency allocation at this time: the frequency band 1400-1427 MHz containing the 21-cm line of neutral atomic hydrogen (HI). This was the first frequency band to be set aside by the ITU purely for passive use. A new footnote was added stating that "all emissions in the band 1400-1427 MHz are prohibited". Nowadays the radio astronomy service has many frequency allocations, totalling 2% of the radio spectrum below 50 GHz (the part that is used for telecommunications) and 24% of the (largely unexploited) spectrum from 50 to 275 GHz. Radio astronomy interest in further frequency bands from 275 to 400 GHz is also noted.

The regulatory status of our allocations is far from ideal, however. What we want to have is frequency allocations with primary status shared only with other passive services. One third of our allocations are of this type, but the other two thirds are shared with active services. Furthermore, there is no mandatory protection for radio astronomy as there is for most other services.

Technical criteria for protecting radio astronomy measurements against interference from transmitters used by other radio services are developed within study groups and task groups of the ITU-R. Working Party 7D (radio astronomy) has produced a Handbook on Radio Astronomy and a set of Recommendations which are described in the following section. As yet, none of these recommendations has regulatory status. The Radio Regulations specify in precise detail how an earth station for telecommunciations should be protected, but not a radio astronomy station. The protection levels for radio astronomy do not appear anywhere in the Radio Regulations. In fact the very concept of harmful interference, on which regulatory protection would be based, is not even defined for the radio astronomy service.

Nevertheless the ITU system could accommodate some of our hopes and aspirations if the political will existed. For example, radio-quiet zones could be set up right now to protect major facilities from terrestrial transmitters. The Radio Regulations are not completely prescriptive: a country has the right to use the radio spectrum how it likes within its own borders, provided that it does not cause interference in other countries where the Radio Regulations are being obeyed. This means that a geographically large country could, if it chose, establish a radio-quiet zone for radio astronomy right now. This would cause no interference to any other country. A measure like this could protect a major terrestrial radio observatory from all transmitters except those on satellites and space stations: it would represent a major safeguard for the future of our science.

4. ITU-R Recommendations on Radio Astronomy

At the present time there are eight ITU-R Recommendations on Radio Astronomy and one in press:

Rec. ITU-R RA.314-8 Preferred frequency bands for radioastronomical measurements

Rec. ITU-R RA.1031-1 Protection of the radio astronomy service in frequency bands shared with other services

Rec. ITU-R RA.517-2 Protection of the radio astronomy service from transmitters in adjacent bands

Rec. ITU-R RA.611-2 Protection of the radio astronomy service from spurious emissions

Rec. ITU-R RA.1237 Protection of the radio astronomy service from unwanted emissions resulting from applications of wideband digital modulation

Rec. ITU-R RA.769-1 Protection criteria used for radioastronomical measurements

Rec. ITU-R RA.1272 Protection of radio astronomy measurements above 60 GHz from ground based interference

Rec. ITU-R RA.479-4 Protection of frequencies for radioastronomical measurements in the shielded zone of the Moon

Rec. ITU-R RA.*in press* A radio-quiet zone in the vicinity of the L_2 Sun-Earth Lagrange Point

Rec. ITU-R RA.314-8 concerns frequency bands for radio astronomy. There are three important tables in the annex to this recommendation, listing frequency bands of greatest astrophysical interest for spectral line measurements below 275 GHz, for spectral lines from 275 to 811 GHz, and for continuum measurements. The IAU periodically reviews these tables through a working group of Commission 40, Division X (Radio Astronomy) and communicates the proposed revisions to Working Party 7D via IUCAF (the Scientific Committee on the Allocation of Frequencies for Radio Astronomy and Space Science, which is described by Robinson elsewhere in these proceedings). The ITU really does pay attention to the tables. For example, as soon as the methanol line at 6.668 GHz was discovered to be a powerful maser (Menten 1991), it was added to the Table of Astrophysically Important Spectral Lines by the IAU in Buenos Aires, July 1991, passed to the ITU-R at the next meeting of the ITU-R Working Party on Radio Astronomy, and it is now listed in the Radio Regulations in Footnote S5.149. *Recommends 5* of Rec. ITU-R RA.314-8 also highlights the issue of spectral lines outside allocated bands and recommends that administrations be asked to help coordinate radio astronomy observations of spectral lines outside the allocated bands.

Rec. ITU-R RA.769-1 provides general protection criteria for the radio astronomy service. The specific recommendations give the substance to my earlier remarks about let-out clauses (my emphasis added):

1. *Radio astronomers should be encouraged to choose sites as free as possible from interference;*

2. *Administrations should afford **all practicable** protection to frequencies used by radio astronomers in their own and neighbouring countries, taking **due account of** the levels of interference given in Annex 1;*

3. *Adminstrations, in seeking to afford protection to particular radioastro-nomical observations, should* **take all practicable steps** *to reduce to the absolute minimum, all unwanted emissions falling within the band of fre-quencies to be protected for radio astronomy, particularly those emissions from aircraft, spacecraft and balloons;*

4. *When proposing frequency allocations, administrations should* **take into account** *that it is very difficult for the radio astronomy service to share frequencies with any other service in which direct line-of-sight paths from the transmitters to the observatories are involved. ...*

The interference levels given in Annex 1 of Rec. ITU-R RA.769-1 are only advisory. Most people take them seriously, however, and they usually form the starting point for negotiations about protection of radio astronomy. The interference analysis on which the Recommendation is based assumes that the interfering signal enters the radio astronomy system through far sidelobes with 0 dBi gain, i.e. it takes no account of the possibility that the interfering source could be near the main beam of the radio telescope (as might be the case for a satellite).

Rec.1031-1 deals with shared frequency bands. The annex, which you will probably guess by now is the best part, contains a description of coordination zones which can be establised around radio observatories to protect them from terrestrial transmitters. The coordination zone concept could be the seed from which radio-quiet zones will one day spring.

The annex to Rec.1031-1 also gives guidelines on how to calculate the appro-priate size for such a coordination zone, depending on details of the tranmsitters. A typical size is 500 km. One of the parameters needed for the calculation is a percentage of time for which the interference levels of Rec.769-1 should not be exceeded due to variable propagation. The figure of 10% which is given there has come to be reinterpreted by the telecommunication community as an accept-able percentage of data-loss for the radio astronomy service, whether in shared bands or passive bands. Radio astronomers within ITU-R Working Party 7D are currently trying to clarify this misunderstanding.

5.   ITU-R Handbook on Radio Astronomy

The ITU-R Handbook on Radio Astronomy (1995) is a useful source of further information on the radio astronomy service within the ITU-R. Its nine chapters complement the ITU-R Recommendations, often giving the rationale behind them. For example, Chapter 3 on Preferred Frequency Bands for Radio As-tronomy Observations (corresponding to Rec. ITU-R RA.314-8) is preceded by a chapter on Characteristics of the Radio Astronomy Service, which sets out the nature of the emissions we study and the physical reasons for the frequency allocations we seek. Chapter 4 on the Vulnerability of Radio Astronomy Ob-servations to Interference contains, in addition to the interference thresholds, much additional material on the derivation of interference criteria in special cases, for example interferometry. Chapter 5 on Sharing the Radio Spectrum with Other Services includes band-by-band discussions of typical interference scenarios, including typical sharing parameters for terrestrial transmitters and necessary separation distances. Chapter 6 on Interference to Radio Astronomy from Transmitters in Other Bands includes detailed discussion of both sides of the electromagnetic compatibility problem: how radio astronomers can best protect themselves against strong signals in adjacent or nearby frequency bands, and levels to which such transmitters need to reduce their unwanted emissions into the radio astronomy bands. Particular threats which are discussed include transmitters on aircraft or spacecraft and transmitters using wideband modulation schemes.

A chapter on special applications deals with VLBI including space VLBI and geodesy, pulsar observing techniques, solar observations, and radio astronomy from Antarctica and from space. There are also chapters on the Search for Extraterrestrial Intelligence (SETI) and on ground-based radar astronomy.

The first edition of the Handbook was published in 1995. It was based on material taken from Reports which were already several years old at that time. A new edition of the Handbook is urgently needed.

6. Getting Involved

There is one official mechanism to ensure that radio astronomers attend meetings of ITU-R and participate in the work of its study groups and task groups: IUCAF, the Scientific Commission on the Frequency Allocations for Radio Astronomy and Space Science. IUCAF was founded in 1960 as an Inter-Union Commission, with members drawn from IAU, URSI and COSPAR (Robinson, 1999). IUCAF membership is restricted to 15 persons, but many more radio astronomers are actively supporting IUCAF, as email correspondents. IUCAF acts as a network to coordinate the efforts of radio astronomers in many countries.

Within individual countries it is up to radio astronomers to make themselves known to their national adminstrations and up to the administrations to then support the astronomers' participation in ITU-R meetings as national delegates. The work is done on a voluntary basis, usually financed from the radio observatory budget. We need more volunteers for this work and more support from observatory directors. It costs money and time. My own observatory already contributes more than 2% of its budget on this work and the pressure is steadily increasing.

Since 1991 some North American countries have sent employees of, or consultants for, companies like Motorola as national delegates to the meetings of Study Group 7 (radio astronomy). ITU-R working parties must reach concensus on the output they produce. On several occasions this has meant that the wishes of the radio astronomy community have been blocked in WP7D by someone in the pay of a satellite corporation. If this situation is not to develop into a wasteful confrontation, we need more radio astronomers to become involved in the task of explaining our needs to other radio users and convincing them that it is in the best interest of all parties to protect our work.

The telecommunications revolution cannot be stopped. But neither is it automatically a bad thing for radio astronomy. New technology has already given us more sensitive receivers, and in future could deliver cleaner radio transmitters and new signal-processing techniques for dealing with interference.

7.   What Do We Want?

Radio astronomers need to be very clear about their objectives before going into battle in the ITU-R arena. We should only ask the ITU for things the ITU member countries can actually give us. In my view we need some of the frequencies all of the time, as we have been asking since 1959, but we also need access to all of the frequencies some of the time.

Officially allocated passive frequency bands are vital to guarantee successful radio astronomy operations anywhere in the world. The most fundamentally important spectral lines, such as the 21-cm hydrogen line, must be accessible globally, by amateurs as well as professionals. There are centuries of work to be done mapping out the structure of our Galaxy using modest telescopes, even just for the protected spectral lines. We also need global allocations throughout the spectrum for continuum observations. There are no all-sky continuum surveys at high angular resolution above 5 GHz. And of course VLBI would be impossible without global allocations.

But in addition to these fundamental needs we also require access, on a best efforts basis, to the *entire* radio spectrum. No one has yet found the right words to explain this to the ITU. Yet our science will slowly wither if all radio telescopes end up confined to the officially allocated frequency bands. The expansion of the Universe produces redshifts of the spectral lines to lower frequencies. The more distant the source the fainter will be its radio emissions and the further they will be from their rest frequency. We need to develop ways to access the redshifted lines anywhere in the radio spectrum. We also need larger bandwidths than those officially allocated for a second reason: to improve the sensitivity of our continuum measurements. Access to wider bands is not impossible. Individual countries are sometimes prepared to take action within their own borders to protect a large-scale radio facility on a special site. We will need to persuade them to do the same for new-generation radio astronomy instruments.

New satellite services are also moving towards wider bandwidths to offer Internet-in-the-sky and live video connections to mobile users. Satellite are a global threat. They can block out large frequency bands and they can be visible from anywhere on the Earth's surface. Unwanted emissions from satellites often mean that the bands denied are far bigger than what is actually used by the satellite service.

Ideally no frequency band should be closed permanently on the Universe. Scientists of the world should have access to all radio frequencies for at least some of the time from at least some places on the Earth's surface. This is not an unreasonable goal.

8.   World Radio Conference 2000

WRC-2000 will be of paramount importance to radio astronomy (Ruf, these proceedings). Radio astronomy issues appear at many places in the WRC-2000 Agenda, the most important being items 1.2, 1.4, 1.5, 1.11, 1.14, 1.15.1, 1.16, and of course item 7.2, which sets out the agenda for future conferences.

The issue of unwanted emissions from satellites comes up under agenda item 1.2. Task Groups 1/3 and 1/5 have been working for 4 years 'to finalize remaining issues in the review of Appendix **S3** to the Radio Regulations with respect to spurious emissions for space services, taking into account Recommendation **66**'. Task Group 1/5 will recommend the first limits to unwanted emissions from spacecraft. This is the vital first step towards protecting radio astronomy from future flotillas of telecommunication satellites. The limits being proposed at this time do not provide all the protection we need, but they are far better than no limits at all, and they can be tightened in future as technology improves.

There are several items dealing with allocations for satellite downlinks in frequency bands dangerously close to radio astronomy bands, where the issue of unwanted emissions spreading into radio astronomy bands is especially relevant. Agenda item 1.4 concerns, among other things, allocation to the fixed satellite service (space-to-Earth) in the band 41.5-42.5 GHz and protection of the radio astronomy service in the adjacent SiO band 42.5-43.5 GHz. Agenda item 1.5 concerns high-altitude platforms (HAPs). These are essentially stratospheric balloons, comparable in their coverage to satellites, with the potential to cause interference to radio astronomy stations over a wide area. Agenda item 1.11 concerns, among other things, a possible allocation of the band 405-406 MHz for a satellite downlink, close to the radio astronomy band 406.1-410 MHz. Agenda item 1.14 concerns the feasibility of a satellite downlink in the band 15.43-15.63 GHz, very close to the passive band 15.35-15.40 GHz. Agenda item 1.15.1 concerns a possible satellite downlink for a new-generation satellite navigation system, just above the radio astronomy band 4990-5000 MHz,.

Arguably the most important item for radio astronomy, however, is item 1.16, 'to consider allocation of frequency bands above 71 GHz to the earth exploration satellite (passive) and radio astronomy services ... '. The last frequency allocations to radio astronomy at millimetre wavelengths were made in 1979, when mm-wave astronomy was still in its infancy. WRC-2000 gives us our first opportunity in 21 years to obtain new mm-wave allocations and to review the existing allocations to radio astronomy.

We need to send a strong delegation to WRC-2000, at least 14 radio astronomers, to lobby their national administrations and to cover all the parallel sessions where the detailed arguments need to be won.

9. Conclusions

As use of the radio spectrum grows throughout the world, radio astronomers could find themselves increasingly confined to the frequency bands officially allocated by the ITU-R. But the science we are trying to do demands much more than this. We need to use wider frequency bands to search in redshift and to maximize the sensitivity of our continuum measurements. A major goal for the next century is to ensure that the best facilities retain access to large tracts of radio spectrum, at least for some of the time.

We also need to protect new instruments, such as ALMA and the SKA, and we need to find the right way to protect special sites, ranging from radio-quiet zones on the Earth, to those in space such as the Sun-Earth Lagrangian point L_2, where PLANCK and other sensitive instruments are planned to operate. The ITU-R Recommendations on these issues are only just the first steps towards guaranteeing the success of these and other new instruments.

We must also continue efforts to keep our allocated frequency bands clean. In particular, the way in which unwanted emissions from satellites can be reduced is a big field waiting to be explored. The satellite community has so far not shown itself strongly minded to tackle the problem, so the push may need to come from higher up. The long-term goal is to achieve regulatory protection of the passive bands to a stated level of power flux density at the surface of the Earth. The technology to achieve this exists today: the problem we face is how to persuade the telecommunications community to help us.

Today the mm-wave bands are clean. It is not clear to me how to keep them that way. The fact that radio astronomy already has 24% of the mm-wave frequency allocations probably means that the rest of the world will be reluctant to give us any more exclusive allocations. Perhaps there are other ways to keep the mm-wave bands clear at special sites, while exploiting the bands for telecommunications in built-up areas. Perhaps the short range of mm-wave transmitters and the efficient blocking by buildings and walls will suffice to keep the observatories quiet. In any case we must keep access to large tracts of the millimetre spectrum, which is so incredibly rich in molecular lines.

Radio astronomy is inevitably a small player within the ITU-R. We need to win powerful friends to succeed. Education is needed at all levels. We have a big story to tell: the origins and evolution of the Universe. It is a story the public likes. The UNISPACE-III meeting with which our symposium is associated is an important step in raising awareness. For the first time the need to keep our radio window on the Universe clear will be discussed at the highest level possible on our planet.

References

Cohen, R. J., 1999, Astron. & Geophys., Vol. 40, No. 6, p.8

Combrink, W. L., West M. E. & Gaylard, M. J., 1994, PASP, 106, 807

Galt, J., 1990, Nature, 345, 483

Handbook on Radio Astronomy, 1995, Radiocommunication Bureau, International Telecommunication Union, Geneva

IUCAF web pages http://www.mpifr-bonn.mpg.de/staff/kruf/iucaf

ITU web pages http://www.itu.int/

ITU-R Recommendations, Radio Astronomy, Volume 1997, RA Series, Radiocommunication Bureau, International Telecommunication Union, Geneva

Kellerman, K., 1997, Sky & Telescope, 94, No.2 (February 1997), pp.26-33

Menten, K., 1991, ApJ, 380, L75

Ponsonby, J. E. B., 1991, Journal of Navigation, 44, 392

Robinson, B. J., 1999, Ann. Rev. Astron. Astrophys., 37, 65

Preserving the Astronomical Sky
IAU Symposium, Vol. 196, 2001
R. J. Cohen and W. T. Sullivan, III, eds.

World Radio Conference WRC-2000

Klaus Ruf

Max-Planck-Institut für Radioastronomie, Auf dem Hügel 69, 53121 Bonn, Germany

Abstract. The World Radio Conference 2000 must be considered the most important one for radio astronomy since WARC-79. The conference agenda contains about 30 topics of substance, and more than 10 of these have direct impact on radio astronomy frequency allocations. From the perspective of radio astronomy the most important items are: "Allocation of Frequency Bands above 71 GHz to the Earth-Exploration Satellite Service (passive) and Radio Astronomy Service" and the agenda items dealing with Recommendation 66 (Unwanted Emissions). A review of the status of preparations is given.

1. Introduction

The International Telecommunication Union, ITU, has long recognised radio astronomy as a service, which needs access to uncontaminated parts of the radio spectrum. Decisions about the distribution of the radio spectrum among the many competing services are taken at international conferences organized by the ITU. Until 1992 these conferences were convened at irregular intervals on request and were called World Administrative Radio Conferences (WARCs). Then ITU changed its structure, and the conferences are now convened regularly and called World Radiocommunication Conferences (WRCs).

2. The Agenda of WRC-2000

WRC-2000 has a number of items on the agenda that are relevant for radio astronomy, at least 11 items out of a total of 33. Agendas of WRCs are written in a formalized language and are difficult to read, therefore only the interesting items are listed in the tables below. The full text has been published by ITU as Resolution 1130 of the ITU Council.

2.1. Highest Priority Issues

Agenda item 1.2 (Fig. 1) as such is a left-over from the previous conference, but the history of Recommendation 66 dates back as long as the allocations in the mm-wave bands. At WARC-79 Recommendation 66 was adopted in order to study the possibility of radio astronomical observations at the limits of sensitivity in an environment of active (transmitting) use of the radio spectrum

---

1.2 TO FINALIZE REMAINING ISSUES IN THE REVIEW OF APPENDIX S3 TO THE RADIO REGULATIONS WITH RESPECT TO SPURIOUS EMISSIONS FOR SPACE SERVICES, TAKING INTO ACCOUNT RECOMMENDATION 66 (REV.WRC-97) AND THE DECISIONS OF WRC-97 ON ADOPTION OF NEW VALUES, DUE TO TAKE EFFECT AT A FUTURE TIME, OF SPURIOUS EMISSIONS FOR SPACE SERVICES;

1.16 TO CONSIDER ALLOCATION OF FREQUENCY BANDS ABOVE 71 GHZ TO THE EARTH EXPLORATION-SATELLITE (PASSIVE) AND RADIO ASTRONOMY SERVICES, TAKING INTO ACCOUNT RESOLUTION 723 (WRC-97);

7.2 TO RECOMMEND TO THE COUNCIL ITEMS FOR INCLUSION IN THE AGENDA FOR THE NEXT WRC, AND TO GIVE ITS VIEWS ON THE PRELIMINARY AGENDA FOR THE SUBSEQUENT CONFERENCE AND ON POSSIBLE AGENDA ITEMS FOR FUTURE CONFERENCES,

---

Figure 1.   Agenda items with highest priority for radio astronomy

at power levels orders of magnitude above the field strength measured from cosmic sources.

Agenda item 1.16 (Fig. 1) is of paramount importance to radio astronomy. All of the radio spectrum between 71 GHz and 275 GHz, where the table of frequency allocations in the radio regulations of the ITU still ends, is opened for reallocation. Resolution 723 of WRC-97 invites the study groups of ITU-R to take into account in their studies the present use of allocated bands. The current allocations in the mm-wave region were made at WARC-79, and the present use is such that all atmospheric windows, also above 275 GHz, are used for spectral line as well as radiocontinuum observations. Little is known about active services making use of an allocation in the part of the spectrum under consideration, and commercial use is non-existent. Nevertheless, the proposals worked out for this agenda item take the existing allocations into due consideration.

Allocations at very high frequencies and results from the studies initiated by Recommendation 66 are also part of the provisional agendas for the next conferences to follow WRC-2000. WRC-2000 will adopt the agenda for the following conference and make proposals for later ones under **agenda item 7.2** (Fig. 1). Although this is a formality repeated on each WRC agenda, circumstances make it an important point for radio astronomers this time.

2.2.   Second Priority Issues

There are many other items important to radio astronomy, although of second priority. The six resolutions quoted in **agenda item 1.4** (Fig. 2) all have to do with (high density) fixed service and fixed satellite service, sharing between these two services and sharing with other services. The frequency range under consideration is roughly 30–60 GHz, and hence the radio astronomy bands at 32 and 43 GHz are possibly endangered. The preparation for this topic is controversial. Europe clearly favors allocations to *terrestrial* services for a number of the frequency bands in question, because of the greater spectrum efficiency offered by improved frequency re-use due to the much smaller cell sizes in Europe. In other parts of the world, where the relation of high traffic density regions to

1.4 TO CONSIDER ISSUES CONCERNING ALLOCATIONS AND REGULATORY ASPECTS
RELATED TO RESOLUTIONS 126 (WRC-97), 128 (WRC-97), 129 (WRC-97),
133 (WRC-97), 134 (WRC-97) AND 726 (WRC-97);

1.5 TO CONSIDER REGULATORY PROVISIONS AND POSSIBLE ADDITIONAL
FREQUENCY ALLOCATIONS FOR SERVICES USING HIGH ALTITUDE PLATFORM
STATIONS, TAKING INTO ACCOUNT THE RESULTS OF ITU-R STUDIES
CONDUCTED IN RESPONSE TO RESOLUTION 122 (WRC-97);

1.11 TO CONSIDER CONSTRAINTS ON EXISTING ALLOCATIONS AND TO CONSIDER
ADDITIONAL ALLOCATIONS ON A WORLDWIDE BASIS FOR THE
NON-GEOSTATIONARY (NON-GSO) MSS BELOW 1 GHZ, TAKING INTO ACCOUNT
THE RESULTS OF ITU-R STUDIES CONDUCTED IN RESPONSE TO RESOLUTIONS
NO. 214 (REV.WRC-97) AND 219 (WRC-97);

1.14 TO REVIEW THE RESULTS OF THE STUDIES ON THE FEASIBILITY OF
IMPLEMENTING NON-GSO MSS FEEDER LINKS IN THE 15.43--15.63 GHZ IN
ACCORDANCE WITH RESOLUTION 123 (WRC-97);

Figure 2. Agenda items of second priority for radio astronomy

low traffic density regions is different, the immediate availability and coverage offered by *satellite* systems have higher priority.

Agenda item 1.5 (Fig. 2) deals with a mixed case: High Altitude Platforms. Such stations, which are still in an early planning state, will be mounted on high flying balloons or airplanes and serve metropolitan areas with fixed links. In a regulatory sense they are treated as (terrestrial) fixed service stations, but they will transmit from high above and under line-of-sight conditions to a huge area, which may contain radio telescopes. The requirement of protecting radio astronomy in the 43 and 49 GHz bands from HAPs unwanted emissions is within the scope of Resolution 122.

Agenda item, 1.11 (Fig. 2), has been on the agenda of several WRCs in one form or another. It has been very difficult to find frequency bands for the mobile satellite service, MSS, in the crowded region of the spectrum below 1 GHz, and the preparation process for this agenda item shows again that the same difficulties persist. Radio Astronomy has achieved protection in a footnote, attached to all allocations to the MSS below 1 GHz, from interference due to unwanted transmissions. Some of the newer proposals, however, are for frequencies very close to our bands.

Agenda item 1.14 (Fig. 2) deals with an existing allocation to the fixed satellite service, FSS, (space-to-Earth), that is too close to a passive band to be useful. Sharing studies in ITU-R (the radiocommunications arm of ITU) concluded that limits in power flux density, necessary to protect radio astronomy in the band 15.35–15.4 GHz, make the allocation unattractive to FSS, and the deletion of this allocation is proposed. While this agenda item only affects one radio astronomical band, the case promises to become such a nice precedent that it is ranked among second priority agenda items.

1.6.1 REVIEW OF SPECTRUM AND REGULATORY ISSUES FOR ADVANCED MOBILE
APPLICATIONS IN THE CONTEXT OF IMT-2000, NOTING THAT THERE IS AN
URGENT NEED TO PROVIDE MORE SPECTRUM FOR THE TERRESTRIAL
COMPONENT OF SUCH APPLICATIONS AND THAT PRIORITY SHOULD BE GIVEN
TO TERRESTRIAL MOBILE SPECTRUM NEEDS, AND ADJUSTMENTS TO THE
TABLE OF FREQUENCY ALLOCATIONS AS NECESSARY;

1.10 TO CONSIDER RESULTS OF ITU-R STUDIES CARRIED OUT IN ACCORDANCE
WITH RESOLUTION 218 (WRC-97) AND TAKE APPROPRIATE ACTION ON THIS
SUBJECT;

1.13.1 TO REVIEW AND, IF APPROPRIATE, REVISE THE POWER LIMITS
APPEARING IN ARTICLES S21 AND S22 IN RELATION TO THE SHARING
CONDITIONS AMONG NON-GSO FSS, GSO FSS, GSO BROADCASTING-SATELLITE
SERVICE (BSS), SPACE SCIENCES AND TERRESTRIAL SERVICES, TO ENSURE
THE FEASIBILITY OF THESE POWER LIMITS AND THAT THESE LIMITS DO
NOT IMPOSE UNDUE CONSTRAINTS ON THE DEVELOPMENT OF THESE SYSTEMS
AND SERVICES;

1.15.1 TO CONSIDER NEW ALLOCATIONS TO THE RADIONAVIGATION- SATELLITE
SERVICE IN THE RANGE FROM 1 GHZ TO 6 GHZ REQUIRED TO SUPPORT
DEVELOPMENTS;

1.15.3 TO CONSIDER THE STATUS OF ALLOCATIONS TO SERVICES OTHER THAN
THE RADIONAVIGATION-SATELLITE SERVICE (NOS. S5.355 AND S5.359) IN
THE BAND 1559--1610 MHZ;

Figure 3.   Agenda items with potential impact on a single radio astronomy band

2.3. Issues Affecting a Single Radio Astronomy Band

The third generation of mobile phones, now termed IMT2000 in **agenda item
1.6.1** (Fig. 3), has already received attention at earlier conferences. The bands
around 2 GHz foreseen for this application do not conflict with radio astronomy
allocations, but the planned extension bands may need to be coordinated with
radio astronomy stations operating at 1720 or 2700 MHz. The extention bands
are thought to be necessary to increase the capacity in high traffic density areas.
The satellite component of this service will probably be put into existing MSS
allocations, such as 1610–1660.5 MHz, and thus increase the pressure on sharing
or coordinating the two 18 cm radio astronomy bands. This pressure is potentially further increased under **agenda item 1.10** (Fig. 3), where the conference
is asked to make appropriate provisions for the global distress and safety service.
These services have lost their exclusive status in the band 1626.5–1660.5 MHz,
but need a kind of priority access in at least part of the band, if this is used for
generic mobile satellite service.

    Agenda item 1.13.1 (Fig. 3) has not been formulated with radio astron-
omy in mind, but radio astronomy is, one of the terrestrial services concerned.
Sharing in the narrow sense of sharing the same frequency band for the operation
of two or more different radio services is not an issue for radio astronomy and
space-based active services like the fixed satellite service (FSS) and the broadcasting satellite service (BSS). Unwanted emissions from these services in nearby

or harmonically related bands into radio astronomy bands, however, need careful consideration in planning such systems, and therefore the exercise described in agenda item 1.13.1 is of interest to radio astronomy.

The last two **agenda items, 1.15.1 and 1.15.3** (Fig. 3), both deal with the radio navigation satellite service. The most well-known example of this service is the Global Positioning Satellite System (GPS). A new allocation is sought for a future system, which is planned to provide timing and location information reliably and with an accuracy that would make it suitable for instrument landing in civil aviation. Unfortunately, one system is planned for a frequency band immediately adjoining the 5 GHz radio astronomy band. Other proposals want to change the usage of the established systems, GPS and the Russian Global Navigation Satellite System (GLONASS), which could threaten the already very much troubled 1612 MHz OH-band.

While the agenda items listed in Tables 2 and 3 require the usual attention, that radio astronomers involved in the frequency protection process within ITU-R have been paying through a number of conferences, the agenda items in Table 1 are of far-reaching interest. They will therefore be described in more detail here.

3.   Studies Related to Recommendation 66

Recommendation 66 has long been considered a major achievement for radio astronomy, and the revision formulated at WARC-92 clearly points out the threat of spurious emissions from satellite based transmitters. WARC-92 had been convened to find allocations for the new services like MSS and BSS (both digital audio and high definition television). All of these new satellite services were allocated frequency bands in the part of the spectrum where they had asked for it. And the revision of Rec. 66 took appropriate account of this.

So called "Recommends" 1 and 5 of RECOMMENDATION No. 66 (Rev. WARC-92) read

1. *study, as a matter of urgency, the question of spurious emissions resulting from space service transmissions, and, on the basis of those studies, develop Recommendations for maximum permitted levels of spurious emissions in terms of mean power of spurious components supplied by the transmitter to the antenna transmission line;*

5. *submit a report to the next competent conference on the result of its studies with the view to reviewing and including spurious and out-of-band emission limits in Appendix 8 of the Radio Regulations, principally for the protection of the radio astronomy and other passive services.*

These studies started smoothly in ITU-R in a Task Group, TG1-3 of Study Group 1. Radio Astronomy protection requirements turned out to be difficult to meet for a number of services, but European countries were generally advocating more stringent limits for some services than some countries in other parts of the world. At a certain stage the space services joined the studies and categorically denied that it would be possible to implement similar limits as for a number of terrestrial services. Protection of radio astronomy to the desired level, the aim which Rec. 66 asked for, would be missed by several tens of dBs.

By the time of WRC-97 spurious emission limits for most radio services could be written into Appendix S3 of the Radio Regulations, mentioning different categories applicable in some countries, and with exemptions from the general limits for some services, but the space service community still only allowed their much relaxed limits to be qualified as "design objectives". Agenda item 1.2 of WRC-2000 will hopefully overcome this situation and the spurious emission limits for space services will become mandatory limits, although undoubtedly entering into force only after a number of years.

Having finished its work on the limitation of spurious emissions, Task Group 1-3 was dissolved and a new Task Group, TG1-5, was set up with the mandate to study possible limitations of out-of-band emissions. (*Out-of-Band Emissions* and *Spurious Emissions* together form *Unwanted Emissions*.) Recommendation 66 was accordingly revised by WRC-97. The clear words of recommends 5. in the WARC-92 revision were turned into the long phrases of recommends 7 and 8 in the WRC-97 revision:

7. *study those frequency bands and instances where, for technical or operational reasons, more stringent spurious emission limits than the general limits in Appendix S3 may be required to protect safety services and passive services such as radio astronomy, and the impact on all concerned services of implementing or not implementing such limits;*

8. *study those frequency bands and instances where, for technical or operational reasons, out-of-band limits may be required to protect safety services and passive services such as radio astronomy, and the impact on all concerned services of implementing or not implementing such limits;*

General protection of radio astronomy to the levels given and explained in a long recognised ITU-R Recommendation has been declared unpracticable, mainly by the space services. Instead, particular frequency bands and instances are to be studied, to see if more stringent limits may be required. This has been named the band-by-band study and is going on in a number of Working Parties of ITU-R Study Groups and in Task Group 1-5. The results of this study will not be ready for WRC-2000, but, as some studies asked for by other recommends of Rec. 66 are fullfilled, WRC-2000 will certainly revise Rec. 66 again and propose related agenda items for future conferences. We must certainly take great care to protect Rec. 66, because it has become apparent that strong forces want to abuse the band-by-band study to open up the ITU-R Recommendation containing the radio astronomy protection requirements. The same forces are determined to drop the whole issue of protecting radio astronomy, if they are not successful in questioning our interference threshold levels. We will need to be very vigilant that neither of these ploys happens.

4.   New Allocations in the Mm- and Submm-Wave Bands

Agenda item 1.16, even though attempting to redistribute almost three quarters of the allocated radio spectrum has been much less controversial in the WRC preparations. As mentioned above, the current usage of the mm- and submm-bands is mostly passive and for scientific purposes. Nevertheless, in order not to exclude the future use of this part of the radio spectrum by radiocommunication services, care has been taken to allocate the same amount of spectrum to active services as they now have. The guidelines in this planning exercise have been

1. to allocate frequency bands to radio astronomy around the most important spectral line frequencies and the central parts of the atmospheric windows, and

2. to avoid any shared or nearby allocations between radio astronomy and space-to-Earth transmissions of satellite services.

One way to accomodate these diverging requirements has been to encourage sharing whenever possible. Best candidates for sharing frquency bands with radio astronomy are terrestrial fixed and mobile services, which can be coordinated so that they are located outside exclusion zones around mm-wave observatories.

The result of this planning exercise, which resulted in very similar allocation proposals being discussed in different parts of the world, has been an enormous increase in bandwidth allocated to radio astronomy, in good representation of current usage, but a decrease in exclusive passive allocations. As an example the current allocation table for the frequency range 105 - 122.5 GHz is given in Fig. 4, together with proposals discussed in the US and in Europe.

105–116	105–109.8	105–109.5
EARTH EXPLORATION-SATELLITE (passive)	FIXED	FIXED
	MOBILE	MOBILE
RADIO ASTRONOMY	RADIO ASTRONOMY	RADIO ASTRONOMY
SPACE RESEARCH (passive)	SPACE RESEARCH (passive) ADD	SPACE RESEARCH (passive) S5.CCC
S5.340 S5.341	S5.CCC	MOD S5.149 S5.341
	S5.341 MOD S5.149	
	109.8–111.8	**109.5–111.8**
	EARTH EXPLORATION-SATELLITE (passive)	EARTH EXPLORATION-SATELLITE (passive)
	RADIO ASTRONOMY	RADIO ASTRONOMY
	SPACE RESEARCH (passive)	SPACE RESEARCH (passive)
	MOD S5.340 S5.341	S5.340 S5.341
	111.8–114.25	**111.8–114.25**
	FIXED	FIXED
	MOBILE	MOBILE
	RADIO ASTRONOMY	RADIO ASTRONOMY
	SPACE RESEARCH (passive) ADD	SPACE RESEARCH (passive) S5.CCC
	S5.CCC	S5.341 MOD S5.149
	S5.341 MOD S5.149	
	114.25–122.25	**114.25–116**
	EARTH EXPLORATION-SATELLITE (passive)	EARTH EXPLORATION-SATELLITE (passive)
	INTER-SATELLITE S5.XXX	RADIO ASTRONOMY
	RADIO ASTRONOMY	SPACE RESEARCH (passive)
	SPACE RESEARCH (passive)	S5.340 S5.341
	S5.138 MOD S5.340 S5.341	

Figure 4. 105-116 GHz Band: Current Table of Frequency Allocations (1st column) and proposals discussed in the US (2nd column) and Europe (3rd column)

Preserving the Astronomical Sky
IAU Symposium, Vol. 196, 2001
R. J. Cohen and W. T. Sullivan, III, eds.

Radio Astronomy and Recent Telecommunications Trends

Tomas E. Gergely

National Science Foundation, Arlington, VA 22230, USA

Abstract. Radio astronomy was born in the 1930s and matured in the 1950s. The telecommunications environment of those decades was dominated by monopolies (often state monopolies) and a strong regulatory environment. Radio communications were analogue in nature and mostly confined to frequencies below a few hundred megahertz. The worldwide telecommunications environment began to change in the 1960s and has undergone a revolution during the last three decades. Impulse for this revolution, which has not yet ended, was provided by political, as well as by technological developments. The most influential among these were the end of the cold war and the emergence of giant, often multinational telecommunications companies or consortia, that provide services previously reserved to state monopolies. Technically, the new era can be characterized by a host of new types of satellite services, the use of digital communication techniques, the proliferation of low power devices that do not require individual licensing and by a steady move towards higher frequencies. I discuss the evolution of some of these trends and their implications for radio astronomy.

1. Introduction

The first attempt to detect extra-terrestrial radio emissions dates to 1890, when Edison tried to detect radio emission from the Sun. There is no record of the results, suggesting strongly that the experiment was unsuccessful. Four years later Sir O. Lodge attempted to detect solar radio emission and documented the first instance of interference to a radio astronomy experiment. He wrote:

> " I hope to try for long wave radiation from the Sun ..." "I did not succeed in this ... There were evidently too many terrestrial sources of disturbance in a city like Liverpool to make the experiment feasible." (Quoted by J. S. Hey 1973)

Radio astronomy was born with K. G. Jansky's detection of radio emission from the Galaxy (1932), while searching for the origin of weak static that was causing interference to ship-to-shore communications. At first, Jansky attributed the emission to weak, man-made interference. Finally, radio emission from the Sun was also detected, by J.S. Hey during World War II (1942), while searching for the origin of radar jamming. In Hey's words:

> "In both instances, the aim has been to study types of interference limiting the effectiveness of practical systems." (J.S. Hey 1973)

Thus, radio astronomy has been closely linked to man- made radio interference since it's beginnings. Its future continues to depend on how effectively it can be protected by regulations or can overcome man-made radio interference by technical means.

2. Telecommunications Trends and Forces on the Spectrum

Table 1.   The changing telecommunications environment.

1930s - 1980s	1980s - Present
•Telecommunication Services Provided by Monopolies - Lack of competition - Spectrum not valued in economic terms	• Privatization and Decentralization - Competition driven - Economic value of spectrum determined by auctions
• Regulatory Environment - Highly regulated - In public interest - Innovations slow to reach market	• Deregulation - Minimal regulation - In consumer interest - Accelerated introduction of new technologies
• Services Provided Locally or Regionally - National ownership requirements - Equipment manufactured to local (or regional) standards	• Globalization (New Technologies Allow Provision of Worldwide Services - WTO agreement on foreign ownership/access - Roaming, cross border transport of radio equipment
• Satellites - No satellites to GSO satellites	• Satellites - Increasing use of low Earth orbit (LEO) and other NGSO systems
• Technologies - Analogue - Low data rate - Compartmentalized - Move to cm-waves	• New Technologies - Digital - High data rate - Convergence - Move to mm-waves

Early on, radio astronomy observatories avoided man-made interference by locating telescopes in remote areas and this strategy was mostly successful until the 1970s. During the last two decades, however, radio observatories have been increasingly experiencing instances of interference. In addition, satellite transmissions now restrict access to large portions of the radio spectrum, no matter how remote the observatory site.

When radio astronomy was born and matured, the telecommunications environment was dominated by state monopolies and a strong regulatory environment. Since radio astronomy observatories are, as a rule, financed and built by governments, regulatory bodies could also be relied on to provide protection from interference. Radio communications were analogue and confined to frequencies below a few hundred megahertz, a fact perpetuated in officially accepted designations, which refer to, e.g. the 3-30 MHz and 30-300 MHz frequency ranges as the "High Frequency" or "HF" and "Very High Frequency" or "VHF" bands, respectively. As radio technology progressed, higher frequency ranges became commercially attractive. The 1-3 GHz range, favoured for satellite applications, became the most fought over portion of the spectrum in the 1990s. This range, of course, is also of great interest to astronomers, because it contains the HI and OH lines and their redshifted extensions.

Table 2.   Spectrum management principles used to formulate policies of the U.S. Federal Communications Commission (FCC)

- **COMPETITION**
    - Rely on market forces to ensure economically efficient use of spectrum
    - Avoid mandating specific services
    - Minimize regulations that limit competition, obstruct innovation, or impede efficient investment
- **FLEXIBILITY**
    - Establish standards sparingly
- **PUBLIC INTEREST**
    - Where the market is unlikely to produce essential public benefits in adequate quantities, minimum intervention may apply to ensure that these benefits are achieved
    - Spectrum set asides for public services or benefits
- **LICENSING AND FEE POLICIES**
    - Support spectrum value
- **ADMINISTRATIVE CERTAINTY**
    - Establish firm ground rules on interference
- **GLOBAL MARKET CONTEXT**
    - Encourage efficient worldwide spectrum use to ensure spectrum availability
    - Promote seamless, worldwide networks
    - Coordinate with other nations (satellites)

The global telecommunications environment begun to change in the 1960's and has undergone a revolution during the last decades. Impulse for this revolution continues to be provided by political and technological developments, e.g. the end of the cold war and the emergence of giant, often multinational companies or consortia, that provide services previously reserved to state monopolies. Some changing telecommunications trends that affect radio astronomy are summarized in Table 1.

In terms of technical developments, the new era can be characterized by the advent of satellite communications and the increasing use of digital modulation techniques. These changing trends affect the use of the radio spectrum. The principles that guide spectrum management and use, particularly by the private sector in the U.S.A., have been summarized by the Federal Communications Commission (FCC) as shown in Table 2. It is fair to say that to a greater or lesser degree these principles represent current trends, worldwide.

3. The Impact on Radio Astronomy

Clearly, radio astronomy is impacted by these telecommunications trends, particularly by: 1) the large increase in the numbers (and power) of satellite borne transmitters, 2) the failure of satellite transmitters to curb their unwanted (out-of-band and spurious) emissions and 3) the increasing use of the mm-wave spectral range (say, above 30 GHz).

3.1. Increase in the Number of Satellite-Borne Transmitters

The sensitivity of radio astronomy observations can be improved by decreasing the system temperature or by increasing the receiver bandwidth or observing time. For many radio astronomy systems it is difficult to decrease further the system temperature. In some cases, at least for the receiver, theoretical limits have been reached. The sensitivity of an observation is inversely proportional to the square root of the observing time multiplied by the receiver bandwidth. For continuum observations a 10% bandwidth is desirable, but most radio astronomy allocations below 70 GHz are much narrower, in fact, only a couple are close to 2%. For this reason, radio astronomers often make use of all available spectrum and do not confine themselves to allocated radio astronomy bands. Unfortunately, proliferation of satellite transmitters, particularly in the 1 - 3 GHz band makes it increasingly difficult to access bands wider than the narrow bands allocated to radio astronomy. Ultimately, sensitivity that is lost due to a decrease in bandwidth can only be made up by increasing the observing time, thus resulting in higher telescope oversubscription rates on existing telescopes.

The increase in the number of satellites over the last 25 years can be illustrated by the increase in advance published Geostationary Orbit satellite systems (GSOs) at the International Telecommunication Union (ITU), shown in Fig. 1. Many advanced published satellites never make it to the launch pad; they are used as placeholders, to keep someone else from occupying an orbital slot. Nevertheless, the exponential growth is indicative of the desire by many countries to introduce more and more satellite-based services. The problem is further compounded by the numerous Non-Geostationary (NGSO) systems now operating or planned. These systems consist of large numbers of satellites and, as a rule, are designed to provide global services, so several satellites are always visible from any point on Earth.

A somewhat more realistic way of visualizing the diminishing fraction of spectrum to which astronomers have access is shown in Figure 2, that illustrates

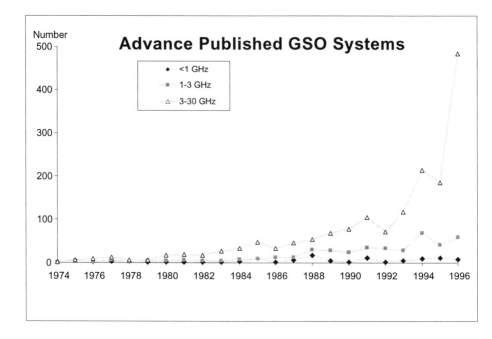

Figure 1.   Advance published GSO systems at the ITU 1974 - 1996.

the variation in the amount of spectrum allocated to the fixed service and to satellite downlinks over the last three decades. Sharing the spectrum between radio astronomy observatories and terrestrial transmitters, particularly those in the fixed service, is relatively easy, through geographical separation. Sometimes other means can also be used, such as placing a null in the transmitting antenna pattern in direction of the affected radio observatory. On the other hand, in-band sharing between a satellite downlink and a radio observatory is not possible, as long as the satellite transmits while above the observatory's horizon (and sometimes even when the satellite is below the horizon, due to troposcatter and other atmospheric effects). 485 MHz of additional spectrum was allocated to satellite downlinks between 1969 and 1992, in the 1 - 3 GHz range. Most of this came from spectrum that was previously allocated to the fixed service, which lost 476 MHz over the same period. Radio astronomers can no longer plan on access to the spectrum allocated to downlinks, possibly from any place on Earth.

3.2.   Unwanted Emissions

Clearly, it is impossible for radio astronomers to observe in frequency bands that are allocated to satellite downlinks. Unwanted emissions of satellite transmitters often spill over into adjacent and neighboring frequency bands, at levels that render them unusable for radio astronomy observations. Recent examples include the GLONASS and Iridium satellite systems. With the proliferation of satellite systems, particularly NGSOs, more problems are likely to occur. Table 3 shows

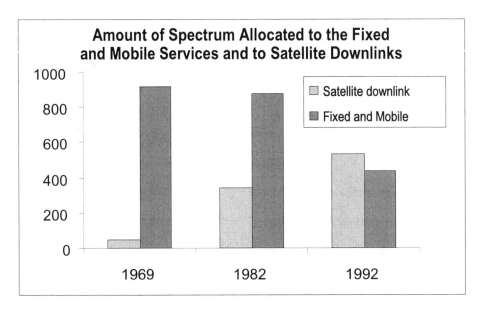

Figure 2. The changing amount of spectrum allocated to fixed and mobile services and to satellite downlinks.

existing and planned satellite service downlinks next to or near primary radio astronomy bands in the 100 MHz to 30 GHz range.

Proposals have also been made recently for downlink allocations next to the 406.1-410 MHz and 1400 - 1427 MHz bands and although they were not adopted, there can be little doubt that such proposals will continue to be advanced. For example, WRC-97 allocated the 40.5 - 42.5 GHz band, adjacent to the 42.5 - 43.5 GHz primary radio astronomy band, to the Fixed Satellite Service (FSS). The allocation is conditioned, however, to finding ways to share the spectrum between radio astronomers and the FSS.

Can the situation be improved? Radio astronomers sought to place limits on unwanted emissions, particularly on those originating from space-based transmitters. Until recently, there were no recommended limits at all on transmitters operating above 17.7 GHz. Spurious emissions of terrestrial transmitters are now subject to limits, but space-based transmitters are still subject only to "design objectives", that are expected to be converted into hard limits at WRC-2000. These limits are, however, far from sufficient to protect nearby radio astronomy bands. Increased use of the efficient modulation methods by satellite designers could improve the situation greatly. Martin, Yan and Lam (1997) compared available modulation methods and showed that at the 60 dB attenuation level, the occupied bandwidth varies between the most and the least efficient methods by a factor of 20. Out-of-band emissions are certainly of the most serious concern to radio astronomers, but use of the more efficient modulation methods would benefit all services in an era of spectrum scarcity and would result in substantial cost savings.

Table 3.   Existing and planned satellite service downlinks next to or near primary radio astronomy bands in the 100 MHz to 30 GHz range.

PRIMARY RADIO ASTRONOMY BAND	ADJACENT SPACE-TO-EARTH BAND	SPACE SERVICE	CLOSEST DOWNLINK
150.05-153 MHz (R 1)	149-150.05 MHz	Radionav Sat	
322-328.6 MHz	none		387-390 MHz mss
406.1-410 MHz	none*		401-402 MHz Space Ops
608-614 MHz	none		460-470 MHz MetSat
1400-1427 MHz	none*		1452-1492 MHz BSS
1610.6-1613.8 MHz	1613.8-1626.5 MHz	mss	
1660-1670 MHz	1670-1710 MHz	Met Sat	
2690-2700 MHz	2670-2690 MHz	MSS (R 2)	
4990-5000 MHz	none*		4500-4800 MHz FSS
10.68-10.7 GHz	10.7-11.7 GHz	FSS	
15.35-15.4 GHz	15.4-15.7 GHz	FSS	
22.21-22.5 GHz	none		21.4-22.0 GHz BSS (R1 and R3)
23.6-24 GHz	24-24.05 GHz	Amateur Sat	

*Downlink recently proposed in nearby band.

3.3.   Increasing Use of the Mm-Wave Spectral Range

Most of the mm-wave spectrum is of interest to astronomers. Over 2,100 spectral lines of 80 chemical compounds have been identified in the 71-275 GHz range and more have been predicted. Many of these lines are listed as being astrophysically important (Rec. ITU-R RA.314-8), as they may yield unique information on various aspects of the Universe. Astronomical observations in the mm-region require as wide receiver bandwidths as possible, to enable continuum observations of faint objects, such as extrasolar planets or very distant galaxies. Very sensitive observations are possible, using bolometer detectors with bandwidths of several tens of GHz and such devices are in use at several observatories. Wide frequency coverage is also needed to observe highly red-shifted objects that provide crucial information on the formation of galaxies and about the early history of the Universe, that may not be obtained at other frequencies.

Mm-telescopes have been in operation for nearly three decades in the Americas, Asia and Europe. Reflecting the growing interest in mm-wave astronomy, the Atacama Large Millimeter Array (ALMA), a giant international project by astronomers from the U.S.A., Europe and possibly other countries that may join the collaboration, is currently under development. Several other large mm-wave telescopes and interferometers are also in the design or construction phases and are expected to begin operations in the next decade.

The spectrum above the 60 GHz oxygen absorption bands is allocated to various radio services, including satellites, but has been scarcely used until now, except by radio astronomers and by the remote sensing community. Spectrum crowding at lower frequencies and recent technological advances have made this range more attractive for commercial use and active radio services are moving towards the millimetre waves. Protection of this spectrum for astronomical observations is therefore timely. Since there are few users yet, reaccommodation of the spectrum allocations is expected to touch fewer entrenched interests than at lower frequencies and would not require costly relocation of existing services. Because of relatively high absorption at ground level at these frequencies, it is estimated that coordination radii of the order of 100 - 150 km are required to protect observatories from terrestrial interference. Assuming, e.g., an absorption coefficient of 0.5 dB/km, spreading and propagation losses over a distance of 100 km amount to 150 dB. The power flux density due to a 1 kW omnidirectional transmitter at a distance of 100 km is then typically -120 dB (W/m^2). Detrimental power flux densities for continuum radio observations in astronomy bands above 71 GHz (listed in Rec. ITU-R RA.769-1, Table 1) range from -125 dB(W/m^2) at 89 GHz, to -113 dB(W/m^2) at 270 GHz. For spectral line observations values range from -144 dB(W/m^2) at 88.6 GHz to -131 dB(W/m^2) at 265 GHz (Table 2 of the same Recommendation). The relatively small area of the coordination regions required, coupled with the limited number of existing and planned mm-observatories and their remote locations, should have only a minimal impact on terrestrial services.

Table 4.   Radio Astronomy access to mm-wave spectrum before and (hopefully) after WRC-2000

Window	Desired ALMA Coverage (GHz)	Current Radio Astronomy Bands (GHz)	%	Desired Radio Astronomy Bands (GHz)	%
3 mm	67 - 116	86 - 92 105 - 116	35	76 - 116	82
2 mm	125 - 211	164 - 168 182 - 185	8	123 - 158.5 164 - 167 200 -211	57.5
1 mm	211 - 275	217 - 231 250 - 252 265 - 275	41	211 - 231.5 235 - 238 241 - 275	88

Line-of-sight sharing between the radio astronomy and satellite downlinks is, as a rule, not possible. Sharing may be possible, however, with some satellite systems and services consisting of a very limited number of satellites, such as Amateur Satellites or the Cloud Profiling Radar, that expect to operate above 71 GHz, eventually. The orbital elements of these systems will be known and their knowledge should permit time sharing in these bands, as radio observatories may schedule observations at times when no satellite is over the horizon, or shut down for the limited periods of time. To protect radio observatories from unwanted emissions of satellites, downlinks could be moved to near the edges of the transparent atmospheric windows, leaving their middle portion available to be shared by terrestrial services and wide band passive receivers.

Proposals to protect radio astronomy at mm-wavelengths along these lines are currently being prepared by spectrum managers of various countries, including the U.S.A., Europe, Japan and others, to be put forward at WRC-2000. Table 4 shows the fraction of mm-spectrum accessible to radio astronomers from a regulatory point of view now and the fraction that is expected to become accessible, if these proposals are adopted by WRC-2000.

4. Summary

Since its beginnings, radio astronomy had to develop and prosper in a noisy environment. The situation worsened considerably with the advent and proliferation of satellite systems, particularly NGSOs, which may render astronomical observations impossible to carry out in large portions of the spectrum. Technical and regulatory solutions should be sought to radio astronomy's interference problems and to preserve mankind's access to this resource. Attempts towards technical solutions are discussed elsewhere in this volume. Regulatory solutions are manpower intensive and costly. Nevertheless, if radio astronomers wish to preserve access to the spectrum, they will need to dedicate a larger share of human resources and their budget to spectrum management activities, so they can compete with the numerous other activities that claim use of the radio spectrum. Finally, education efforts will have to be undertaken or intensified, so that policy makers and the general public appreciate the value of radio astronomy and what it would mean for our understanding of the Universe to lose access to most or all of the radio spectrum. Only by pursuing all these routes are astronomers likely to retain access to crucial segments of the radio spectrum that they need.

References

Hey, J. S. 1973, The Evolution of Radio Astronomy, New York: Science History Pub.

Martin, W.L., Yan, T. and Lam, L.V. 1997, Efficient Modulation Methods Study at NASA/JPL, Space Frequency Coordination Group Doc. SF17-28/D

ITU-R Recommendations 1997, RA Series, Geneva

Preserving the Astronomical Sky
IAU Symposium, Vol. 196, 2001
R. J. Cohen and W. T. Sullivan, III, eds.

Protection of Millimetre-Wave Astronomy

Masatoshi Ohishi

The National Astronomical Observatory, 2-21-1, Osawa, Mitaka, Tokyo, 181-8588, Japan

Abstract. Development of radio technologies will lead to a serious conflict between millimetre-wave astronomy and telecommunication services. I describe characteristics of millimetre-wave astronomy and technical aspects related to radio astronomical observations. Three examples of possible interference to millimetre-wave astronomy are described. It is very important to advertise what millimetre-wave astronomy contributes to human culture and to get support from the non-astronomical community to keep the radio windows open and clean.

1. Introduction

Over the past quarter-century, astronomical studies in the millimetre and submillimetre wavebands have grown into one of the major disciplines of astronomy. These wavebands contain thousands of observable rotational molecular spectral lines, many of which are invaluable physical and chemical probes of the interstellar and circumstellar media. Furthermore, these bands contain the long-wavelength emission of cool dust and the short-wavelength tail of synchrotron emission, both of which are important for establishing the spectrum and physical properties of the emitting source. Low energy phenomena produce emission in the millimetre wave spectral region. Interstellar clouds have kinetic temperatures of a few tens of Kelvins, and can excite the low-energy rotational transitions of gas molecules. The spectroscopic emission from these transitions can, in principle, yield the abundance, temperature and velocity structure of the emitting gas. To achieve unique solutions for these quantities, observations of multiple transitions of the same or different species are often required. For the purposes of this paper, I will define the mm/submm bands as the frequency range from 30 to 1000 GHz (wavelengths 10 mm to 300 μm respectively).

1.1. Science in the Millimetre-Wave Bands

What happened just after the Big Bang and what is the destiny of our Universe? How do stars and galaxies form? How did our solar system form? Do planets like the earth exist around other stars? What role did the cosmos play in the development and evolution of life on earth? How do stars die and how is their material recycled into new stars? These fascinating questions – fundamental to our understanding of the Universe – are but a few that astronomers seek to answer.

Radio astronomy utilizes the very latest radio technology to address such questions and to contribute to the cumulative knowledge of society. Recent research in radio astronomy has provided important information on the birth and death of stars, the detection of planets and proto-planetary systems around other stars, the formation of external galaxies just after the Big Bang, discoveries of black holes, discoveries of molecules and unexpectedly complex chemistry in space, to give but a few examples.

1.2. Millimetre-Wave Radio Astronomy

Millimetre-wave radio astronomy is a special branch of astronomy developed over the past quarter-century. The field is driven by, and often drives, the creation of new technology, including the development of high-precision and thermally-stable mechanical structures and the development of receiving elements approaching the quantum noise limit.

Millimetre-wave radio astronomy provides unique scientific contributions. Since the detection of interstellar formaldehyde (H_2CO) in 1969 by Snyder, Buhl & Zuckerman (1969), approximately 110 interstellar molecules have been identified by their line emissions in the millimetre and submillimetre wavebands (see Table 1), with more than 3,000 spectral lines observed, distributed more or less continuously in frequency except where oxygen absorption bands near 50 GHz, 118 GHz and 183 GHz prevent observation from the ground (Fig. 1).

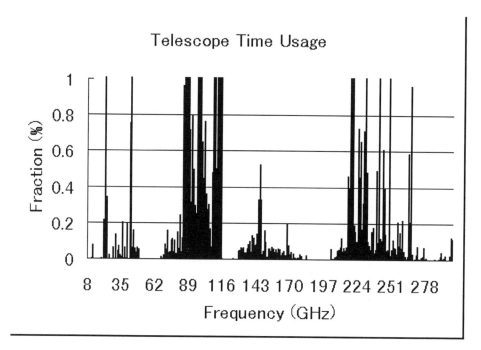

Figure 1.   Statistics of the combined telescope time usage of the NRAO 12-m, Onsala 20-m, NRO 45-m, IRAM 30-m and SEST 15-m millimetre-wave radio telescopes.

Table 1.   Observed Interstellar Molecules as of July, 1999

Simple Hydrides, Oxides, Sulfides, Halogens and related molecules

H_2 (IR)	CO	NH_3	CS	$NaCl^*$
HCl	SiO	SiH_4^* (IR)	SiS	$AlCl^*$
HF (IR)	SO_2	C_2 (IR)	H_2S	KCl^*
H_2O	OCS	CH_4^* (IR)	PN	AlF^*
N_2O				

Nitriles and Acetylene derivatives

C_3^* (IR)	HCN	CH_3CN	HNC	$C_2H_4^*$ (IR)
C_5^* (IR)	HC_3N	CH_3C_3N	HNCO	$C_2H_2^*$ (IR)
C_3O	HC_5N	CH_3C_5N ?	HNCS	
C_3S	HC_7N	CH_3C_2H	HNCCC	
C_4Si^*	HC_9N	CH_3C_4H	CH_3NC	
	$HC_{11}N$	CH_3CH_2CN	HCCNC	
	HC_2CHO	CH_2CHCN		

Aldehydes, Alcohols, Ethers, Ketones, Amides and related molecules

H_2CO	CH_3OH	HCOOH	CH_2NH	CH_2CC
H_2CS	CH_3CH_2OH	$HCOOCH_3$	CH_3NH_2	CH_2CCC
CH_3CHO	CH_3SH	$(CH_3)_2O$	NH_2CN	
NH_2CHO	$(CH_3)_2CO$	H_2CCO		
		CH_3COOH		

Cyclic Molecules

c-C_3H_2	SiC_2	c-SiC_3	c-C_3H	c-C_2H_4O

Molecular Ions

CH^+ (OPT)	HCO^+	$HCNH^+$	H_3O^+	HN_2^+
HCS^+	$HOCO^+$	HC_3NH^+	HOC^+	H_3^+
CO^+	H_2COH^+	SO^+		

Radicals

OH	C_2H	CN	C_2O	C_2S
CH	C_3H	C_3N	NO	NS
CH_2	C_4H	$HCCN^*$	SO	SiC^*
NH (UV)	C_5H	CH_2CN	HCO	SiN^*
NH_2	C_6H	CH_2N	MgNC	CP^*
HNO	C_7H	NaCN	MgCN	
	C_8H			

For molecules observed in other frequency regions than radio, the wavebands are shown in parentheses : *IR* – infrared, *OPT* – optical and *UV* – ultraviolet.

* – detected only in the envelope around the evolved star IRC+10216.

? – claimed but not yet confirmed.

Such line emissions are used to study the star-formation process, the structure and formation of galaxies, and the process of star death. In general, this information is not available by any other means. In addition, broadband continuum emission arises in the millimetre wavebands from several physical mechanisms including free-free (bremsstrahlung) emission, synchrotron emission and the thermal emission from dust. The millimetre wavebands are critical for determining the spectral energy distribution of this emission, which in turn identifies the physical conditions of the emitting source.

Millimetre-wave radio astronomy places special requirements on the availability and usage of the frequency spectrum. These requirements fall in two broad categories: sensitivity and broad frequency bands. In general, radio astronomy at all frequencies requires extreme sensitivities, achieved by large, high-accuracy antennas, cryogenically-cooled electronics and long observations in which random noise is averaged to allow the detection of very weak cosmic signals. Millimetre-wave astronomy is certainly no exception to this requirement. Millimetre-wave astronomy has a nearly unique need for wide, interference-free bands. This requirement derives from four scientific and technical factors:

1. The ensemble of molecular rotational spectral line emission is distributed uniformly over the frequency spectrum. A few lines are observed more often than others, but many transitions of many species are required for the full and unique utilization of the science.

2. One of the most important applications of astronomical millimetre-wave spectroscopy is the observation of emission from gas in extremely distant, early-epoch galaxies. These studies are crucial to the understanding the evolution of galaxies and the early universe. Because of the expansion of the universe, the frequency at which this emission is received is reduced by factors ranging from a few percent (the local universe) to 80% (the currently-known earliest epochs of galaxy formation). Because of the continuous range of these redshifts, emission from important spectral lines can occur at any frequency in the millimetre wavebands.

3. Millimetre-wave continuum emission from galactic and extragalactic sources is typically exceedingly weak, but very important. Often, this emission can be detected only by using extremely wide instrumental bandwidths (up to ∼40 GHz wide). In particular, incoherent bolometre detectors, currently the most sensitive detectors for mm and submm continuum work, require these very broad bandwidths to achieve their sensitivity.

4. The most sensitive instruments for mm and submm spectroscopy are SIS mixer receivers (Section 3.1), which respond to signals over bandwidths as large as 30% of their operating centre frequency. These exquisitely sensitive devices can be saturated or destroyed by a strong interfering signal anywhere in this bandwidth. Although it is possible to identify certain narrow frequency bands of very high importance to radio astronomy, the utilization of the full millimetre-wave spectrum is fundamental to the science.

2. Possible Radio Interference in the Near Future

In the millimetre-wave regions several active services plan to utilize frequencies that may interfere with millimetre-wave observations. I show three examples below, but I note there are also other planned uses of the millimetre-wave bands by the military, etc.

2.1. Collision Avoidance Radar at 76 GHz

Vehicular radars are coming into operation in the band 76-81 GHz, which is allocated to the Radio Location Service on a primary basis. The third harmonic of the radar lies within the band 217-231 GHz which is allocated to radio astronomy and other passive services on a primary basis. ITU-R Footnote S5.340 states that all emissions are prohibited within this band.

The proposed FCC standards on unwanted emissions above 200 GHz are such that a coordination zone nearly 4 km in radius would be needed to avoid harmful interference to radio astronomy from just one single radar near an observatory. Most mm-wave observatories are not in a position to regulate vehicular traffic over such a large area. Indeed some are in areas of great interest to tourists and skiers. The radio astronomy community is concerned that the proliferation of these devices, coupled with possible degradation of their performance with age, will lead to an electromagnetic smog in the band 217-231 GHz analogous to the light pollution which now compromises many once great optical observatories.

The radio astronomy community would like to propose that an on/off capability be incorporated into these devices so that, at the very least, mm-wave observatories could place signs along nearby roads requesting that vehicular radar be turned off in the vicinity of the observatory. A better solution to the difficulty would be possible if industry could accept tighter specifications on unwanted emissions such that the coordination zone were less than 1 km in radius (within which most observatories have some control over traffic).

2.2. HAPS near 31 GHz and 47 GHz

High Altitude Platform Systems (HAPS) have been proposed by the USA and Japan. The US system is proposed by a commercial company, Sky Station International (http://www.skystation.com/), and uses the band 47.2-47.5 GHz for downlinks and the band 47.9-48.2 GHz for uplinks. Airships are to be located at an altitude of $\sim$21 km and more than 250 airships are planned, to cover every major city in the world. There are several important radio astronomy spectral lines near the 47-GHz band : CS at 48.991 GHz, CH_3OH at 48.372 and 48.377 GHz, and SiO at 43.424, 43.122 GHz and so on, of which the CS and SiO bands are allocated to the radio astronomy service. Although ITU studies suggest that there will be no harmful interference from HAPS into the radio astronomy bands listed above, we need to be very careful to monitor developments.

Furthermore, Japan proposes to launch another HAPS system using bands at 19, 29 and 31 GHz. Japan plans to launch more than 90 airships to cover almost all of Japan. Several airships will be visible from the Nobeyama Radio Observatory, where the 45-m radio telescope and the Nobeyama Millimetre Array are located. The 31-GHz HAPS band is 31.0 - 31.3 GHz, and the frequency

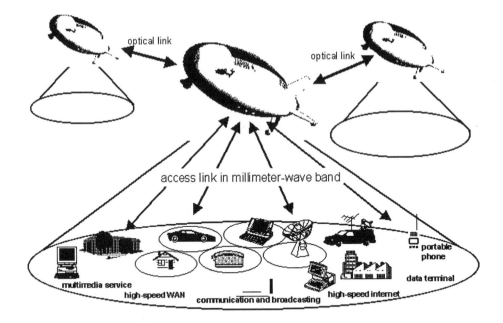

Figure 2.   The HAPS system proposed by Japan

range is adjacent to an exclusive passive co-primary radio astronomy band, 31.3 - 31.5 GHz, and a co-primary radio astronomy band, 31.5 - 31.8 GHz. These bands are used to measure cosmic microwave background radiation.

The Japanese system is more poisonous for us, because main beam - main beam couplings may occur between the radio telescope and the HAPS. In this case the received signal is estimated to be more than 150 dB above the harmful interference level when the airship is over the radio telescope! Even if the airship is over Tokyo the 45-m telscope can detect the HAPS signal about 56 dB stronger than the harmful interference level. Therefore a guard band of ~200 MHz bandwidth is necessary in the HAPS band to protect millimetre-wave astronomy.

2.3.   Cloud Profiling Radar at 94 GHz

WRC-97 allocated the band 94.0-94.1 GHz to the Earth Exploration Satellite (active) and Space Research (active) services and added Footnote S5.562: "The use of the band 94-94.1 GHz by the earth exploration-satellite (active) and space research (active) services is limited to space-borne cloud radars." The space-borne cloud radars are downward-looking and will provide vertical sounding of cloud-layer structure down to the earth's surface.

The band 94.0 - 94.1 GHz is close to several bands allocated to mm-wave radio astronomy:

   86 - 92 GHz (Footnote S5.340, all emissions are prohibited in this band)
   93.07 - 93.27 GHz (Footnote S5.149)
   97.88 - 98.08 GHz (Footnote S5.149)

Several potential interference mechanisms have been investigated. Unwanted emissions from the radar into the radio astronomy bands are not the only difficulty: indeed calculations suggest that with care the unwanted emissions can be kept below the radio astronomy interference thresholds. The main problem is how to protect the sensitive radio astronomy receiver from the powerful adjacent band radar signals. In practice the most common type of receiver used in mm-wave observatories is the SIS mixer, which can saturate at very low power levels. Cloud radar transmissions could affect the operation of such receivers over their total frequency range, whenever the radio astronomy antenna is pointing within 14° of the satellite, according to calculations based on preliminary antenna patterns. Furthermore, calculations suggest that an SIS mixer could actually be destroyed by a radar transmission directly into the main beam of the radio astronomy antenna.

Possible remedies to these problems have been investigated. It is technically very difficult to provide adequate filtering to radio astronomy receivers, to reject the cloud radar transmissions, while having very low loss so as not to compromise the receiver noise figure. The other main option is time sharing: either cover the radio telescope feed horn or switch off the radar transmissions whenever the satellite is above the observatory.

3. Protection of Millimetre-Wave Astronomy

3.1. Technical Issues on Radio Astronomy Receivers

1. Radio Astronomy Sensitivity Thresholds

   General sensitivity thresholds for the Radio Astronomy Service are described in detail in ITU-R Recommendation RA. 769-1 and in the ITU-R Handbook on Radio Astronomy (Sections 4.3 - 4.6). The input power levels for detrimental interference are $\sim$ -185 dBW for broadband continuum observations and $\sim$ -204 dBW for narrow band spectral line observations.

2. SIS Receiver Broadband Response

   Heterodyne mixer receivers utilizing SIS (Superconductor - Insulator - Superconductor) tunnel junctions are the devices of choice for modern mm- and submm-wave spectroscopy. These devices are capable of noise performance approaching the quantum limit. The spectral bandwidth of such receivers can be up to a few GHz wide, but this width is usually limited by isolators and amplifiers following the mixer. A modern, tunerless SIS mixer can have efficient RF coupling over more than 30% of the band centre frequency as set by the mixer's local oscillator. A strong narrow-band system can overload an SIS mixer to the point of saturation or burnout (Pan et al., 1989).

3. Continuum Bolometers

   Incoherent bolometers are the most sensitive detectors for broadband continuum emission in the millimetre and submillimetre bands. They are sensitive to photons with frequencies ranging from the infrared to the long millimetre wavelengths. To eliminate atmospheric noise and to define the wavelength band for detection, bolometers use bandpass filters. However, to maximize the bolometer sensitivity, the passbands typically cover the full width of an atmospheric window. Such wide frequency bandwidths can make bolometers vulnerable to RF interference. Weak, narrowband interference signals may be of little consequence because they will be diluted by the wide bandwidths of the bolometers. A strong CW signal may saturate a bolometer, however.

4. Sidelobe Levels

Interference in the main diffraction beam of an antenna sets the most stringent constraints on interference thresholds. Interference entering through the antenna sidelobes can also be a serious problem, as the sidelobes cover a much larger solid angle than the main lobe. Sidelobe response is also relevant to interference management schemes such as geographical sharing of the spectrum. A recommended sidelobe pattern of large parabolic antennas is given in Recommendation ITU-R SA. 509-1, which is applicable for frequencies between 2 GHz and 30 GHz. The model for the envelope of the gain (G) of the sidelobes is given by

$$G = (\ 32 - 25 \log \phi)\ \text{dBi}, \quad 1° < \phi < 47.8°;$$
$$G = \text{-10 dBi}, \qquad\qquad 47.8° < \phi < 180°;$$

for angle ϕ measured from the axis of the main beam. It should be noted that the sidelobe pattern of the millimetre-wave antenna is similar to that of the centimetre-wave antenna.

5. Atmospheric Propagation and Transparency

Although molecular rotational emission is one of the great tools of millimetre-wave astronomy, some of the same kinds of transitions also cause absorption of cosmic signals as they traverse the earth's atmosphere. Rotational transitions of water vapour and spin-rotational transitions of molecular oxygen absorb signals to some extent at all millimetre-wave frequencies and completely stop the signal at certain frequencies. These absorption effects drive millimetre-wave observatories to high mountain-top locations where they are above as much of the atmospheric water vapour as possible. At some frequencies and in much of the submillimetre band, observatories must be located entirely above the earth's atmosphere in space-borne platforms. Indeed, some of the lines of O_2 and H_2O that form the strongest atmospheric absorption are also of great interest in cosmic sources. Although atmospheric absorption attenuates cosmic signals, it can also be used to advantage in shielding observatories from terrestrial interference. For the details of the atmospheric propagation, see, for example, Erickson & Merino (1997).

3.2. Possible Methods to Protect Millimetre-Wave Astronomy

In millimetre and submillimetre-wave astronomy, a few frequency bands are of critical importance, but access to wide frequency bands is also fundamental to much of the science and technology. Frequency allocations above 30 GHz must take this into account, or this field of astronomy will be seriously jeopardized. The radio astronomy community believes it possible to satisfy the needs of the radio astronomy service and the active services. The recommendations and strategies for future new allocations or reallocations are described below.

1. Reallocation of satellite down link frequencies

   Currently the frequency bands allocated to the radio astronomy service are sometimes surrounded by satellite downlink frequencies. This is a potential threat to the radio astronomy service owing to unintentional interference to sensitive wideband radio astronomy receivers. Reallocating satellite downlink bands to the edges of the atmospheric windows should be a compromise acceptable to the radio astronomy service: it would leave wide bands available for spectroscopy throughout the centre of the atmospheric windows, it would not affect the centre frequencies of the absorbing atmospheric lines, which may be of considerable importance to space-borne observatories and it may make it more feasible to construct high-pass or low-pass filters for receivers (as opposed to narrow notch filters). It should be noted that this arrangement will not come without cost to the radio astronomy service: some important spectral lines are near the edges of atmospheric windows and the construction of high-pass and low-pass filters may not be trivial.

2. Co-Primary with other Passive Services

   Because all of the radio astronomy mm-wave bands are passive, it is possible to share such bands with other passive services such as the Earth Exploration Satellite Services (passive). It is necessary to allocate these bands in the co-primary status to avoid potential interference from active services.

3. Geographical and Time Sharing

   A further possibility for protection of the radio astronomy service is through use of geographical coordination zones and time-sharing schemes. There are relatively few millimetre and submillimetre observatories. Most are located at isolated, mountain-top sites. Consequently, geographical sharing of the spectrum is a possibility. In such a scheme, interfering ground-based transmissions would be prohibited in the vicinity of the observatory. Satellite downlinks could, in principle, be blanked as the transmission beam passed over the observatory. Ground-based transmissions in the vicinity of an observatory can be regulated by national administration, or possibly jointly with the administrations of the neighbouring countries, and do not necessarily require ITU agreements.

   In time-sharing schemes, interfering transmitters would, by prior agreement, be turned off at certain times. This option is much less appealing than other allocation means because it restricts the time available for astronomy. Furthermore, it may be impractical for communication services to turn off their systems once the public becomes reliant on them. However, as a last resort, it could be used to protect mm and submm astronomy in certain geographical regions and/or for certain times of the day, or times of the year.

4. Conclusion

The peaceful era for millimetre- and submillimetre-wave astronomy is now over. The radio astronomy service has been regarded as an eyesore for the active services, because radio astronomy needs very low interference threshold levels for its protection. How can we change this perception? I believe that we need to do more ourselves to advertise that radio astronomy is a very useful field of science that provides a lot of attractive new knowledge to human society. This is a very important activity that all radio astromoners can do, to help protect radio astronomy in the future.

Acknowledgments. I appreciate the work of all the members of the Millimetre-Wave Working Group (MMWG) of IUCAF, which was co-chaired by Phil Jewell of the Joint Astronomy Centre (JAC) and myself. This paper is based on the MMWG report presented at the ITU-R, adding several new topics.

References

Eriksson, J. E. P., Merino, F., 1997, "On Simulating Passive Observations of the Middle Atmosphere in the Range 1 - 1000 GHz", Research Report No. 179, Department of Radio and Space Science with Onsala Space Observatory, Chalmers University of Technology, Göteborg, Sweden

Pan, S.-K., Kerr, A. R., Feldman, M. J., Kleinsasser, A., Stasiak, J., Sandstrom, R. L., and Gallagher, W. J., 1989, IEEE Trans. Microwave Theory Tech., vol. MTT-37, no. 3, 580

Snyder, L.E., Buhl, D., & Zuckerman, B. 1969, Phys.Rev.Lett, 22, 679

Preserving the Astronomical Sky
IAU Symposium, Vol. 196, 2001
R. J. Cohen and W. T. Sullivan, III, eds.

Utilization of the Radiofrequency Spectrum above 1 GHz by Passive Services

Juan R. Pardo[1] and Pierre J. Encrenaz

DEMIRM - Observatoire de Paris, 75014 Paris, France

Daniel Breton

Formerly at Centre National d'Etudes spatiales, Toulouse, France

Abstract. Microwave atmospheric radiometry and radio, mm and sub-mm astronomy are "passive" services, i.e. not involved in any man-made transmission but only concerned with the reception of naturally occurring radio waves. The intensity of the radiation received is not subject to human control, unlike the situation for active services. All active services operate in bands occupied by natural signals of atmospheric and cosmic origin and the active service tranmissions may be powerful enough to noticeably interfere with reception of those signals by scientific services. A conflict exists for the coexistence of active and passive services in many frequency bands, which leads to a need for regulating how to share the electromagnetic spectrum. This document gives an overview of the problems of frequency sharing in the longwave region of the electromagnetic spectrum (radio to submillimetre waves).

1. Introduction: General Capabilities of Passive Microwave Atmospheric Radiometry and Radio Astronomy

Passive microwave radiometry is a tool of fundamental importance for the Earth Exploration-Satellite Service (EESS). The EESS operates passive sensors that are designed to receive and to measure natural emissions produced by the Earth's surface and its atmosphere. The frequency and the strength of these natural emissions characterize the type and the status of a number of important geophysical, atmospheric and surface parameters (land, sea, and ice-caps), which describe the status of the Earth/Atmosphere/Oceans System, and its mechanisms :

1. Earth surface parameters such as soil moisture, sea surface temperature, ocean wind stress, ice extension and age, snow cover, rainfall over land, etc.;

---

[1]Present address: Division of Physics, Mathematics and Astronomy, California Institute of Technology, MS 320-47, Pasadena, CA 91125, USA

2. Three-dimensional atmospheric parameters (low, medium and upper atmosphere) such as temperature profiles, water vapour content and concentration profiles of radiatively and chemically important trace gases (for instance O_3, SO_2 and ClO).

Microwave observation techniques below 100 GHz allow us to study the Earth's surface and its atmosphere from space-borne instruments even in the presence of clouds, because the clouds are almost transparent at these frequencies. This "all-weather" capability has considerable interest for the EESS, because more than 60% of the surface of our planet is overcast with clouds. Passive microwave sensing is an important tool widely used for meteorological, climatological and environmental monitoring and survey (operational and scientific applications), for which reliable repetitive global coverage is mandatory.

Observation of the longwave electromagnetic spectrum is also extremely important for the understanding of our universe. Whereas "optical astronomy" detects the radiation from hot objects such as stars, radio waves in the universe often come from cooler objects, such as interstellar gas or electrons in ordered motion in a magnetic field. Radio waves thus open our knowledge of the Universe to a series of physical phenomena that optical waves alone would never reveal.

The spectrum of cosmic radio waves consists of a broad continuum covering the whole range of frequencies that can penetrate the Earth's atmosphere and a large number of spectral lines (atomic and molecular resonances), each of which is confined to a more or less narrow frequency range. The fact that some of these lines are keys for studying our universe (21-cm neutral hydrogen line, carbon monoxide rotational resonances, the lines of the OH radical) makes the protection of their frequencies highly important.

2. Spectrum Requirements

2.1. EESS Requirements

Several parameters may generally contribute, at varying levels, to the natural emission which can be observed at a given frequency. Therefore, measurements at several frequencies in the longwave spectrum must be (quasi)simultaneously implemented, in order to isolate and to retrieve each individual contribution. This is true for both geophysical and astrophysical measurements, although the required degree of simultaneity in recording the information can be very different.

The absorption characteristics of the atmosphere, shown in Figure 1 (Pardo et al, 2001), are characterized by absorption peaks due to rotational resonances of atmospheric molecules (of which H_2O and O_2 are the most important), "dry"continuum-like absorption (collision-induced absorption involving O_2 and N_2) and water vapour continuum-like absorption (far wings of IR lines most probably).

The selection of the frequencies best suited for passive microwave sensing depends closely on the characteristics of the absorption:

- Frequencies for observation of surface parameters are selected below 100 GHz, where atmospheric absorption is the weakest. One frequency band per octave, on average, is necessary.

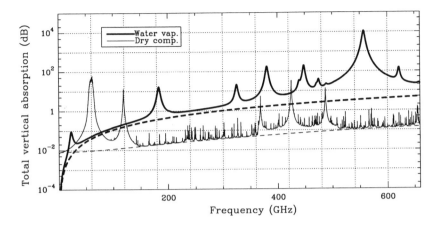

Figure 1. Spectrum of the atmospheric absorption. The dashed lines represent the contributions of the collision-induced "dry" absorption and the water vapour pseudocontinuum (water opacity not related to lines below 10 THz).

- Frequencies for observation of atmospheric parameters are very carefully selected mostly above 50 GHz within, and in the vicinity of, the absorption peaks of atmospheric gases.

The frequencies and bandwidths required for Earth remote sensing up to 200 GHz are listed in the Table 1. Most frequency allocations above 200 GHz contain absorption lines of important atmospheric trace species but have been in general much less used to date.

2.2. Radio Astronomy Requirements

The electromagnetic radiation detected in radio astronomy is either emission from atoms or molecules at very specific frequencies (line emission), or very broadband thermal or non-thermal radiation (continuum emission). In both cases the polarization characteristics of the signal are also important as an indicator of the physical conditions of the emitting object. Thus, the following requirements are essential for radio astronomical research:

- Good frequency coverage.

- High spectral resolution.

- High spatial resolution.

- High time resolution.

Table 2 provides information on some of the main targets of radio astronomy at frequencies between 1 and 200 GHz. We stress in bold the extremely important bands.

Table 1.   Frequency bands and bandwidths used for satellite passive sensing of the atmosphere below 200 GHz

Frequency (GHz)	Necessary BW (MHz)	Main measurements
Near 1.4	100	Soil moisture, salinity, sea temperature, vegetation index
Near 2.7	60	Salinity, soil moisture
4.2-4.4	200	Ocean surface temperature (back-up for 6.9 GHz, with reduced performance)
6.7-7.1	400	Ocean surface temperature
10.6-10.7	100	Rain, snow, ice, sea state, ocean wind
15.35-15.4	200	Water vapour, rain
18.6-18.8	200	Rain, sea state, ocean ice, water vapour
23.6-24	400	Water vapour, liquid water
31.3-31.8	500	Window channel associated with temp. measurements
36.5-37	500	Rain, snow, ocean ice, water vapour
50.2-50.4	200	O_2 (Temperature profiling, magnetic field)
52.6-59.3	6700 (1)	O_2 (Temperature profiling, magnetic field)
86-92	6000	Clouds, oil spills, ice, snow
100-102	2000	N_2O
109.8-111.8	2000	O_3
115.25-122.25	7000 (1)	O_2 (Temperature profiling, magnetic field), CO
174.8-191.8	17000 (1)	H_2O (Moisture profiling), N_2O, O_3

Table 2.   Current and requested radio astronomy frequency allocations which are shared with active services.

Frequency Band	Main measurements
1330-1400 MHz	Doppler-shifted radiation from hydrogen
1400-1427 MHz	**21-cm line of neutral atomic hydrogen**
1610.6-1613.8 MHz	**Important OH line**
1660-1670	OH lines and continuum
1718.8-1722.2 MHz	Study of the OH radical
2655.0-2690.0	Continuum emission of radio sources
3100.0-3400.0 MHz	Lines of CH
4800.0-4990.0 MHz	Interstellar ionized H clouds and supernova remnants
10-15 GHz	Non-thermal synchrotron sources
22.01-22.21 GHz	Red-shifted H_2O
22.21-22.5 GHz	**H_2O line (accessible from the ground)**
22.81-22.86 GHz	Studies of non-metastable ammonia and methyl formate
23.07-24.0 GHz	Ammonia lines
31.5-31.8 GHz	Continuum band
36.43-36.5 GHz	HC_3N and OH lines
42.5-50.2 GHz	Two important diatomic molecules: SiO and CS
86.0-95.0 GHz	Continuum and various lines
95.0-100.0 GHz	J=2-1 CS line
105.0-116.0 GHz	**Carbon monoxide and its isotopes**
121.0-182.0 GHz	Continuum and various lines
182.0-185.0 GHz	**Important water vapour line**
185.0-200.0 GHz	Various lines

3. Performance Parameters and Constraints

Passive sensors are characterized by their radiometric sensitivity and their spectral and geometric resolutions.

3.1. Radiometer Sensitivity

This parameter is generally expressed as the smallest temperature differential, ΔT_e, that the sensor is able to detect (σ level). ΔT_e is is given by:

$$\Delta T_e = \frac{\alpha T_s}{\sqrt{B\tau}} \quad \text{(K)} \tag{1}$$

where:

- B = receiver bandwidth (Hz);

- τ = integration time (s);

- α = receiver system constant (depends on the configuration);

- T_s = receiver system noise temperature (K).

The radiometer threshold ΔP is the smallest power change that the passive sensor is able to detect. ΔP is given by :

$$\Delta P = \Delta T_e B \quad \text{(W)} \tag{2}$$

where $k = 1.38 \times 10^{-23}$ (J/K) is Boltzmann's constant.

3.2. Geometrical Resolution

In case of two-dimensional measurements of surface parameters (see Section 4), it is generally considered that the transverse resolution is determined by the -3 dB point of the antenna pattern projected onto the ground. In case of three-dimensional measurements of atmospheric parameters (see Section 5), the longitudinal resolution along the antenna axis is also to be considered. This longitudinal resolution is a complex function (generally known as a weighting function) of the frequency-dependent characteristics of the atmosphere and the receiver performance characteristics (noise and bandwidth).

3.3. Integration Time

The integration time is also an important parameter which results from a complex trade-off taking into account in particular the desired geometrical resolution, the scanning configuration of the sensor and its velocity with respect to the scene observed.

3.4. Frequency Coverage and Spectral Resolution

Some sensors carry out a contiguous spectroscopic coverage along the line profile. Their total coverage $\Delta\nu$ and frequency resolution $\delta\nu$ determine the range of pressures that is covered.

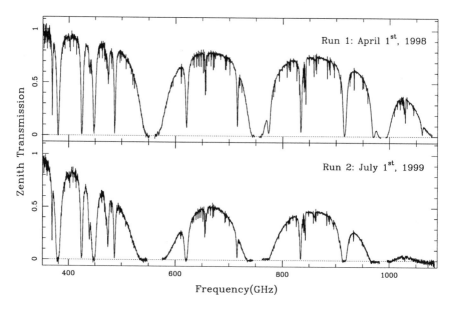

Figure 2.   Fourier Transform Spectroscopy measurements of the atmospheric transmission at Mauna Kea (Hawaii), 4100 m above sea level, showing outstanding conditions for submillimetre-wave astronomy. Data from Pardo et al. (2001).

4.   Typical Operating Conditions of Passive Sensors

Passive sensors can be ground-based or space-borne, the first type being dominant nowadays for radio astronomy and the second for the EESS.

The evolution of radio astronomy towards the submillimetre domain has required high and dry sites in order to allow for an at least partially transparent at such wavelengths. Figure 2 shows the measured atmospheric spectrum at one such submillimetre site.

Satellite-borne facilities are deployed essentially on two complementary types of satellite system:

- Low altitude (polar) orbiting satellites.

  Passive radiometers operating at frequencies below 100 GHz are currently flown only on low-orbiting satellites. This is essentially due to the difficulty of obtaining adequate geometric resolution at relatively low frequencies, which may change in the future. Those systems, based on satellites in low sun-synchronous polar orbits, are used to acquire high resolution environmental data on a global scale. The repeat rate of measurements is limited by the orbital characteristics and only a maximum of two global coverages can be obtained daily, with a single satellite.

- Geostationary satellites.

Systems involving satellites in geostationary orbits are used to gather low-to medium-resolution data on a regional scale. The repeat rate of measurements is limited only by hardware technology and is typically one regional coverage every 30 minutes or less.

5. Interference Criteria and Recommendations

Passive sensors integrate all natural (wanted) and man-made (unwanted) emissions. They cannot, in general, differentiate between these two kinds of signals because the atmosphere is a highly unstable medium with fast changing characteristics, spatially and temporally. The sensors are therefore extremely vulnerable to interference which may have extremely detrimental consequences:

- It was demonstrated that as few as 0.1 % of contaminated satellite data could be sufficient to generate unacceptable errors in Numerical Weather Prediction forecasts, thus destroying confidence in these unique all-weather passive measurements.

- The systematic deletion of data where interference is likely to occur may prevent recognition of new developing weather systems and vital indications of rapidly developing potentially dangerous storms or other phenomena may be missed.

- For climatological studies and particularly "global climate change" monitoring, interference may lead to mis-interpretation of climate signals.

- If the exclusive use for the Radio Astronomy Service of key bands of the electromagnetic spectrum is not guaranteed, our studies of the Universe will suffer to the extent of preventing understanding of some of its basic physical processes.

In those frequency bands shared by active services and passive EESS, it is necessary to establish recommendations on the maximum accepted interference level. The International Telecommunication Union has given the following such recommendations:

1. For EESS performing surface measurements, the typical accepted interference level is around -163 dBW for a reference bandwidth of 100 MHz (spectral region from 1.4 to 90 GHz). The limits can be exceeded by less than 5% of all measurement cells within a sensor's service area in the case of a random loss and by less than 1% of measurement cells in the case of a systematic loss.

2. In the 50 - 66 GHz frequency band, the required radiometric resolutions are 0.3 K and 0.1 K for scanning sensors and for push-broom sensors respectively. The resulting interference thresholds are -161 dBW for a scanning sensor and -166 dBW for a push-broom sensor, in a reference bandwidth of 100 MHz. These levels are equivalent to brightness temperature increases of 0.06 K and 0.02 K respectively and can be considered as a normal contribution to the error budget of the instrument.

3. Above 100 GHz the required radiometric resolution is currently 0.2 K at all frequencies, leading to an interference threshold of -160 dBW in a reference bandwidth of 200 MHz. However, these figures need to be revised in light of the most recent achievements in atmospheric sciences.

The interference criterion applicable to all three-dimensional measurements in the atmosphere is the following (from an ITU recommendation): "The interference levels given above can be exceeded for less than 0.01% of the pixels in the sensor's service area for three-dimensional measurements of atmospheric temperature or gas concentration in the absorption bands including those in the range 50.2-59.3 GHz and bands near 118 GHz and 183 GHz."

In addition to these quantitative recommendations established by the ITU for the EESS, the Committee on Radio Astronomy Frequencies has given other general recommendations:

- To define in the ITU Radio Regulations the term "level of harmful interference to radio astronomy".

- To adopt a definition of a passive service in the ITU Radio Regulations.

- Improve current definitions to make passive frequency use better understood.

- To improve communication/contact between the radio astronomical bodies and administrations on the one hand and "industry" on the other hand.

- To pay attention in frequency allocation procedures that existing passive bands should not be touched.

- To avoid that "passive" bands should be shared with "active" services.

- To create more "Primary exclusive Passive" bands.

6.   Future Developments that will need Protection

Technical advances during the past years have opened the doors for the access of passive services to the submillimetre domain of the electromagnetic spectrum. In fact, the technical improvements to allow access to this frequency region (ν >300 GHz) have been the results of efforts by the passive services. As a result, practically no active service operates today in those bands, but their protection should now be secured to ensure the quality of future research.

Here we provide a list of the most significant passive submillimetre projects that will operate in the coming years.

6.1.   EESS Submillimetre-Wave Projects

- **Microwave Limb Sounder** (EOS - MLS) NASA/JPL, with various channels from 118 GHz to 640 GHz, with possibly other channels at 1.23 and 2.5 THz.

- **ODIN** (Swedish Space Corporation), with various channels from 118 GHz to 576 GHz. This satellite will also perform passive astronomical observations in the same bands.

- **Sub-mm Observation of Processes in the Atmosphere Noteworthy for Ozone** (SOPRANO) - ESA. The submm bands of this instrument will range from 497 to 955 GHz.

- **Millimetre-wave Acquisitions for Stratosphere-Troposphere Exchanges Research** (MASTER) - ESA, with mm and submm bands from 199 to 505 GHz.

6.2. Submillimetre-Wave Astronomy Projects

- **Far InfraRed and Submillimetre Telescope** (FIRST), will perform photometry and spectroscopy in the 60-670 μm range. It will have a radiatively-cooled telescope and carry a science payload complement of three instruments housed inside a superfluid helium cryostat. It is hoped that FIRST will be operated as an observatory for a minimum of three years following launch and transit into an orbit around the Lagrangian point L2 in the year 2007.

- **Atacama Large Millimetre Array** (ALMA), will be a 64-element interferometer (12-m antennas) located at an elevation of 5300 m in Llano de Chajnantor (Chile) that will image the Universe at frequencies in all atmospheric windows between 10 mm and 350 μm.

- **Stratospheric Observatory for Infrared Astronomy** (SOFIA), will be a Boeing 747SP aircraft modified to accommodate a 2.5-m reflecting telescope. It will be the largest airborne telescope in the world. It is expected to carry submillimetre receivers as well as infrared instruments.

Acknowledgments. J.R. Pardo acknowledges the financial support for his work provided by Observatoire de Paris, CNES and Météo-France under the *décision d'aide à la recherche 795/98/CNES/7492*, the Goddard Institute for Space Studies and Columbia University, and by American NSF grants # ATM-9616766 and # AST-9980846.

References

"CRAF Handbook for Radio Astronomy" second ed. (1997), (European Science Foundation, Strasbourg, France; also available from CRAF Secretariat, P.O. Box 2, 7990 Dwingeloo, The Netherlands.)

International Telecommunication Union. Working Party 7C reports 1-205 (http://www.itu.int/itudoc/itu-r/sg7/docs/wp7c/1998-00/contrib).

Pardo, J.R., Serabyn E., and Cernicharo J., 2001, J. Quant. Spect. Radiat. Transfer, 68/4, 419-433.

Preserving the Astronomical Sky
IAU Symposium, Vol. 196, 2001
R. J. Cohen and W. T. Sullivan, III, eds.

Radio Astronomy in the European Regulatory Environment

R. J. Cohen

University of Manchester, Jodrell Bank Observatory, Macclesfield, Cheshire SK11 9DL, UK

Abstract. European radio astronomy has major world-class facilites which operate successfully in a hostile electromagnetic and economic environment. In 1988 the Committee on Radio Astronomy Frequencies (CRAF) was established under the auspices of the European Science Foundation, 'to keep the frequency bands used for radio astronomical observations free from interference'. Coordination of the European efforts through CRAF adds value through the sharing of expertise and information. Having one recognized voice for European radio astronomy also gives us strength. For example, the agreement concluded with Iridium LLC offered radio astronomy significant concessions compared with agreements reached elsewhere in the world. As Europe moves towards harmonized use of the radio spectrum, CRAF members participate in the discussions alongside representatives of governments and industry, to ensure that radio astronomy will have a secure future in Europe. This paper gives an overview of the European regulatory environment and the ways in which CRAF is working to protect radio astronomy.

1. Introduction

This paper describes some of the interference issues for radio astronomy in Europe and the work of the Committee on Radio Astronomy Frequencies (CRAF), of which I am Chairman. In Europe we have a very diverse collection of radio telescopes. Altogether there are 40 radio observatories in 17 countries. Figure 1 shows the locations of most of them.

All types of antenna and array configuration are here, and all radio astronomy frequency bands are covered, from 10 MHz up to 275 GHz, beyond which there are no international frequency allocations, either to radio astronomy or to any other radio service.

The European facitities are characterized by large collecting area and high sensitivity at all radio wavelengths. There are decametric and decimetric arrays at Nançay, Medicina and Cambridge, large reflectors for centimetre wavelengths at Effelsberg, Jodrell Bank, and Nançay, the large millimetre wave dish at Pico Valeta, the MERLIN and Westerbork arrays for aperture synthesis imaging at centimetre waves, the Plateau de Bure interferometer for millimetre-wave aperture synthesis, and the European Very Long Baseline Interferometer Network (EVN). The Joint Institute for VLBI in Europe (JIVE) operates the world's fastest VLBI correlator, designed for data rates of up to one gigabit per second.

Figure 1.   Locations of radio observatories in Europe.

Radio telescopes and people are crowded together in Europe as nowhere else on Earth. This presents us with particular regional problems and challenges. A good example is the use of television channel 38, at 606-614 MHz. Some European countries such as the UK use the frequency band for radio astronomy. For example, at Jodrell Bank Observatory the band is used continuously for pulsar timing, and periodically for VLBI. Other countries use the band for television broadcasting, most recently digital TV. Sharing studies show that 600-km separation is needed between a typical TV transmitter and a radio telescope if interference to radio astronomy is to be at an acceptable level. In European terms this is a big coordination zone. For example, a 600-km radius around Vienna includes 10 countries outside of Austria (namely Hungary, Italy, Czechia, Slovakia, Ukraine, Romania, Yugoslavia, Switzerland, Poland, and Leichenstein), each with its own laws and uses of the radio spectrum. How can radio astronomers deal with this and keep channel 38 clear for astronomical use?

CRAF was founded in 1987 because radio astronomers saw the need for a body to defend radio astronomy in regional interference issues such as that of TV channel 38. The first chairman was Hans Kahlmann of the Netherlands Foundation for Research in Astronomy (NFRA). The CRAF Secretariat and 'clearing house' were also established at NFRA. From these beginnings CRAF has developed to become the voice of radio astronomy within the European radio regulatory arena.

2. The European Environment

The countries of Europe are small and politically diverse. In order to coexist and to make efficient use of the radio spectrum they are having to harmonize their use of radio frequencies and set tight standards for electromagnetic compatibility. The European Common Market policy means that there should be free circulation of equipment and mutual recognition of standards and licences. There is rapid technical and political change. There are many pan-European bodies, each with different powers.

The European Commission (EC) sets policy based mainly on political and commercial considerations. Directives and decisions of the European Union are binding on member states. Radio astronomy is not a big player in this market-driven forum. The recent EC Green Paper on Radio Spectrum Policy sets out broad strategy for commercially exploiting the radio spectrum. The radio spectrum has become a scarce resource, and pressures are growing to manage it in a way which reflects its market value. As part of the consultation process, CRAF has drawn the attention of the EC to the value of non-commercial use of radio and has asked for a policy on defective space systems.

The Conferénce Européen des Postes et des Télécommunications (CEPT) is the body of 43 member countries which has to work through the technical issues flowing from policy decisions. This is the forum where radio astronomers can make an impact. The CEPT has a complex hierarchy, with a permanent secretariat, the European Radiocommunications Office (ERO), and three permanent committees of which the European Radiocommunications Committee (ERC) is the most important for radio astronomy. In particular, the ERC prepares common European positions for the World Radio Conferences (WRCs). The ITU rule of "one member one vote" gives the CEPT considerable leverage at WRCs. European Common Proposals to a WRC coming from a block of 43 countries carry great weight.

Working groups and project teams within the ERC carry out detailed technical studies and produce reports which lead to ERC decisions. Radio astronomers are active in Working Groups SE (spectrum engineering) and FM (frequency management). Project team SE21, for example, deals with spurious emissions and prepares European contributions to the work of the ITU task group TG1/5. Project team SE28 deals with mobile satellite issues, such as the introduction of Iridium and Globalstar into Europe. Project Team FM33 has been looking at new allocations for passive services at frequencies above 71 GHz, in preparation for WRC-2000. Radio astronomy is represented in these project teams by CRAF members who work alongside representatives of government, industry, broadcasting, and science. ERC decisions and recommendations, however, are not mandatory for the CEPT countries.

The European Telecommunications Standards Institute (ETSI) sets European standards for new equipment. ETSI is an open-forum organization with about 700 members from 50 countries, representing administrations, network operators, manufacturers, service providers, and users. CRAF needs to be aware of developments within ETSI, and may comment on the impact for radio astronomy through national administrations. Coordinated action through CRAF stands a much bigger chance of influencing standards than actions taken by individual astronomers in isolation.

3. CRAF Activities

CRAF was established as the regional body to deal with European interference issues, and to coordinate the efforts of individuals at European observatories. The first meeting of CRAF was held in Paris in 1987. This was an open meeting of radio astronomers who were concerned at growing levels of interference to their observations, and who wanted to do something about the situation. Interference problems can be local (such as a nearby radio link), regional (such as television broadcasting in channel 38), or global (such as global satellite systems). Each problem needs to be tackled at the appropriate level.

In 1988 CRAF became an associated committee of the European Science Foundation (ESF). CRAF's mandate is 'to keep the frequency bands used for radio astronomical observations free from interference'. Under the ESF umbrella CRAF has published a Handbook for Radio Astronomy, already in its second edition (1997). The ESF has ensured wide publicity and a wide circulation for the CRAF Handbook (which is available free on request).

Since 1997 CRAF has had a full-time Frequency Manager for European Radio Astronomy, Dr. Titus Spoelstra of NFRA, who has also served as Secretary since CRAF began. The position is funded by CRAF member countries and observatories. The availability of a full-time professional has greatly strengthened our efforts, and has allowed CRAF's voice to be heard in many new fora.

Today, CRAF has 24 members from 17 countries. We meet two or three times per year, and also have teleconferences. Most of our work is carried out between meetings, communicating by email and correspondence. There is a CRAF web-site (http://www.astron.nl/craf), with an electronic newsletter. The CRAF Newsletter is read by people from government and industry, as well as by radio astronomers. Via the Newsletter CRAF is able to publicize issues such as that of the TEX Satellite, which was for many years an unknown source of interference to radio astronomy at 328 MHz, a frequency not even allocated to satellites (CRAF Newsletters 1996/3 and 1998/2, on the CRAF web site).

CRAF's remit has broadened beyond radio astronomy to include aeronomy, geodesy and the European Incoherent Scatter Scientific Association EISCAT, which can all face similar interference problems in reception. The CRAF network allows us to share information and coordinate our efforts. Having a single voice for radio astronomy in Europe gives us a strong bargaining position, which is recognized within the CEPT and its working groups.

CRAF also maintains close working relations with the other regional body, the Committee on Radio Frequencies (CORF) in North America, and with the global "action group" IUCAF. Like IUCAF, CRAF now participates in ITU activities as a sector member, entitled to submit papers and speak at conferences, but not to vote. CRAF has taken the lead on several key issues within ITU, such as the protection requirements for millimetre-wave astronomy. This work included a collaboration with industry (European Space Agency and Oerlikon-Contraves) on techniques for fabricating millimetre-wave filters (Natale et al. 1998). CRAF has also held workshops to discuss the issue of defining an acceptable percentage of time lost to radio astronomy through interference, and has produced input documents to the ITU on this matter. Finally, and most importantly, CRAF takes a strong lead on presenting the radio astronomy position for WRCs to the European administrations.

4. CRAF and Iridium

A topical illustration of CRAF's work is the negotiations recently concluded with Iridium LLC (Abbott 1998, 1999). Iridium is the first of a new generation of mobile satellite communication systems. Unfortunately the frequency band chosen for the Iridium downlink to subscriber terminals is very close to the frequency of the hydroxyl (OH) spectral line at 1612 MHz. This poses the threat of interference to radio astronomy from the unwanted emissions of the Iridium satellites, so-called radio smog (Schenker 1999). The potential danger was recognized as early as 1991, when Iridium and Motorola personnel visited Jodrell Bank. They were referred to CRAF and the CEPT, as this was clearly not a local UK problem.

Technical studies on the impact of Iridium in Europe were conducted within the ERC project teams SE17 and SE28, which looked at many sharing issues: sharing with military radars in Sweden, with tactical point-to-point communications links used by NATO in Germany, with the Russian Global Navigation Satellite System GLONASS, with the INMARSAT satellite communications via geostationary satellite, and of course with radio astronomy. The technical studies reached deadlock over the issue of protecting radio astronomy. Iridium could not guarantee to keep the unwanted emission levels of its satellites below the interference thresholds for radio astronomy given in the ITU-R Recommendation RA.769-1. Furthermore, they claimed that radio astronomy did not really need this protection level. They proposed a series of mitigation factors and techniques, all of which were rejected by CRAF. Iridium clearly felt in a strong position at this time, having reached a time-sharing deal with US astronomers. However, the 4 hours per night agreed in the USA was totally inadequate to the needs of European radio astronomy, where some facilities spend up to half their observing time at 1.6 GHz.

In 1997 direct negotiations began between CRAF and Iridium, under the auspices of the Milestone Review Committee of the ERC, to try to reach a compromise deal. The discussions were difficult and often hostile. The tactics of big business were brought to bear, along with corporate lawyers, and CRAF was glad to be able to call on the services of the ESF legal advisor. Under great pressure agreement was reached to give radio astronomy full protection for part of the time. Until 2006 there would be full protection to ITU levels for 7 hours per day and 2 whole weekend days per month, plus protection to an intermediate level for all other weekend days. Furthermore from 2006 there would be no need for operational restrictions on either side. In the interim period there was a work plan to study the use of mitigation factors at the radio astronomy observatories, as well as suppression or prevention of unwanted emissions at the satellite transmitters. The complete text of the agreement is published on the CRAF web pages.

The Iridium case shows the need for radio astronmers to fight together. Iridium tried many ways and many times to destroy our unity, making approaches to individual observatories and administrations in order to circumvent CRAF. Our persistence was rewarded in that achieved a deal which was less damaging to radio astronomy than the deals struck elsewhere in the world. It is also legally binding in Europe. My hope now is that the mitigation factors don't come back to haunt us in the future.

5. Summary

At the turn of the millenium CRAF has unique opportunities to influence events to secure a good future for radio astronomy in Europe. As Europe works towards harmonized frequency use and standards we are making a place for radio astronomy. We are registering the radio observatories and their frequency requirements in the European database of spectrum usage. We are explaining, which usually means defending, our protection requirements and interference thresholds, and we are carrying out technical studies and interference assessments within the CEPT working groups and project teams and elsewhere, based on the ITU protection criteria. Now is the time to state our requirements clearly, or active services will simply dictate to us what we are getting. Our experiences with Iridium have amply demonstrated the advantages of having a single voice for European radio astronomy.

It is very important that CRAF participates directly in the European preparations for WRCs. We can thereby influence the CEPT position, which counts for 43 votes at the conference. Incidentally, WRC-2000 will be the first WRC at which European radio astronomers can register as CRAF delegates. We have already had CRAF delegates to ITU Working Party 7D (radio astronomy) and Task Group 1/5 (unwanted emissions).

Finally, through our umbrella organization the ESF, we have new links to explore outside the telecommunications community. It is in the wider arena that our arguments need to be carried, not just within the confines of the specialist working groups and committees, but with the entire scientific community, as well as with the educated public.

References

Abbott, A. 1998, Nature, 394, 607

Abbott, A. 1999, Nature, 399, 513

CRAF Handbook for Radio Astronomy, 2nd edition 1997, European Science Foundation: Strasbourg

CRAF web pages http://www.astron.nl/craf

Natale, V, Tofani, G., Tomassetti, G., Dainelli, V. & van't Klooster, C. G. M., 1998, in Proceedings of the 12th International Wroclaw Symposium on Electromagnetic Compatability, J. Janiszewski, W. Moron & W. Sega, pp.621-624

Schenker, J. L. 1999, in TIME Magazine International, vol. 154, No. 2, p.112

Preserving the Astronomical Sky
IAU Symposium, Vol. 196, 2001
R. J. Cohen and W. T. Sullivan, III, eds.

Preserving Radio Astronomy in Developing Nations

G. Swarup and C. R. Subramanya

National Centre for Radio Astrophysics,
Tata Institute of Fundamental Research Poona University Campus,
Pune - 411007 India.
gswarup@gmrt.ernet.in

Raman Research Institute, C. V. Raman Avenue Sadashivnaga,
Bangalore, Karnataka 560 080, India.

Abstract. Due to the very weak nature of signals from cosmic radio sources, the sensitivity of a radio telescope and receiver is about 40-60 dB higher than those of communications receivers. Hence, radio telescopes are generally located in relatively radio-quiet locations and operate in frequency bands that are protected against radio interference through frequency planning by national governments. Taking advantage of the much lower degree of radio interference in developing countries and the relatively labour-intensive nature of metre-wave radio telescopes, several such radio telescopes have been built and are planned in Argentina, Brazil, China, India, Mauritius and South Africa. Radio telescopes operating at cm-wavelengths are also planned in Egypt and Mexico.

A particularly severe problem arises for the radio astronomy service and other passive services below 2 GHz from the possibility of unacceptable emissions from satellites in unwanted bands (out-of-band and spurious emissions), due to the specific modulation schemes used in satellite transmitters. It is noted that this can be circumvented within the existing technologies if the satellite transmitters employ suitable bit-shaping or filtering techniques or use modulation schemes like Gaussian-filtered Minimum-Shift Keying (GMSK) which produce very little out-of-band emission. Although radio astronomy started in the western world at low frequencies, much low frequency radio astronomy is now planned or operational in developing countries. In order to protect the interests of these and other passive services within developing nations, it is important that suitable regulations be recommended to UNISPACE-III to provide appropriate protection.

Preserving the Astronomical Sky
IAU Symposium, Vol. 196, 2001
R. J. Cohen and W. T. Sullivan, III, eds.

Steps to Establish International Radio Quiet Zones

Harvey Butcher

Netherlands Foundation for Research in Astronomy,
P.O. Box 2, 7990 AA Dwingeloo, The Netherlands

Abstract. Future radio telescopes will feature large increases in sensitivity, not only to celestial sources but also to man-made interference. In addition, they will need to measure spectral lines as a function of redshift and hence observe at arbitrary frequencies. Current international regulations governing the use of the radio spectrum are in principle inappropriate to these science-driven needs. There are still a few places on Earth that are relatively interference-free, but the coming generation of telecommunications satellites in low orbit will compromise even those sites. The Organization for Economic Cooperation and Development (OECD) has recently sponsored a working group on the future of radio astronomy, which recommended that the possibility of establishing one or more formally recognized International Radio Quiet Zones (IRQZ) be studied seriously. First results of that effort will be discussed.

Preserving the Astronomical Sky
IAU Symposium, Vol. 196, 2001
R. J. Cohen and W. T. Sullivan, III, eds.

A Potential Site for the World's Largest Single Dish, FAST

B. Peng

Beijing Astronomical Observatory, National Astronomical Observatories, Chinese Academy of Sciences, Beijing 100012, China

R. G. Strom

NFRA and University of Amsterdam, Postbus 2, 7990 AA Dwingeloo, The Netherlands

R. Nan

Beijing Astronomical Observatory, National Astronomical Observatories, Chinese Academy of Sciences, Beijing 100012, China

Abstract. We have carried out a series of measurements at some locations in Guizhou Province and one additional site at the Urumqi Astronomical Station in the Xinjiang autonomous region, to check on their suitability, from the point of view of interference, for the construction of a Five-hundred-meter Aperture Spherical Telescope (FAST). This large facility will in some sense act as a prototype for the Square Kilometer Array (SKA). Measurements were made using a commercial receiver in the frequency range 25–1500 MHz. The results in Guizhou look quite promising. A protected radio quiet zone in Guizhou would make it an ideal location for an international radio astronomical facility, and would establish the FAST site as a natural SKA location.

1. Introduction

One way to realize the Square Kilometer Array (SKA), with a total collecting area of about 1 km^2 to achieve $\simeq 1\,\mu$Jy flux density sensitivity, is to construct a spherical reflector array of about 30 individual telescopes, each roughly 200 m diameter. Ideally, there would be continuous frequency coverage between about 0.2 and 2 GHz, with possible capability up to 5 GHz or even 8 GHz depending upon the cost. As the first step, a Five-hundred-meter Aperture Spherical Telescope (FAST) acting as a pilot for SKA has been proposed (Peng 1998).

FAST is not simply a copy of the existing Arecibo telescope, but has a number of innovations. Firstly, the proposed main spherical reflector (Qiu 1998), to fit a paraboloid of revolution in real time by active control, enables the realization of broad bandwidth and full polarization capability by using standard feed design. Secondly, a feed support system that integrates optical, mechanical and electronic technologies (Duan et al. 1996) will effectively reduce the cost. With an overall diameter of 500 m and radius of its spherical surface of 300 m, FAST will be the world's largest single dish.

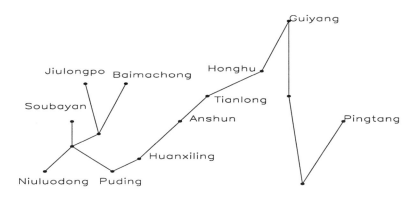

Figure 1. A schematic map of test sites measured in March 1995.

There are a number of important considerations affecting the choice of any telescope site. It has been realized that the right terrain, such as the Karst formations of Guizhou Province in southwestern China, would be critical to a successful implementation of an Arecibo–style spherical reflector (Braun 1994). The Chinese FAST/KARST[1] project team has been searching for suitable sites and testing them for an Arecibo–style telescope. A database of about 400 Karst depressions in Pingtang and Puding counties in the province has been set up. It is well-known that a critical consideration in choosing the best telescope site is the interference environment. As part of the international effort to build such a new telescope, the Chinese Academy of Sciences and the Royal Dutch Academy of Sciences have funded a bilateral exchange program to enable radio interference measurements to be carried out at a number of potential sites in China. In the following, we present preliminary results from a series of interference monitoring tests at various locations in Guizhou and Xinjiang.

2. Preliminary Results

In November 1994, the first measurements were made at 8 Karst depression sites located at an altitude of about 1,000 m above sea level in Pingtang and Puding counties in Guizhou. In addition, we monitored a site in the centre of Guiyang, the capital of Guizhou, and as a further test, made measurements at the site of the 25 m telescope of the Urumqi Astronomical Station in the Xinjiang autonomous region in northwestern China. Guiyang was monitored just to compare the interference situation in a populated area with that of the remote sites. Four months later in March 1995, additional observations were made at a number of remote sites for comparison. These included monitoring at four Karst depressions in Puding for a period of one month in order to obtain

---

[1]Kilometer-square Area Radio Synthesis Telescope, a Chinese concept for the SKA project.

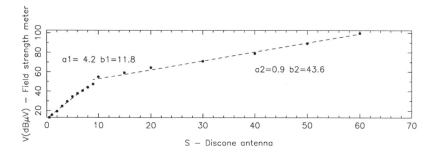

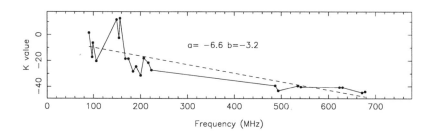

Figure 2.   *Upper plot:* Relative voltage S versus real voltage V_r; *Lower plot:* K coefficient of discone antenna versus frequency.

more complete statistics; and an investigation of interference at several locations between sites in Puding county, Anshun and Guiyang cities, and Pingtang county in Guizhou in an attempt to understand distance effects. A schematic route of this trip, not shown to scale, is plotted in Figure 1

The measurements were carried out with an Icom IC-R7100 receiver coupled to a Diamond Antenna Corp. D-130N omnidirectional discone antenna. This combination can receive over the range 25 – 1500 MHz (the receiver is continuously tunable up to 2 GHz, and a check showed that the antenna can receive there, but presumably with reduced gain). The R7100 can receive and detect single side band (SSB), both upper and lower side bands (USB and LSB), and AM and FM (in wide and narrow-band mode) signals. For the purpose of our observations, we scanned in SSB (USB) mode, but most signals found were then checked in the various modes, particularly to look for voiced modulation. The measurements were done in automatic scanning mode, with the squelch threshhold set just above the background noise level (about 1 on the S-meter) so that any signal above noise would be detected. A few scans were made by hand to search for signals in the noise, and a number were found. The scan step was usually 10 kHz below 200 MHz, and 25 kHz above. About 2.5 hr was typically required to monitor each site over the entire 25 – 1500 MHz range.

To put our values on an absolute intensity scale, we calibrated both the receiver and the antenna. The field measurements of intensity give the relative voltage (S) of signals picked up at each interfering frequency. Using a signal

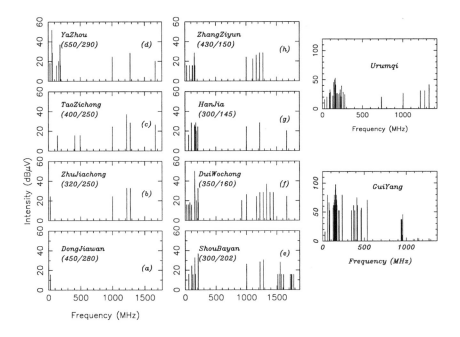

Figure 3.   The first measurements of interference levels E, made in November 1994.

generator (MG615B) we have determined S and the real voltage V_r at frequencies of 30, 100, 200, 400 and 800 MHz. The measured S (of the IC-R7100) versus V_r (dBμV) (of the ML518 field strength meter), based on average values of V_r, is shown in Figure 2 (0dBμV = -60dBmV), the upper plot. By least-squares fit we have, for $S \leq 9.0$,

$$V_r(\text{dB}\mu\text{V}) = 4.2 \times S + 11.8 \tag{1}$$

otherwise,

$$V_r(\text{dB}\mu\text{V}) = 0.9 \times S + 43.6 \tag{2}$$

The K coefficient of the D-130N omnidirectional discone antenna, defined as $V_r - E$ (the field strength), can be approximated by

$$K = -6.6 \times f(\text{MHz}) - 3.2, \tag{3}$$

and is plotted in Figure 2 at the bottom. It has been measured over a frequency range of 50 MHz $\sim$ 700 MHz using a log-periodic antenna MP635A coupled to the ML518 field strength meter. This gives us a general idea of the absolute interference strength E at the surveyed sites.

The results of our measurements are presented in Figures 3, 4 and 5. In the left 8 graphs of Figure 3, the name of each Karst depression is labelled at the top left: the diameter D and the depth H were roughly estimated (in meters) and are given just under each name in the form D/H. Four depressions,

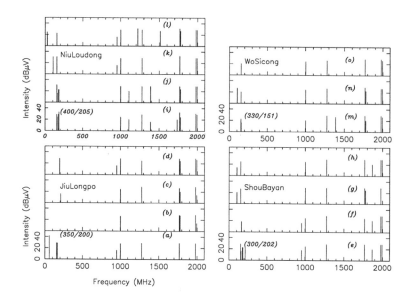

Figure 4.   Weekly monitoring at four locations in Puding county, March 1995.

a, b, c and d, lie in Pingtang county, and the others in Puding county. Some signals at frequencies of 256, 512, 768, 1024, 1500 MHz and their harmonics were found to be produced by the quartz oscillators of the receiver itself; they have not been reproduced in the figures. The lower right plot in Figure 3 shows the interference situation in Guiyang city, which was measured from the top of an 11-storey building of the Science and Technology Commision of Guizhou Province, the highest structure in that part of the city. The upper right plot shows the situation at the 25 m telescope station, about 50 km southwest of Urumqi city, the capital of Xinjiang.

Figure 4 contains 15 graphs, showing the interference monitoring results for four Karst depression sites during one month in Puding, with an interval of one week between measurements at each location. It seems that the interference environment there was quite stable during this period. Results obtained in March 1995 are presented in Figure 5 by means of graphs that show interference measurements of eleven sites ordered according to their geographic location (see Fig. 1). These include four Karst depressions in Puding and sites going from Puding via Anshun and Guiyang to Pingtang. They give us some indication of variations with distance. As one can clearly see, at frequencies below 1000 MHz a few sources of radio interference are strong, and some wide band interference is also present in Guiyang City.

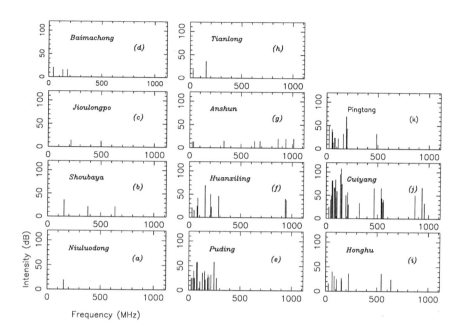

Figure 5.   Interference measurements ordered by geographic location (see Fig. 1), March 1995.

3. Concluding Remarks

The results of these measurements provide information about the frequency, strength and characteristics of the interfering signals. Most of the interfering signals found appear to be narrow-band (<10 kHz) beacon signals of unknown origin. There are also a few FM radio stations, mobile communication bands and television channels. Moreover, additional observations were made at remote sites for comparison, a longer period of one month's monitoring at four sites in Puding for more complete statistics, and an investigation of interference at several locations between sites from Puding via Anshun and Guiyang cities to Pingtang in Guizhou to understand distance effects.

Qualitatively and semi-quantitatively, the data are about what one might expect. As one leaves Guiyang (Fig. 5j) going west, the interference level generally and number of interfering signals drops in Honghu (Fig. 5i). Continuing to Tianlong (Fig. 5h) it decreases further, but as one approaches Anshun (Fig. 5g, which was outside of the city itself) the numbers increase again. As for changes with time, Fig. 4 shows that the situation at four sites was quite stable over a month. For one of these, Shou Bayan (Fig. 4e–h), the picture is quite similar to that four months earlier (Fig. 3e).

All in all, we conclude that, due to the remoteness of this region and terrain shielding, the results in Guizhou Province look quite promising, with relatively little interference found in the region 220 – 920 MHz.

A State Standard, "Protection Criterion for the Radio Astronomy Service" has been submitted to the State Supervisory Bureau of Technologies, and is expected to be issued as one of the laws for radio management in China. Recently we made proposals for designating a radio quiet zone (RQZ) in Guizhou province for a potential site of FAST.

Guizhou regional Bureau of Radio Management agreed to work and collaborate on designating a RQZ as soon as the final site for FAST/KARST was chosen and the project FAST approved as a National Megascience Project of China. Guizhou provincial government would like to make a plan to modestly develop their radio industry and communications. We have resumed monitoring the radio interference environment since January 1999, and believe that a protected RQZ in Guizhou would make it an ideal location for an international radio astronomical facility, and would establish the FAST site as a natural SKA location.

Acknowledgments. B. Peng thanks the MPIfR in Germany and the IAU travel grants for supporting his attendance at this IAU symposium 196. We are grateful to the Chinese Academy of Sciences and the Royal Netherlands Academy of Sciences for travel grants in the context of a bilateral exchange programme (grant number 96CDP042) which made much of this research possible. We wish to thank all the members of the FAST/KARST project team for their efforts in such a site survey, the S&T Commission of Guizhou Province, the Institute of Remote Sensing Application, the officials of counties Pingtang and Puding, and the Karst Comprehensive Research and Experimental Station for their practical help. This FAST project has now become a key project in the Chinese Academy of Sciences, and has also been supported by the Ministry of Science and Technology of China.

References

Braun, R. 1994, Draft Minutes of URSI LTWG-1-minutes, Jodrell Bank

Qiu Y.H. 1998, MNRAS, 301, 827-830

Peng B. 1998, The Observatory, 118, p.261

Preserving the Astronomical Sky
IAU Symposium, Vol. 196, 2001
R. J. Cohen and W. T. Sullivan, III, eds.

Techniques for Coping with Radio Frequency Interference

J. R. Fisher

NRAO, P.O. Box 2, Green Bank, WV, USA

Abstract. As a complement to spectrum management efforts by radio astronomers a number of observatories and research groups around the world have begun looking into technical solutions to the problem of separating weak cosmic radiation from man-made radio signals. Some of the technical research now getting underway includes: high dynamic range receivers, low-noise superconducting filters, passive digital filtering, adaptive filters, adaptive sidelobe nulling, multi-feed correlation of RFI, and various techniques for signal blanking.

Increased technical support to spectrum management can also be provided in the form of accurate and statistically significant characterization of the radio environment, empirical and theoretical improvement of over-the-horizon propagation models, and timely measurements of spurious radiation falling in the protected radio astronomy bands.

Finally, credibility of our spectrum management effort can only be maintained by making sure that local radiation under the control of our radio observatories is in compliance with the field strength limits of Recommendation ITU-R R.A.769. This requires sensitive radiation measurements and often shielding of digital equipment, microwave ovens, test equipment, local oscillators, etc.

1. Introduction

The title of this paper was chosen to emphasize radio frequency interference (RFI) excision and cancellation techniques, but let me expand the topic a bit to include technical support for spectrum management and the measures that we can take to maintain a clean environment in the vicinity of our radio observatories. To sustain a credible presence in the national and international spectrum management agencies the radio astronomy community must show that it is allocating significant resources to quantitatively monitor the radio spectrum and that it is keeping up with the state of the art in RFI reduction. Our current effort in these areas is too small.

Radio astronomy and passive remote sensing are unique amongst the radio services in several ways. They are mostly receive-only; their received signals are typically from three to six orders of magnitude weaker that those of the communications services; and their signals from natural sources are spread throughout the entire electromagnetic spectrum. These last two aspects of our science make the task of monitoring the radio spectrum a daunting but not impossible one.

The allocation of resources for RFI control can be divided into two categories: organization of current technologies and research and development. The first includes an adequate number of monitoring stations to maintain cleanliness of the exclusive radio astronomy bands, to verify spectrum sharing agreements, to enforce quiet zone and coordination zone agreements, and to evaluate new sites for radio observatories. It also includes thorough shielding and filtering of RFI sources in the vicinities of radio telescopes and the engineering of receivers that can tolerate signals from other radio services.

Research and development of RFI excision techniques are also essential. The protected radio astronomy bands cannot possibly cover all frequencies of importance to the science. Radio astronomers have been resourceful in finding remote sites and observing at times and frequencies that are not fully occupied by the allocated radio services, but the increasing density of spectrum use is making this more problematic. We need to more efficiently separate signals in the spatial domain with sidelobe cancelling and null steering techniques that have already been developed by radar and acoustic engineers. The extremely low signal-to-noise requirements of radio astronomy demand, however, a significant extension of the current state of the art.

2. Support for Spectrum Management

I think that it is fair to say that radio astronomy's spectrum managers have been inadequately supported by field measurements of the radio environment around our observatories. Reports of interference to observations in the protected bands tend to be anecdotal, qualitative, and very sporadic. Although we can fully justify the harmful interference levels set out in Recommendation ITU-R R.A.769, the case is seen as a largely hypothetical one by many of our competitors for spectrum space. If radiation in the protected bands were reported in a more timely and systematic manner, our spectrum managers could make a stronger argument for better protection to the rule-making agencies.

The enormous increase in satellite services for navigation and communication has created a strong demand for downlink spectrum allocations near protected radio astronomy bands. Because satellites are line-of-sight to radio telescopes, their signals can be quite strong. This puts stringent requirements on the satellite transmitters to avoid producing harmful spurious radiation in the radio astronomy bands. To verify that the satellite transmitters comply with these requirements, radio astronomers have had to measure the spurious radiation during the satellite's test or commissioning phase. In the case of the Iridium satellite system this took many man-months of engineering effort to refit receivers, and the tests occupied a number of days of radio telescope time that would otherwise have been used for astronomy. We can expect a demand for more of these measurements in the future. It will behove us to invest in several dishes in the 10-15 meter class with receivers and signal processing back-ends to avoid using valuable time on our best radio telescopes. Furthermore, if such measurements were made routine, they would serve as a monitor of compliance throughout the lifetime of many satellites.

In recent years the National Radio Quiet Zone around NRAO's Green Bank observatory has been challenged several times on the basis of dubious propagation calculations used to request limits on the radiation from transmitters installed in the Quiet Zone. Different propagation models give quite different answers in some instances. The only way to conclusively resolve each challenge is with statistically significant measurements over a range of foliation and atmospheric conditions. We are just now designing measurement facilities for this purpose, and it will be several years before we have a good understanding of the accuracy of our propagation model.

3. Sensitivity

One occasionally hears the lament that an RFI monitoring station cannot hope to reach the sensitivity of a radio telescope because of the latter's very low noise receivers and integrating radiometers. That is true if all one has is a simple antenna and a common spectrum analyzer, but we need to do much better.

The harmful limits set by Recommendation ITU-R R.A.769 assume an isotropic gain for the far sidelobes of a radio telescope with a typical system temperature of 20 K. A monitor antenna and uncooled preamp necessarily has a noise temperature of about 300 K, but this can be compensated by a reasonably modest 12 dB of antenna gain in the direction of the RFI source.

There is no reason that we cannot use the signal processing power of radio astronomy spectrometers and continuum radiometers on the monitor antenna signal to realize the same integration times and spectral and time resolutions as are used for astronomical measurements. Ideally, one would like to build a small, dedicated integrating spectrometer for RFI detection and monitoring, but various time or channel sharing schemes can also be made to work with existing equipment. Digital spectrometer bandwidths of several tens of MHz are now relatively inexpensive.

Our monitor station plans for the new 100-m Green Bank Telescope (GBT) include a set of antennas near the top of the feed support arm which will share the alternate LO synthesizer and one or more of the eight fiber optic IF channels to the control room. An old 2x40 MHz bandwidth FFT spectrometer will be available most of the time for RFI measurements while the telescope is using a new 800 MHz correlator. A second monitor station near the lab will have a similar set of antennas and three optical fiber links to the GBT control room and the FFT spectrometer.

4. Control of Our Own Radiation

A radio observatory requires thousands of pieces of electronic and computing equipment, each with real potential for interference to astronomical measurements. Many of us have experienced this painful fact. As a practical matter and to demonstrate that we are serious about the radiation limits that we ask of other radio services, we are obligated to suppress the radiation from our own equipment to the levels set by Recommendation ITU-R R.A.769, at least in the radio astronomy bands but, ideally, at all of our observing frequencies.

Retrofitting an observatory for RFI suppression is an expensive and time-consuming business. The DRAO observatory at Penticton is a notable success story. They have gone to considerable lengths to shield computers, fax machines, correlators, and even a postage scale to clean up their environment around 21 cm wavelength, where they are making a large scale survey of the sky. Similarly, the VLA went to considerable expense and effort to suppress radiation from equipment in their antennas to meet the requirements of new 330 and 74 MHz receivers. The lesson to be learned is that RFI control must be an integral part of an observatory design from the very beginning.

At Green Bank we have the fortunate opportunity to partially start from scratch with the new GBT and its control room. An important part of this new construction has been a program to measure most of the new electronic equipment that will be installed on or near the GBT. These measurements are done in a shielded anechoic chamber. The isolated environment and occasional use of an FFT spectrometer and specially programmed continuum signal processing have allowed us to measure RFI emissions that are commensurate with the limits of Recommendation ITU-R R.A.769. The field strength measurements for each piece of equipment are then converted to a power density at the GBT feed, using free space loss and any shielding external to the unit. Where necessary the equipment is sent back to the lab for further shielding and filtering. Very few pieces have escaped the need for additional suppression, and even apparently innocuous items like linear power supplies have been found to be a source of substantial RFI.

Other measures that have been implemented in connection with GBT construction are a 60 dB-shielded secondary receiver room, a 60 dB-shielded equipment and control room removed about two kilometers from the telescope, and transfer of computers and other equipment for the other telescopes to the shielded control room. The biggest problems that remain are many pieces of computing and test equipment in an unshielded lab on site. These will be slowly addressed on a room by room and unit by unit basis.

5.   Site Characterization

To my knowledge there is very little quantitative information that allows us to compare the RFI environments of the many radio observatories around the world. Hence, we have essentially no basis from which to begin a site evaluation for future radio telescopes, most notably the Square Kilometer Array (SKA).

Part of the problem is that we have never established a standard measurement procedure, primarily because making accurate and statistically representative field strength measurements over several decades of frequency is a large task. An informal group that met at the SKA Workshop in Dwingeloo, the Netherlands, in April 1999 is making a modest beginning at establishing measurement guidelines or standards. A number of observatories have most of the facilities needed to begin a comparative measurement program, but it remains to be seen how much support this effort will gather. It is important to the radio astronomical community that substantial resources should now be devoted to this task as an investment in future RFI control and spectrum management.

6.   RFI Excision

We commonly think of RFI in the frequency domain, but from an observer's viewpoint the location of interference is better described in at least four dimensions: frequency, time, and two directional coordinates. If a source of cosmic radiation does not coincide in all four coordinates with any RFI sources, we can, in principle, isolate the cosmic radiation for measurement. Of course, in practice this is a challenging and imperfect task.

In this four-coordinate phase space the electromagnetic spectrum is very sparsely populated. At a given observing location one frequency may be occupied by only one RFI source in a specific direction, or the RFI source may cover a continuum of frequencies but have a small duty cycle. In many cases the cosmic radiation may be isolated by placing a single null on the interfering source in only one of the four coordinates. All forms of RFI excision do just that in one way or another.

For the following brief discussion of excision techniques we assume that the frequency of the cosmic radiation is occupied by an interfering signal so that we must separate the two in one of the time and spatial coordinates.

6.1.   Blanking (Temporal Excision)

The simplest method of RFI excision is to divide the observational data into a modest number of time intervals and throw away, based on visual inspection, data that is corrupted by interference. When the number of time bins gets to be more than a few hundred, this process must be automated, but matching the ability of the human eye at picking out anomalies in data is a difficult task. The trick is to find a property of the interference that differs mathematically from random noise and from the intended astronomical signal.

If the interference has a sufficiently low duty cycle it can be detected in the data by comparing the instantaneous power to a running mean or median of surrounding time samples. This has been successfully demonstrated by Peter Fridman at the Special Astrophysical Observatory and NFRA on RATAN-600 data and by Morgan and Fisher (1977) on data from the late NRAO 91-m telescope. To be effective the data sampling interval must be short enough to resolve the interference temporal structure, and the mean or median interval must be long enough to get a good sample of the data in the absence of interference. Hence, the sampling and averaging time scales must be carefully matched to the temporal properties of the RFI, and one must be careful that the exciser does not affect the astronomical measurements. The necessity for manual optimization has been a deterrent to implementation of this type of RFI excision. A good user interface with sufficient display of the results of the excision process would probably help. A more robust excision algorithm which used as many as 10^5 time samples in real-time detection of outliers in a 22 MHz interferometer was reported by Kasper *et al* (1982).

A very different approach to detection and blanking has been simulated for a signal which is correlated in the outputs of a synthesis array such as the Westerbork radio telescope (Leshem and van der Veen 1999). When the correlated signal rises to a level that is above anything expected from an astronomical source, the data integration process is suspended until the correlated signal falls below threshold. This detection scheme has the considerable advantage of a known zero-interference reference level (zero correlated signal), but, like all blanking methods, a successful astronomical measurement depends on having a significant fraction of interference-free time.

6.2.   Spatial Nulling

Since total integration time is part of the sensitivity equation we would prefer not to discard any data. Also, we need an excision technique that works when RFI is present all or most of the time at the frequencies of the cosmic signals. To accomplish this we must isolate and discard the interference in the spatial domain.

The ideal spatial isolation would be a radio telescope with no sidelobes, but this is a practical impossibility. The next best thing is to create nulls in the sidelobe response in the directions of interfering signals while being careful not to affect the gain of the antenna's main beam. In the signal processing literature (e.g., Widrow and Stearns 1985) the technique of spatial nulling is often divided into two categories, interference cancellation and null steering, but they really amount to the same thing. The first is usually found in connection with acoustics and single antennas, and the latter is found in the antenna array literature.

The basic idea is that a sample of the interfering signal is added to the RF path of a radio telescope, but in opposite phase and equal amplitude to the original RFI signal that entered through the telescope's sidelobes. In the case of a paraboloid antenna the interference sample can be obtained by an auxiliary antenna whose gain is somewhat higher than the sidelobe gain. In an antenna array the cancelling signal is generated with a slight modification of the complex weights of the array element signals that make up the main-beam-forming network. An auxiliary antenna could also be used with an array. One advantage of an auxiliary antenna is that it can have very little gain in the direction of the astronomical source, and hence its signal will have little effect on the main beam gain. It is not clear at this point whether this is an important advantage in practice.

Because the radio telescope and possibly the interfering source are not stationary, the phase and amplitude of the cancelling signal must be made to track the changing amplitude and phase of the interference entering through the sidelobes. This requires a servo loop that maintains cancellation under all expected conditions, hence the names *adaptive* cancellation and *adaptive* null steering. Also, because most interfering signals are not monochromatic, and the delays and filter characteristics in the main and auxiliary signal paths are necessarily different, the auxiliary signal must have its phase and amplitude conditioned as a function of frequency. This requires an adaptive filter, usually in the form of a tapped delay network, in the auxiliary signal path.

Figure 1, which is adapted from a figure in Barnbaum and Bradley (1998), shows the block diagram of a simple adaptive cancelling arrangement. The servo criterion for the adaptive algorithm is to minimize the output of the summing network by adjusting the weights of the tapped delay network. More than one interfering signal can be cancelled with this simple arrangement as long as the signals are at different frequencies and there are a sufficient number of taps in the delay network. Lab measurements have demonstrated at least 70 dB of

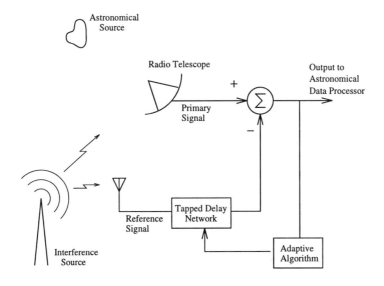

Figure 1.   Basic adaptive cancellation scheme.

signal cancellation with this scheme, and preliminary tests on the NRAO 43-m telescope showed suppressions on FM broadcast signals of up to 20 dB. A number of improvements in the adaptive hardware are required before attempting to understand the fundamental limits of this technique for astronomical observations. A second reference antenna and adaptive filter will be needed to cancel signals in the orthogonal polarization.

One question that remains to be answered is how effective the adaptive cancellation will be on weak RFI signals that are still strong enough to affect astronomical measurements. Barnbaum and Bradley showed that the signal-to-noise ratio of the interfering signal in the auxiliary channel must be 10 dB for every 20 dB of suppression required. This is due to the small DC offset in the error signal caused by white noise in the auxiliary channel. My intuition tells me that there is another limit imposed by the noise in the main channel in the sense that the adaptive servo cannot set the delay tap coefficients any more accurately than is allowed by the integration time of the adaptive loop. If the astronomical integration time is much longer than the inverse of the loop bandwidth, we may see a residual RFI signal that cannot be cancelled by the adaptive filter. We plan to test this with the current hardware.

The adaptive array null steering technique is being investigated by the Square Kilometer Array development group at the NFRA in Holland. They have prototyped an eight element array and demonstrated the basic concept in their antenna range (Goris 1997). This work has posed a number of interesting questions that need to be answered before we see a practical implementation of null steering in a radio astronomy array.

Another important question that needs to be answered for any RFI cancelling scheme is how much does the adaptive servo affect the *stability* of the astronomical measurement? We know that the auxiliary signal channel will add noise with frequency structure and that adaptive nulling in an array has the potential for affecting the main beam gain and shape. Can these be sufficiently controlled with constraints on the adaptive algorithm or by suitable calibration methods?

6.3.   Correlation Removal

Ekers and Sault (1997) have suggested a method for subtracting interference from single dish data using the cross correlation of signals from elements in a focal plane array. If one assumes that the astronomical signals are essentially uncorrelated between the array elements and that the number N of interfering sources and the number M of array elements are related as $N \leq 2M - 1$, then the gain of each element in the direction of the interfering sources can be computed using a method analogous to self-calibration of synthesis array data. Using these computed gains, the interfering signals may be subtracted from the autocorrelation (total power) of each element. Initial trials of this technique with the Parkes 13-beam, 21-cm receiver were encouraging, but, as with so many interference subtraction attempts, one is left with a puzzle about why it works less well than expected.

Ekers and Sault pointed out that a more accurate interference subtraction might be obtained if some of the receiver array elements were deliberately pointed at the sources of RFI. This is somewhat analogous to using several reference antennas to sample the interference for the purpose of nulling with an adaptive filter as described above.

Interference subtraction with correlation data is a power subtraction. Hence, it requires an accurate gain calibration to be effective. This may be a disadvantage when compared to the adaptive filter technique, which does its subtraction in the voltage domain for which there is a definitive null criterion for success. Conversely, the correlation technique might lend itself to a global solution of interference subtraction, using all of the data in an astronomical integration to overcome the noise limitation of the adaptive filter loop bandwidth. At this point these are just speculations on my part.

6.4.   Other Experiments

Jon Bell (ATNF), Rick Smegal (SETI Institute), and collaborators have recorded baseband data of interfering sources at the Australia Telescope Compact Array (ATCA) and are in the process of recording correlation data of interference on the Parkes 64-meter antenna. The Compact Array data are now available on CDROM for use by anyone who would like to experiment with RFI subtraction algorithms. The Parkes data will be similarly published.

One distinguishing feature of all man-made signals is that they are fully polarized at their point of origin. Single path propagation effects (reflection, diffraction, and refraction) and the response of a radio telescope's sidelobes will change this polarization to a different elliptical polarization, but the polarization fraction should remain 100%. To test this I tried recording the four Stokes parameters of the many spectral features of an 80 MHz commercial TV signal with the NRAO 43-m telescope. Subtraction of the root sum of the polarized Stokes parameters from the total intensity, $I - \sqrt{Q^2 + U^2 + V^2}$, should eliminate the TV signal. The measured results were that the TV signal was reduced between 0 and 20 dB, depending on the spectral feature of the signal. This leads me to suspect that multipath propagation coupled with the wide TV signal bandwidth is causing depolarization of the signal. This could be tested by doing a similar experiment on narrow band signals at higher frequencies, for which depolarization should be less severe. If multipath propagation is a prevalent phenomenon, any adaptive cancelling of correlation subtraction techniques will need to take this into account.

My guess is that we must understand propagation effects on RFI signals before our cancelling and null steering techniques can reach their full potential. Some of what we need is in the engineering literature, but the unique nature of radio astronomy measurement will likely require advances in the general understanding of propagation. This may very well become an interesting cross-disciplinary research area.

7. Communication of Research

My apology to anyone in radio astronomy whose work in the areas of RFI monitoring and excision I have not mentioned. Part of this is due to my incomplete survey of the literature, but a bigger problem is that many efforts are in internal reports or possibly never written up. Much of this work deserves a wider distribution, even if not at the level of a journal article. To this end we are starting a Internet web site and email distribution list to which we invite contributions and subscription. You can contact me at *rfisher@nrao.edu* or Jon Bell at *Jon.Bell@atnf.csiro.au* for more information.

References

Barnbaum, C. and Bradley, R. F. 1998, AJ, 115, 2598

Ekers, R. and Sault, B. 1997, "Notes on estimation of interference using arrays", ATNF Internal Memo, 4 April 1997

Goris, M. J., 1997, "The adaptive beamformer of the SKAI adaptive antenna demonstrator", NFRA Technical Report - 459/MG/CB/V2.2

Kasper, B. L., Chute, F. S. and Routledge, D. 1982, MNRAS, 199, 345

Leshem, A. and van der Veen, A-J., 1999, "Detection and blanking of GSM interference in radio-astronomical observations", IEEE Workshop on Signal Processing Advances in Wireless Communication (SPAWC), Annapolis, Maryland, May 1999

Morgan J. V. and Fisher, J. R. 1977, "An impulse noise suppressor for continuum radiometry", NRAO Electronics Division Internal Report, No. 178

Widrow, B. and Stearns, S. D. 1985, Adaptive Signal Processing (Englewood Cliffs, NJ; Prentice Hall)

Preserving the Astronomical Sky
IAU Symposium, Vol. 196, 2001
R. J. Cohen and W. T. Sullivan, III, eds.

Radio Interference and Ejecting Techniques at Beijing Astronomical Observatory[1]

X. Zhang[2], T. Piao, B. Peng[2], X. Wang

Beijing Astronomical Observatory, CAS, Beijing 100012, China
National Astronomical Observatories, CAS, Beijing 100012, China
E-mail: zxz@class1.bao.ac.cn

Abstract. In this paper, we first describe the situation of radio interference at the Miyun station in Beijing Astronomical Observatory, and then new developments in both hardware and software techniques of interference rejection for the Miyun Synthesis Radio Telescope (MSRT) are described.

1. Introduction

The Miyun Synthesis Radio Telescope (MSRT) was built in 1967 on the north bank of the Miyun reservoir, about 120 km from the Beijing city. There was a very quiet radio environment at that time, at its operating frequency of 232 MHz. But radio interference has become very serious as the economy has developed. In the 1980s radar and taxi calling telephones were the main interference sources. Later on, TV transfer stations, cellular telephones, water and electricity power monitoring systems, PCs, and so on became new manmade radio interference sources. In 1994, due to the very serious interference, the MSRT had to stop its daily observations. While the MSRT staff appealed to the National Radio Management Committee of China to solve the problem, some anit-interference hardware and software were developed. One attempt is that 28 surface-acoustic-wave filters (SAWFs) were added to the telescope to reduce the effects of bad interference at 234.5 MHz. Although it is just out of the protected frequency band, it is so strong that it produced big trouble in the telescope receiver. The properties of the SAW filter and some examples of interference are demonstrated in this paper.

For interference within the MSRT band, software was developed to eject the interference from the observational data so that a much better map can be obtained. The software was designed based on the differences of phase and amplitude properties between interference and astronomical objects. Some results before and after using the software are presented.

---

[1]The project is supported by the National Natural Science Foundation of China and the National Climbing programme of China.

[2]Visiting Astronomer, MPI Bonn, Germany

2. New Technical Developments on Interference Rejection

The working frequency of the MSRT is 232 MHz and its bandwidth at the 3-dB level is 1.5 MHz, i.e. from 231.25 to 232.75 MHz. In the 1990s after the problems from radar and car-calling-telephone systems were solved, strong interference then was caused by telemetry services for flood, earthquake and electricity monitoring. Among them the worst interference, from the electricity monitoring system, is about 30 dB stronger than the receiver system level at 234.5 MHz. In addition to this, other interferences are from inside the telescope control room: (a) the antenna-control-computer gave interference about 10 dB higher than the system background noise at 236.28 MHz; (b) the correlation shell of the receiver radiated on average 7-8 dB high interference in the frequency range from 235 to 236 MHz; (c) from the GPS clock system at frequencies 230.63 and 235.5 MHz, about 10 dB high interference signals were found. To eliminate these interference effects, new hardware and software techniques were developed.

2.1. Hardware Techniques

We have two kinds of hardware techniques to constrain the interference. 1) For interference from the PCs, we simply jump the PC-bus-frequency so that its harmonic frequencies will be far from the telescope band. 2) Facing the more serious interference problems, new SAWFs were developed and added to the telescope. The good properties of SAWFs, especially the very steep passband, helps to constrain the strongest interference at 234.5 MHz which is just out of the telescope working band. The left-hand plot in Fig. 1 shows the passband of a SAWF and in contrast the passband of the original filter is shown in the right-hand plot in Fig. 1. Table 1 lists the related numbers and shows clearly the superior anti-interference capability of the new SAWF. The main parameters of the SAWF include: flat passband of ~2 MHz, shape factor ~1.7, passband ripple ~0.2 dB, noise factor ~6.5 dB and output dynamic range ~ 0 dBm.

Table 1. Interference at Miyun station: a comparison of the anti-interference capacity of the new and old systems.

Frequency(MHz)	229.10	230.63	232.0	234.5	235-236	236.28
Interf. source	PC	PC	?	moni. sys.	GPS clock	PC
NEW	-41.7	-20	0	-43	< −45	< −45
OLD	-5.0	-1.0	0	-2.0	-10/-20	-20

2.2. Software Techniques

For interference originating just outside our operating band, the hardware technique mentioned above is very effective. But hardware techniques cannot deal with interference within our band. To solve this problem two software techniques have been developed since 1989 (Yang et al. 1990), but new interference appeared after 1989, and new functions and features have been added to the original software in recent years.

Our new software design is based on understanding the observational data. The relevant data properties are:

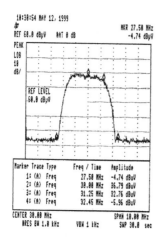

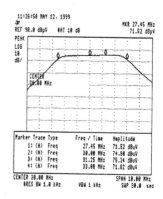

Figure 1.   The left-hand plot shows the passband of a SAWF and the
right the original passband

1) It appears like random noise.

2) For most sky area, signal is much lower than the telescope system noise.

3) For a signal from a radio source, phase and amplitude variation can be calculated.

4) For interference with high intensity and short duration, its effects on the map
plane are long, narrow and straight lines. The interference can be simply edited
out.

5) An important case is that of low level interference of long duration and varying in phase and amplitude. Its effects on the map plane could be very complex.

6) Receiver zero-offset is a special interference. It causes a series of circles of
varying amplitude and radius, centred near the North Celestial Pole.

Based on the knowldge mentioned above, the new software was designed
and then was used sucessfully for the Miyun Radio Sources Survey (X. Zhang
el al. 1997). Fig. 2 gives an example of the recorded data before and after the
interference was rejected.

3.   Results

These new developements in both hardware and software make the MSRT run
successfully. Fig. 3 shows the sky maps before and after the new software has
been applied to the observing data. Its improvement to the map quality can be
easily seen.

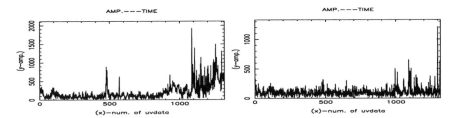

Figure 2. The two diagrams show examples of MSRT data before (left) and after (right) interference was rejected using software techniques.

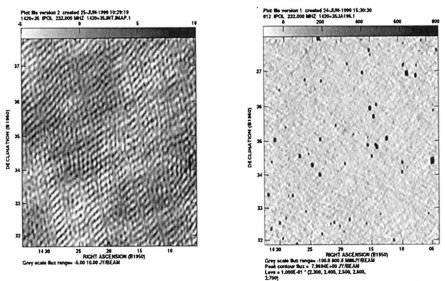

Figure 3. The left and right are sky maps before and after interference was rejected.

References

Yang, Y.P., Zheng, Y.J., Zhang, X.Z., et al., 1990, Proceedings of IAU/URSI Symp. on *Radio Astronomical Seeing*, editors: J.E. Baldwin and S. G. Wang, (Pergamon Press, Oxford), pp.249-254

X. Zhang, Y. Zheng, H. Chen, et al., 1997, A&AS, 121, 59

Preserving the Astronomical Sky
IAU Symposium, Vol. 196, 2001
R. J. Cohen and W. T. Sullivan, III, eds.

Radio Interference Monitoring and Databases

W. van Driel

Unité Scientifique Nançay, Observatoire de Paris, France

Abstract. For astronomers, monitoring of the radio spectrum is an obvious necessity in order to do observations which are free of interfering signals. Recent practice has shown that the installation of dedicated RFI surveillance antennas at radio astronomy sites has virtually become a necessity. Data obtained with such antennas have carried significant weight in discussions with other radio spectrum users. Furthermore, the growing number of dedicated antennas for RFI surveillance at radio observatories has opened the possibility for the establishment of common RFI databases. Such databases can provide objective numbers showing the degradation of the effective use of protected frequencies due to interference, to be used by astronomers as well as by their partners in frequency protection. As it is all too easy to drown in such a sea of data, discussions between astronomers themselves and with their partners are clearly necessary to define the form, implementation and use of such databases.

1. RFI Monitoring

Frequency bands allocated to the Radio Astronomy Service (RAS) - even primary allocations - clearly are increasingly threatened. This is seen in the everyday experiences of astronomers world-wide and in the proposals by active Services for further frequency allocations in bands shared with astronomy or in adjacent bands. Clearly, radio astronomy needs to protect itself, to keep its allocated bands clean and useable for the scientific research for which they were allocated. A crucial step in the complex process of frequency protection is the organised and objective monitoring of our allocated bands for interfering signals (see also R. Fisher, this Volume).

In order to see how much astronomical observations are suffering from radio interference (RFI, or EMI), and to study its increase in time, we first need to define the "pain" in an objective way so we know what to measure. The International Telecommunication Union (ITU) has provided a handle on this aspect with its definition of the level of interference considered "harmful" to radio astronomy observations as "... the level which causes an increase of 10% in the measurement errors, relative to the errors due to system noise alone" (ITU-R Handbook on Radio Astronomy 1995). Note that this means in no way that "radio astronomers should accept the loss of 10% of their data", as it is sometimes misquoted by other spectrum users.

Radio astronomers need measures of the degradation of their instruments' performance made in a consistent, quantitative and objective way on a regular basis. Such RFI data can provide vital evidence in case of conflicts with other spectrum users, when rulings or arbitration are called for. In a strictly legal sense, only observations made at official monitoring stations are acceptable in spectrum regulation matters; in practice, however, observations made at radio astronomy sites of violations of protected bands have been proven to be very useful as well.

In practice, RFI monitoring data can be obtained in several ways:

• Monitoring by monitoring stations of regulatory authorities, like those in Leeheim (Germany) and Baldock (UK) in Europe. These stations can be used to monitor both the main, side-band and spurious emissions of targetted emitters, including satellites. Disadvantages of the presently operating official tracking antennas are their size (about 10 meter diameter), small compared to the size of radio astronomy telescope dishes, and the heavy demand made on them for monitoring all sorts of emitting sources all over the spectrum, limiting the time they can be allocated to monitor RFI in threatened RAS bands. An advantage is their fast tracking capability, allowing on-source integrations of fast-moving Low Earth Orbiting (LEO) satellites. Also, their use is not free of charge, so in case of conflict the question is who pays, the polluter or the victim service?

• Large radiotelescopes themselves can also be used for RFI monitoring. Their advantage is their sensitivity, as these are very large antennas - the largest, Arecibo (Puerto Rico), has a diameter of 300 meter while 100-meter class instruments are in operation at Effelsberg (Germany), Green Bank (Virginia, U.S.A.) and Nançay (France). Their disadvantage is their inability to measure the location and main frequency of an interfering signal, crucial for the determination of the polluter. In fact, their sensitivity makes them virtually omnidirectional RFI detectors, as even their far sidelobes can pick up interfering signals. Also, they are very sensitive to side-band and spurious emissions of emitters in adjacent frequency bands, in which they are not observing, so much so that sometimes they cannot measure the main frequency of the interfering signal (because the sensitive receivers saturate). On the observer's monitoring console these sidelobe-detected, out of band interfering signals are simply seen mixed with the main beam signal of the targetted astronomical source.

• Dedicated RFI monitoring antennas have already been put into operation by a number of radio astronomy observatories. The disadvantage of the systems presently in use is their small size (a few meters in diameter at most) and their inability to track fast LEO satellites, but their advantage is the operation around the clock by the victim service, radio astronomy itself. They can be used for regular monitoring of known sources of interference (such as radars, satellites) as well as searches for interfering signals, either as a general RFI surveillance programme or as a rapid reaction response to an alert given by the site's main radio telescopes.

An example is the Nançay Surveillance Antenna (NSA) operating in central France (see Figure 1). Since 1988, after the launch of the notorious GLONASS satellites, regular RFI monitoring has been carried out at Nançay. The system was mounted on a 22-m high tower in 1992 to allow its unattenuated operation above treetop level, as the Nançay Observatory site lies in the great Sologne forest. The NSA system has two steerable antennas, covering the 100-500 MHz

Figure 1.   An example of a dedicated RFI monitoring antenna at a radio astronomy observatory, the Nançay Surveillance Antenna (NSA), mounted on a 22-m high tower. The 1.8-m diameter parabola is used for RFI monitoring in the 1–3.5 GHz range; the log-periodic antenna covers the 100–500 MHz range.

and 1-3.5 GHz frequency ranges corresponding to, respectively, the frequency coverage of the site's radioheliograph and decimetric telescope.

For RFI monitoring at radio observatories, whether performed with a dedicated monitoring system or picked up with the main telescopes, the registration of the occurrence of RFI is a task for the personnel responsible for the execution of the observations (in general technical observatory staff) rather than for the astronomers whose scientific programs are being executed, as the latter often recover their data later at their own Institute through the Internet.

As mentioned above, the principal goal of RFI monitoring is the measurement of the degradation of the data quality, for which the ITU Regulations provide guidelines. For an effective handling of a potential avalanche of data, automatized RFI recognition and characterisation algorithms listing the relevant parameters of the detected RFI (see below) need to be developed and made available to the astronomical community, recognizing the need to provide homogeneous data to be accumulated into international databases.

2. RFI Databases

Once the Radio Astronomy Service has access to a regular stream of homogeneous RFI monitoring data, whether through official monitoring stations or through its own operations, it is extremely useful to assemble this input into databases. An obvious caveat is that one can easily drown in such a sea of data, while no useful product or conclusion will emerge from our efforts to help protect our allocated frequencies. Our partners in protection, namely the Agencies charged with national frequency allocation, international regulatory organisations and mediators, need clear, objective measured numbers to be able to assist us and to decide on our future access to the Universe.

Though an RFI database can be extremely useful for an observatory in its struggle to keep its local environment clean, gathering data from sites on a national or international level is an obvious goal, as it provides a powerful tool for the Service in frequency protection. In a European framework, a database concept has been established by CRAF (the Committee on Radio Astronomy Frequencies of the European Science Foundation) in its effort to gather information on the RFI situation and its evolution. Recent negotiations on the continued use of the 1610.6-1613.8 MHz band by CRAF, representing all European observatories, for the protection of one of radio astronomy's primary allocations, have shown the strength of international solidarity among observatories and the added value of sharing RFI monitoring data.

The format of the CRAF database is given in Table 1; it contains basic elements needed to quantify a recorded RFI event. The first version of an accompanying software package has also been developped for interrogating the database, which can provide data on

- the frequency of occurrence as function of time (year, month, weekday, hour)

- the percentage of data lost

- the periodicity of a particular interference

- the comparison between different antennas

For further information, see the CRAF Website
http://www.astron.nl/craf/

Table 1.   CRAF RFI database parameters

Field	Field_name	Width (in bytes)	Format description
1	DATE	8	yy-mm-dd
2	STATION	10	up to 10 characters
3	START	5	hh:mm (UT)
4	END	5	hh:mm (UT)
5	ANTENNA	4	(according coding)
6	RFIFREQ	10	EMI bandwidth in MHz (accuracy 0.001 MHz) - format is ffffff.fff
7	BANDWIDTH	10	EMI bandwidth in MHz (accuracy 0.001 MHz) - format is ffffff.fff
8	REP_INTERVAL	4	Repetition interval of pulses (for radars): seconds
9	INTENSITY	6	Intensity of interference
10	INT_UNIT	2	Intensity unit: KE=Kelvin, JY=Jansky
11	RFI_AZ	3	Azimuth of EMI source (if available)
12	RFI_EL	3	Elevation of EMI source (if available)
13	TYPE	2	kind of observation: BR=broadband SP=spectral
14	ANT_AZ	3	Azimuth of observations in degrees
15	ANT_EL	3	Elevation of observations in degrees
16	DEG	3	Degree of degradation in percent
20	Reserved	1	
	Total	80	bytes

In the course of gathering information on RFI in its own protected bands, the radio astronomy service needs to observe outside these bands as its telescope are very sensitive to out of band emission. Thus, information can also be accumulated on spectrum occupancy in other bands. It is clear that such data are liable to non-disclosure, as they cover military and commercial uses. Therefore, access to databases on RFI for radio astronomy, such as the one operated by CRAF, needs to remain restricted.

References

ITU-R Handbook on Radio Astronomy, 1995, Radiocommunications Bureau, International Telecommunication Union, Geneva

Preserving the Astronomical Sky
IAU Symposium, Vol. 196, 2001
R. J. Cohen and W. T. Sullivan, III, eds.

Fixed and Mobile RFI Search Facilities at Medicina

S. Montebugnoli, G. Tomassetti, C. Bortolotti and M. Roma

Institute of Radio Astronomy, CNR, Bologna, Italy

Abstract. Fixed and mobile RFI search facilities at the Medicina radio astronomy station are described. A complex system of wide-band antennas that can be steered, has been designed and installed on the top of a 25-m high tower and connected with a wide-band sweeping receiver system. At present it works from 0.08 GHz up to 2.5 GHz in continuous mode. Low-noise front ends are used in the radio astronomy bands. A mobile system, equipped with a similar receiving system, has been designed to facilitate the interfering transmitter's position localization.

The Medicina radio telescope station comprises two radio telescopes: the Northern Cross 610-m×564-m "T"-shaped antenna and the 32-m VLBI dish (Figure 1). Due to both the high sensitivity and the pollution situation in the radio bands, these systems are very vulnerable to radio frequency interference (RFI). In order to control the frequency bands allocated to radio astronomy, a 25-m high tower, with a proper antenna system and a completely equipped van have been set up since the beginning of the 1980s. The RFI situation control is performed through both the tower and the van (Figure 2).

The antenna system is installed on a tower (25-m height) and it is completely adjustable, over a 360° range in the horizontal plane, from a remote control room. The antenna position is controlled via a 12-bit absolute optical encoder. Both antennas and front ends are user-selectable and depend on the used band. In Figure 2 the overall multi-band antenna system is shown. It is remotely adjustable at a programmable speed with a steering error of about 0.1°. The received signals are sent to a 40-m distant control room (Figure 3). Very low-loss coaxial cable has been widely employed.

The operating frequencies of the tower system are:

- 80/500 MHz: both linear polarizations are available with two log-periodic antennas. The FM is attenuated with a proper filter band.

- 500/1000 MHz: both linear polarizations are available with two log-periodic antennas.

- 325±8 MHz : double yagi system for each polarization. Low-noise frontends are used.

- 408±8 MHz : double yagi system for each polarization. Low-noise frontends are used.

- 608±4 MHz : double yagi system for each polarization. Low-noise frontends are used.

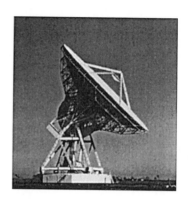

Figure 1.   The Medicina radio telescopes. *Left:* The Cross antenna; *Right:* The 32-m VLBI antenna

Figure 2.   *Left:* The 25-m RFI tower at Medicina; *Right:* The multi-band system installed on the RFI tower.

Figure 3.   RFI control room at Medicina radio astronomy station.

- 1/2.5 GHz : 1.2-m dish with 45° linear polarized feed. A low-noise frontend is used. A future upgrade up to 12 GHz is planned.

The van is a fully equipped mobile RFI monitoring system and a remotely controlled mast extendable up to 12 m is installed on it (for antenna support and orientation). The antenna position is controlled via an incremental optical encoder with programmable offset. A completely autonomous power supply source is installed to allow operations in any situation. A GPS receiver and communication facilities have been installed on board. The operating frequencies of the mobile facility are:

- 310/430 MHz : Log-periodic antenna (linear polarization)

- 603/611 MHz : Log-periodic antenna (linear polarization)

- 1/2.5 GHz : Double-ridged horn or Log periodic antenna (linear polarization)

- 2/18 GHz (four bands): Double-ridged horn (linear polarization)

In Figures 4 and 5, rear views of the van are shown. In the interior view the HP8562A 2 6.5-GHz Spectrum Analyzer, the pneumatic extendable mast, the receiver and the Pentium II PC are visible. The rear sides of the instruments are accessible for connections and/or expansion. These instruments are installed in special shock-absorber crates. All the frontend blocks (tower and van) are fully equipped with low-noise amplifiers and good shaped filters while a coaxial relay set allows the operator to chose the proper antenna.

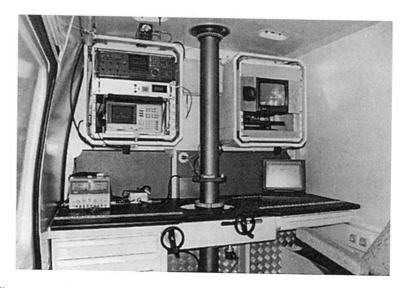

Figure 4.   Medicina RFI van, interior view.

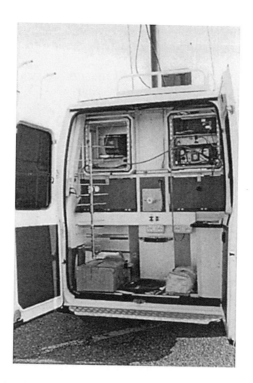

Figure 5.   Medicina RFI van, rear view.

Preserving the Astronomical Sky
IAU Symposium, Vol. 196, 2001
R. J. Cohen and W. T. Sullivan, III, eds.

RFI Sentinel 2

S. Montebugnoli, M. Cecchi, C. Bortolotti, M. Roma and S. Mariotti

Radio Astronomy Institute, National Research Council, Bologna, Italy

Abstract. Nowadays we have a massively increasing use of radio techniques in a wide variety of application fields. Meanwhile state-of-the-art receiver technology dramatically increases the sensitivity of modern radio telescopes. This situation produces a worrying vulnerability of ground-based radio telescopes to Radio Frequency Interference (RFI). In order to monitor the RFI scenario within the frequency bands reserved for radio astronomy activities, a monitoring system, based on a quite new approach, has been developed and is presented here.

1. The System

The main characteristic of the "Sentinel 2" system is to perform an FFT analysis on a down-converted band up to 20 MHz wide, step-tuned under programmed control over a broader frequency range. This approach, unlike that of high-cost commercial sweeping receivers, allows the system to achieve fast scanning while maintaining high frequency resolution.

The system architecture, along with the designed post-processing tools, offers the possibility to deal with different kinds of interfering weak signals independently of whether they are stationary, moving or burst-like. A hypothetical commercial receiver (i.e 80-kHz frequency resolution) might take 23,750 steps to sweep a 0.1-2.0 GHz band. Since each step takes time, the RFI scenario obtained could be very different from the real situation. Both rapidly variable interference and spread spectrum signals are, generally, missed (Figure 1). Scanning the same range with a wideband window (20 MHz) it is possible to compute a power spectrum over a programmable number of channels (i.e. 256) maintaining a similar resolution but drastically decreasing the number of steps (in this case by a factor of 250). If the FFT computation is fast enough, a good RFI scenario, much closer to the real one, is obtained (Figure 2).

The signal coming from the sweeping receiver is digitized with a 12-bit analogue-to-digital (A/D) converter after a proper downconversion to the video band (theoretically 12-bits should allow a 72-dB dynamic range). Data are acquired in burst mode and transferred (via PCI bus) to the PC memory (500-MHz Pentium III powered), then they are block-processed for the spectral analysis. Since the power spectral computation takes time, the analysis is not in real time. In fact, while it is in progress, the A/D converter is stopped, introducing a "hole" in the acquisition beside every working cycle. The resulting duty cycle is, anyway, acceptable because the system is not requested to work in real time.

Figure 1.   Commercial receiver's working mode.

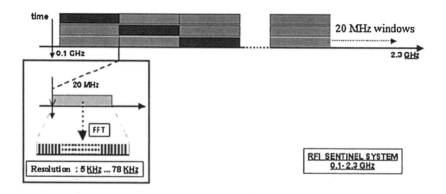

Figure 2.   Sentinel 2 working mode.

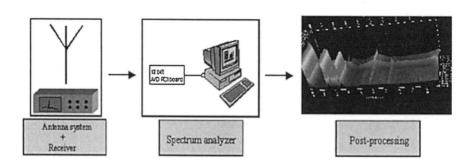

Figure 3.   Sentinel 2 System Block Diagram.

The system architecture is shown in Figure 3. Each *working cycle* is composed, in more detail, of the following steps:

- data acquisition in a burst;

- data transfer from the ADC board buffer to the PC memory;

- complex spectrum computation (Radix 2) for every burst series;

- power spectrum and average computation;

- data storage on a SCSI 2 hard disc;

- local oscillator and antenna selection control (IEEE 488 bus);

- data-base management and display.

The time taken to perform the above steps in the case of bursts of 256, 128, 64 spectra is given in the following table[1]:

Channels	Resolution	Mode	256	128	64
256	78kHz	1 work-cycle (ms)	2,57	3,91	6,34
256	*78kHz*	*Sweep time (s)*	*66*	*52*	*37*
512	39 kHz	1 work-cycle (ms)	3,84	5,25	7,81
512	*39 kHz*	*Sweep time (s)*	*96*	*68*	*53*
1024	19.5 kHz	1 work-cycle (ms)	6,71	7,94	10,5
1024	*19.5 kHz*	*Sweep time (s)*	*162*	*98*	*68*
2048	10 kHz	1 work-cycle (ms)	13,79	14,28	17,09
2048	*10 kHz*	*Sweep time (s)*	*325*	*172*	*106*
4096	4.9 kHz	1 work-cycle (ms)	X	25,88	28,56
4096	*4.9 kHz*	*Sweep time (s)*	*X*	*305*	*172*

Processed data are finally stored in an ASCII database and handled by off-line software to produce an RFI map.

Two monitoring modes are possible:

- Scanning of the whole band 0.1- 2.35 GHz at low system time efficiency;

- Scanning of different radio bands of interest for radio astronomy at a very high system time efficiency.

In the first case a complete overall RFI scenario is obtained. In the second one an RFI scenario within the programmed bands is given. The System is able to detect a very wide variety of RFI with levels spanning from -51 dBm up to 14 dBm (measured at the IF output). Weak RFI needs long integration time, while military spread spectrum signals need wide bands and very short integration time.

---

[1] The higher the burst-size, the faster the system throughput (execution time for a single work-cycle), because there is a fixed time for board management and control that loses relevance as the board memory is exploited.

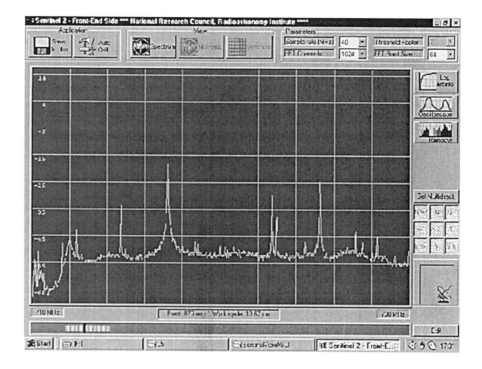

Figure 4.   On-line Processing and Control Graphical User Interface.

Here are the main characteristics of the Sentinel 2 system:

Scanning window size (tuning step)	20 MHz
Dynamic Range:	[-52,+14] dBm
Total frequency range	[100,2350] MHz

- User-friendly on-line control and flexible off-line analysis software

- Modularity

- Less expensive than the commercial monitoring systems

2.   The Software

Software for the Sentinel 2 control and off-line processing was designed and developed *ad hoc* under the Microsoft Visual Studio 6.0 environment. Care was taken to develop a flexible, fast and user-friendly package. We have two applications:

- On-line Processing & Control;

- Off-line Spectrometer.

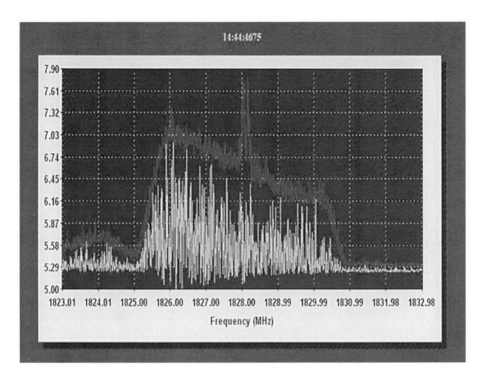

Figure 5. Spread-Spectrum Military Interference: The red (upper) plot represents the maximum channel value over the time period while the yellow plot represents a single spectrum (at 14:44:4675).

The first one is an on-line application in charge of the frontend/backend control, computing power spectra, averaging and data saving. It works under the Windows NT 4.0 operating system and was developed in the Microsoft Visual C^{++} 6.0 environment. The graphical interface (see Figure 4) allows the user both to easily programme the working parameters and to control the system activity.

The second application works off-line. It reads spectra from the database, providing graphical tools for better display. It operates under Windows 95/98 or NT 4.0 and it was developed under Microsoft Visual Basic 6.0 using the Olectra Chart 6.0 ocx library.

A miscellany of different kinds of RFI processed by S2 is illustrated in Figures 5, 6 and 7.

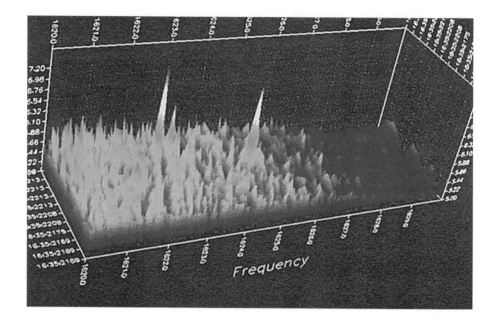

Figure 6.   Iridium Interference at ~1620 MHz over 320 msec of acquisition (3-D view).

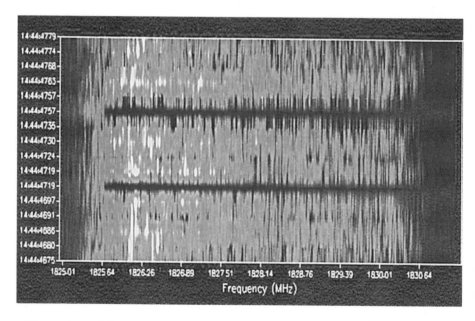

Figure 7.   The same interference displayed in a Time-Frequency chart. The colour represents the intensity of the signal.

Preserving the Astronomical Sky
IAU Symposium, Vol. 196, 2001
R. J. Cohen and W. T. Sullivan, III, eds.

Radio Interference in Astronomical Observatories of China

B. Peng[1], R. Nan[1], T. Piao[1], D. Jiang[2] and Y. Su[1]

*National Astronomical Observatories, Chinese Academy of Sciences,
Beijing 100012, China*

R. G. Strom

*NFRA and University of Amsterdam, Postbus 2, 7990 AA Dwingeloo,
The Netherlands*

S. Wu[1], X. Zhang[1], L. Zhu[1] and X. Liu[3]

*National Astronomical Observatories, Chinese Academy of Sciences,
Beijing 100012, China*

Abstract. We first very briefly introduce the major radio facilities for astronomical research in China, and then report on the present interference situation at major radio observatories. Some of the radio interference problems are caused by paging services, mobile phone satellites, telemetry services for power supply, waterpower and earthquake activity, or radar systems, but some causes are unknown. In the worst case, harmful to radio astronomy, the Sesan VLBI station has not been able to do any observations at 92 cm due to serious radio interference problems since 1992. Still more serious interference coming from satellites can be expected in the next decade. International efforts on frequency protection should be urgently pursued if ground-based radio astronomy is to survive.

1. Introduction

Radio astronomy in China started with solar research about 40 years ago. At present, radio telescopes covering a wavelength range from 3 mm to 1.3 m for astronomical research are in operation at Beijing Astronomical Observatory (BAO), Purple Mountain Observatory (PMO), Shanghai Astronomical Observatory (SHAO), Yunnan Astronomical Observatory (YAO) and Urumqi Astronomical Observing Station (UAO). The Bureau of Radio Management (BRM) in the Ministry of Information Industry is currently responsible for policy-making and implementation of radio frequency allocation in China.

---

[1]Beijing Astronomical Observatory, Chinese Academy of Sciences, Beijing 100012, China

[2]Shanghai Astronomical Observatory, Chinese Academy of Sciences, Shanghai 200030, China

[3]Urumqi Astronomical Station, Chinese Academy of Sciences, Xinjiang 830011, China

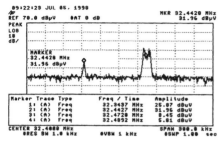 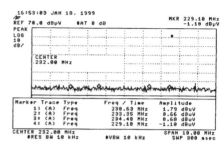

Figure 1.   *Left-hand plot:* Interference at MSRT in July 1998 caused by a telemetry service for power supply, accompanied by mobile communication at 234 MHz
in Beijing, LO=202.00 MHz; *right:* Spectrum in January 1999 after the telemetry
service for power supply, accompanied by mobile communication, was gone.

2.   Major Radio Astronomical Facilities

- The Miyun Synthesis Radio Telescope (MSRT) at BAO consists of an
  East-West array of 28 parabolical dishes, each of 9 m diameter, with a
  longest baseline of 1164 m working at both 232 and 327 MHz since 1984.

- A 13.7 m, mm-wave telescope run by PMO is equipped with acousto-
  optical and multi-channel spectrometer backends. It has been in service
  since 1990 at two frequency bands, 22 and 85-115 GHz.

- Two 25 m radio telescopes, the first built at SHAO in 1987 and the second
  at UAO in 1994, were put into operation, equipped with VLBA and MK3A
  data acquisition systems, respectively, at 5 bands – 92, 18, 13.6, 6 and 3.6
  cm - and with an additional one (1.3 cm) at SHAO.

- Instrumentation for solar radio astronomy consists of

  1. A total power radio telescope working at 2.84 GHz, and two spectrom-
     eters on 7 and 3 m diameter dishes receiving circular polarization with
     multi-channel and frequency sweeping over $1.0 - 2.0$, and $2.6 - 3.8$
     GHz, respectively, at BAO;

  2. Two total power radio telescopes at 2.7 and 9.375 GHz at PMO;

  3. A $230 - 300$ MHz spectrometer on a 10 m diameter dish, which also
     has continuum receivers at 1.42, 2.13, 2.84 and 4.26 GHz, at YAO.

The Solar Radio Broadband Dynamic Spectrometer (SRBDS) project has been
underway since 1994 and will eventually cover the frequency range 0.7—7.6 GHz.

3.   Current Interference Situation at Radio Observatories

There are other irregular and illegal radio services within frequency bands specified for radio astronomy, situated close to observatories.

  1. MSRT at BAO: two kinds of interfering signals were identified as to whether
     they were due to instrumental effects inside the observing room or not.

     Firstly, there is some interference coming from outside the observing room,
     such as (a) a telemetry service for power supply accompanied by mobile

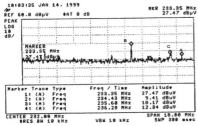 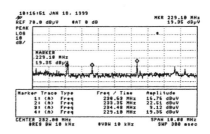

Figure 2. Interference in Jan. 1999 due to bad EMC problems generated by some subsidiary instruments inside the MSRT observing room.

phone communication at 234 MHz, and causing serious interference problems (Figure 1, left-hand plot), although the band 232 ± 4 MHz was said to be protected 20 years ago. After our strong and reasonable arguments with the BRM, this interference has been eliminated according to the regulations of the state management and allocation of radio frequencies, as shown in Figure 1(right-hand plot); (b) some interference around 232 MHz, caused by a military radar system and telemetry services for waterpower and earthquake activity, has been successfully removed in the same way; (c) some interference at 623.37 and 630.0 MHz of 36.12 dBμV and above 21 dBμV respectively, can be well suppressed by taking defensive measures.

Secondly, some internal interference (Figure 2) is actually due to problems generated by subsidiary instrumental effects inside the observing room, such as a PC controller at 236.28 and 229.10 MHz (signal levels of 14.18 and 15.50 dBμV), a GPS clock at 230.63 and 235-236 MHz (15.50 and 18.17 (maximum) dBμV), unshielded correlators at 233.35 MHz (26 dBμV) and at 235-236 MHz an average noise increase of $\sim$ 7.5 dBμV. Altogether this led to a total system noise increase of $\sim$ 9 dBμV.

2. VLBI station at SHAO: there is occasionally unknown interference in some VLBI bands. For instance, some harmful interference at L band in 1997 was recorded in Figure 3 (left-hand plot). After taking some defensive measures in 1998, such as sending the amplified RF signal directly to the observing room, and doing mixing between the RF and LO in the observing room with extra filters, the interference was clearly eliminated (Figure 3, right-hand plot). However in the worst case, which has been very harmful to radio astronomy, the Seshan VLBI station has not been able to do any observations at 327 MHz since 1992 due to serious radio interference mainly produced by paging transmitters.

3. VLBI station at UAO: Some interference of unknown origin shows up occasionally in some directions at 327 MHz. Recent interference monitoring in March 1999 demonstrates that there is no problem at other wavebands.

4. 13.7 m telescope at PMO: there has been no harmful interference.

5. Solar radio telescopes: there are several wide-band paging services around 150 MHz in Kunming city, leading to interference problems at YAO. In addition, we have built wideband solar radio spectrographs, covering most

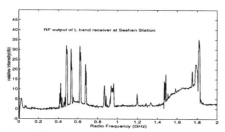 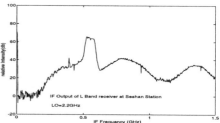

Figure 3.   *Left-hand plot:* Radio interference at 0—2 GHz in 1997 at SHAO. The high continuous spectrum around 1.67 GHz is due to the operation of the preamplifier; *right:* Spectrum on Feb. 6, 1999 was plotted after taking some defensive measures (LO=2.2 GHz). The peak is the passband at L band for VLBI observing.

frequencies from 1 to 12 GHz. It is not reasonable to ask for protection for all of these from BRM.

6. Interference from Space: Satellite TV programme broadcasts cover the whole territory almost 24 hours a day. The Communication and Broadcasting Satellite Company of China was established in 1996. New ground-based stations for satellite communications have been built in Beijing and Shanghai, operating in the frequency range of 1610 — 1626.5 MHz. Iridium satellites have been and will continue to be launched regularly in China for the Motorola Company. More serious interference coming from space is expected in the next decade, especially in one of the most important radio astronomy bands – the OH band close to 1.62 GHz.

4.   Summary of Frequency Protection Efforts

A State Standard, "Protection Criterion for the Radio Astronomy Service", has been submitted to be one of the laws for radio management in China.

Some efforts have also been made by the Chinese Astronomical Society, such as making proposals for designating a radio quiet zone in Guizhou province for a potential site of a Five-hundred-meter Aperture Spherical Telescope (FAST); appealing to the public, news media and concerned institutions to improve frequency protection for radio astronomy use; organizing national meetings to discuss issues or measures on protecting astronomical frequencies. On the other hand, new techniques have been proposed to develop telemetry or remote control instruments via power feeders instead of radio waves.

We have maintained frequent cooperation and communication with CRAF and IUCAF, exchanging information and participating in some international conferences. Cooperation with CORF takes place occasionally in some special cases. We have actively joined discussions concerning interference from satellite systems in the OECD Megascience Forum Working Group on Radio Astronomy in the past few years. We also co-proposed and signed the radio observatory directors' "Kyoto Declaration". Finally, the Asia-Pacific Commission of Radio Astronomy Frequency is to be established to strengthen efforts and collaboration on frequency protection for radio astronomy in this region.

Acknowledgments.   B. Peng thanks the MPIfR in Germany and the IAU travel grants programme for supporting his attendance at this symposium.

Preserving the Astronomical Sky
IAU Symposium, Vol. 196, 2001
R. J. Cohen and W. T. Sullivan, III, eds.

Measurements of Radio Interference at Solar Radio Stations in Beijing

Yihua Yan[1,2], Qijun Fu[1], Yuying Liu[1] and Zhijun Chen[1]

Beijing Astronomical Observatory, Chinese Academy of Sciences, Beijing 100012, China

Abstract. Shahe Station of our Solar Radio Group has suffered from radio interference in recent years, so we decided to move our solar radio telescopes to Huairou Station of BAO. We measured radio interference at both sites recently and found that the radio interference is more serious in Shahe than in Huairou. Although the interference is low at the single working frequency, we do find some radio interference within the working band at Shahe. It is comparatively radio quiet at Huairou and suitable for placement of the solar radio instruments there.

1. Introduction

The study of solar radio astronomy at Beijing Astronomical Observatory (BAO) began in 1959 when a solar radio telescope at 3.2 cm wavelength made in the former Soviet Union was sent to BAO after the observing campaign at Hainan Island in south China by a Joint China-Soviet Union Eclipse Team. Later the antenna was returned to the Soviet Union and in the early 1960s, the first solar radio telescope made at home with a 3.2 cm wavelength calibration system was developed and put into operation at Shahe Station for predictions of solar activity. A solar radio flux observing system at 10 cm wavelength was established in 1970 and began measurements at Shahe Station. Then the Solar Radio Group began to study rapid variations of solar radio emission and in 1981, a solar radio fast recording system at 2840 MHz with a temporal resolution of 1 ms was put into operation (Zhao & Jin 1982, Jin et al. 1986). Since 1983 it has served as an important element in a network all over the country for observing solar radio emission simultaneously with high temporal resolution, and considerable achievements have been made in the 22nd solar cycle. In January 1993, a solar radio observing system was established at Zhongshan Station in Antarctica.

Since 1994, we have been developing a broadband solar radio spectrometer with a frequency coverage of 0.7-7.6 GHz, a frequency resolution of 1-10 MHz, and a temporal resolution of 1-10 ms (Fu et al. 1995). This instrument is composed of 5 spectrometers, covering 0.7-1.4 GHz, 1.0-2.0 GHz, 2.6-3.8 GHz,

---

[1]National Astronomical Observatories, Chinese Academy of Sciences, Beijing 100012, China

[2]Chinese Academy of Sciences-Peking University joint Beijing Astrophysics Center, Beijing 100871, China

4.0-5.2 GHz, and 5.2-7.6 GHz. The three spectrometers at 1.0-2.0 GHz, 2.6-3.8 GHz, and 5.2-7.6 GHz are located at Shahe Station. The other two are located in two cities: Kunming and Nanjing, about 2000 km and 1000 km away from Beijing, respectively. The 1.0-2.0 GHz spectrometer has been in operation since January 1994 and the one at 2.6-3.8 GHz since September 1996. The other parts are scheduled to finish this year. At present, we routinely provide solar radio observations at 2840 MHz for inclusion in *Chinese Solar-Geophysical Data* published by Beijing Astronomical Observatory, and in part II of *Solar-Geophysical Data* published by National Geophysical Data Center at Boulder, Colorado.

Shahe Station has suffered from radio interference in recent years and we decided to move our solar radio telescopes to Huairou Solar Observing Station of BAO, about 50 km away in a northern suburb of Beijing. We measured radio interference at both sites recently and found that the radio interference is more serious in Shahe than in Huairou. Although the interference is low at our single working frequency, we do find some radio interference within our working band. It is comparatively radio quiet in Huairou and suitable for operation of the solar radio instruments there. Here we report the measurements and analyses of the radio interference at Shahe and Huairou Stations. In Section 2 we describe the measurements and introduce the instruments and their performance, and in Section 3 we analyze the radio interference at both sites and discuss its influence.

2.   Measurements and Instruments

2.1.   Instruments

The measurements of radio interference at Huairou and Shahe Stations of the BAO were conducted on 9 and 10 December, 1998, respectively. Figure 1 shows the location of both stations and the diagram of the testing system. The characteristics of the testing system are listed in Table 1.

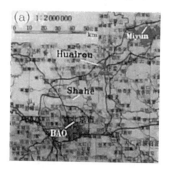
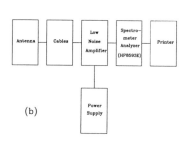

Figure 1.   (a) Locations of the Headquarters of BAO, as well as Shahe and Huairou Stations. (b) Block diagram of the testing system.

2.2.   Interference Measurements

Firstly, all instruments were connected according to the system diagram shown in Figure 1. At Huairou Station, the antenna was mounted on the top flat of

Table 1.   Performance of the testing system

Item	Type	Frequency range	Performance
Log-periodic dipole antenna	LPDA-9531	0.2-1.0 GHz	gain: -18 dB to +6 dB
Parabolic reflector antenna	1m diameter	1.0-2.0 GHz 2.0-4.0 GHz 4.0-8.0 GHz	gain: 20 dB gain: 28 dB gain: 30 dB
Cable losses		all band	∼ 3 dB
Low Noise Amplifier (LNA)		0.2-1.0 GHz 1.0-2.0 GHz 2.0-4.0 GHz 4.0-8.0 GHz	gain: 50 dB; noise: 1 dB gain: 50 dB; noise: 1 dB gain: 50 dB; noise: 1 dB gain: 50 dB; noise: 1 dB
Spectrometer Analyzer	HP8593E	Sensitivity: -109 dBm (at 10 KHz resolution)	

the tower of the Solar Magnetic Field Telescope, whereas at Shahe Station, it was mounted on the roof of a two-story building. Before measuring the radio interference the self-testing procedure of the system was executed to ensure that the system was in good order. Then the elevation of the antenna was adjusted to 0 degrees. For each frequency band the antenna was scanned at 5 degree intervals in azimuth, and the maximum radio interference was carefully identified and recorded. The horizontal and vertical polarizations were measured repeatedly at these maximum values of the interference.

3.   Analysis of the Radio Interference

More than 60 plots of the original radio spectral distributions were thus obtained and are collected in a technical report (1998). Figure 2 shows the radio interference measured at Shahe and Huairou stations. Interference mainly comes from cellular communications below 1 GHz, the Multichannel Microwave Distribution System (MMDS) around 2.6 GHz, and other microwave interference from Beijing and Huairou county's directions. Fortunately the MMDS for cable TV in Beijing is to be replaced by a fibre-optic system, which might improve the situation in this band. It is comparatively radio-quiet at Huairou and suitable for operation of the solar radio instruments there. The influence of radio interference on the solar radio observations and the measures to reject or eliminate it are discussed in Yan et al. (2001).

A document entitled "Interference Protection Criteria for Frequencies used by the Radio-astronomy Service" has been drafted for several years, but it is still not approved to become a national standard in China. We hope that the present concern to preserve the astronomical sky may help China to have its first national standard on frequency protection for radio astronomy.

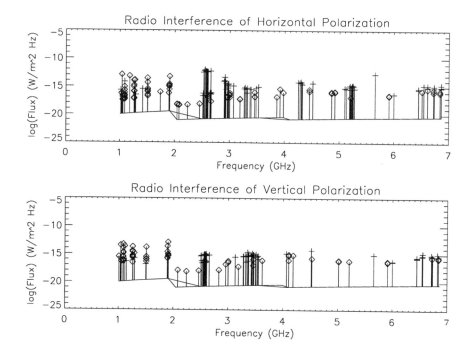

Figure 2.   Flux of vertical and horizontal polarizations of radio interference measured at Shahe (+) and Huairou (◇) Station.

Acknowledgments. The Solar Radio Broadband Dynamic Spectrometer is supported by CAS and NNSF of China. We acknowledge Profs. Chen H., Ji H. and Piao T. for kind help. The measurement of radio interference was pursued by the Technical Service of NRMC of China under a Contract with the Solar Radio Group of BAO.

References

Fu, Q., Qin, Z., Ji, H. and Pei, L. 1995, Solar Physics, 160, 97.

Jin, S., Zhao, R. and Fu, Q. 1986, Solar Physics, 104, 391.

Report of the Radio Interference Measurements, National Radio Monitor Center of China, December, 1998

Yan, Y., Ji, H., Fu, Q., Liu, Y. and Chen, Z. 2001, these proceedings.

Zhao, R. and Jin, S. 1982, Scientia Sinica (Series A), 25, 422.

Preserving the Astronomical Sky
IAU Symposium, Vol. 196, 2001
R. J. Cohen and W. T. Sullivan, III, eds.

Analysis of Solar Radio Observations and the Influence of Interference

Yihua Yan[1,2], Huirong Ji[1], Qijun Fu[1], Yuying Liu[1] and Zhijun Chen[1]

Beijing Astronomical Observatory, Chinese Academy of Sciences, Beijing 100012, China

Abstract. Based on the data observed by a solar radio spectrometer at 2.6-3.8 GHz and other measurements at Shahe Station, we analyze the radio interference and its influence on observations. We have identified three different types of interference: (a) from antenna; (b) from IFs; and (c) internal signals, etc. Corresponding measures can be taken to reject or correct these errors and to improve the observations.

1. Introduction

Solar radio spectrometers covering 1.0-2.0, and 2.6-3.8 GHz have been in operation since 1994 and 1996 respectively at Shahe Station of Beijing Astronomical Observatory (Fu et al. 1995). Shahe Station has suffered from radio interference in recent years. In this paper we analyze the radio interference and its influence on observations, and suggest measures to reject or correct for interference and so improve the observations.

2. Data Description

The data employed in the present work include the following materials taken at Shahe Station:

(i) Measurement of the radio interference in the 0.1-7 GHz range (Yan et al. 2001).

(ii) Long-term observational data at 2.6-3.8 GHz taken using the radio spectrometer with 0.2 s temporal resolution.

(iii) Measurements of radio interference at intermediate frequency (IF) range.

(iv) Data at 2.6-3.8 GHz taken by the radio spectrometer with 0.2 s temporal resolution under the following conditions:

---

[1]National Astronomical Observatories, Chinese Academy of Sciences, Beijing 100012, China

[2]Chinese Academy of Sciences-Peking University joint Beijing Astrophysics Center, Beijing 100871, China

(a) one end of the IF cable was connected to the indoor receiver but the other end was disconnected from the outdoor device and in open circuit;

(b) one end of the IF cable was connected to the indoor receiver but the other end was disconnected from the outdoor device and connected with a matched load;

(c) no IF cable connected to the receiver at all.

From these data we can identify the sources and characteristics of radio interference, which are discussed in the next section.

3.   Sources and Characteristics of Radio Interference

The performance of the 2.6-3.8 GHz radio spectrometer is shown in Table 1 and its block diagram is shown in Figure 1. The sources of radio interference may be summarized as follows: (a) from the antenna; (b) from transmission lines that connect the outdoor device and the indoor receiver; and (c) internal interference signals.

Table 1.   The 2.6-3.8 GHz Solar Radio Spectrometer

Frequency range	2.6-3.8 GHz (operated since Sept. 1996)
Temporal resolution	8 ms
Frequency resolution	10 MHz (120 channels)
Sensitivity	2% $S_{\text{quiet Sun}}$
Polarization	LHCP, RHCP
Observing time	22-10^h UT(Summer), 0-8^h UT(Winter)

We identified the category of the interference by checking: (a) whether the intensity of the interference varied when rotating the antenna; (b) whether the sun, the sky background, the calibration noise source, or the terminal signal, suffered interference; and (c) whether the interference had repeatability or similarities under different conditions.

We found that interference introduced from the antenna frequently occurs, either when observing the sun or the background, or rotating the antenna, but it does not have regularities over long-term observations. When the instrument is connected to the noise source or terminal load, this kind of interference does not occur.

Interference introduced from IF transmission lines is not influenced by the rotation of the antenna and it may occur whenever observing the sun, the background, or connecting to either the noise source or the terminal load. It may exhibit repeatability, but this does not frequently occur.

Internal interference is not influenced by the rotation of the antenna and occurs occasionally. It can occur when observing the sun or the background, or when connecting to either the noise source or the terminal load.

The effects of the radio interference on the observations are as follows:

(1) Microwave interference introduced from the antenna influences data from 0-10 channels.

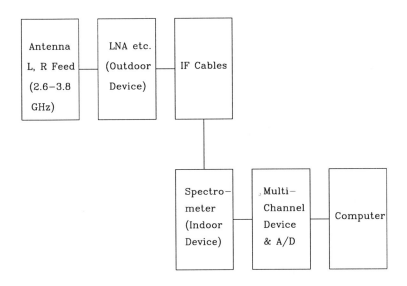

Figure 1.   The block diagram of the solar radio spectrometer at 2.6-3.8 GHz.

(2) Interference introduced from the IFs influences data of 28-29 channels. For example, the environment noise levels are respectively -39.7 dBmW at 489.3 MHz and -48.2 dBmW at 496 MHz.

(3) The internal interference, e.g., due to the computer and/or interfaces influences data in one particular channel, 106.

4.   Measures to Eliminate Interference

Hardware Methods:

(1) Measuring the environment interference and then choosing the frequency range of the radio instrument to escape from the worst interference band. Using an antenna with low sidelobes and backlobes, and adding resistance filters when necessary.

(2) Selecting a suitable intermediate frequency to get rid of IF interference. Compressing the IF band as small as possible. Increasing IF transmission power. Making the IF cable as short as possible with good sheath, ground and connectors.

(3) Using absorbing and separation techniques to eliminate leakage signals, if any, and eliminating interference from the computer.

Software Methods:

(1) Using data processing techniques to eliminate known fixed interference.
(2) Using wavelet transform for noise depression and edge enhancements.

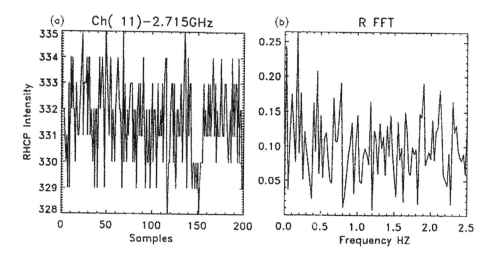

Figure 2.   Signal when pointing to the background shows typical white noise feature and no significant structure due to interference. (a) recorded signal of right-hand circular polarization at channel 11. (b) its FFT spectra (arbitrary unit).

Having taken into account the above measures to eliminate errors and to improve the observations, the Solar Radio Broadband Dynamic Spectrometer works quite well. Figure 2 shows an example of a normal signal that was not affected by any interference when pointing to the background. The data shows typical white noise features. Therefore in this channel there is no significant structure due to interference.

Acknowledgments.   The Solar Radio Broadband Dynamic Spectrometer is supported by CAS and NNSF of China. We acknowledge Profs. Chen H. and Piao T. for kind help.

References

Fu, Q., Qin, Z., Ji, H. and Pei, L. 1995, Solar Physics, 160, 97.
Report of the Radio Interference Measurements, National Radio Monitor Center of China, 1998
Yan, Y., Fu, Q., Liu, Y. and Chen, Z. 2001, these proceedings.

Preserving the Astronomical Sky
IAU Symposium, Vol. 196, 2001
R. J. Cohen and W. T. Sullivan, III, eds.

GPS Satellite Interference in Hungary

T. Borza and I. Fejes

FÖMI Satellite Geodetic Observatory, H-1373 Budapest, Pf. 546, Hungary

Abstract. Civil users of the NAVSTAR Global Positioning System (GPS) in Hungary occasionally experience interference at the 1575.42-MHz GPS signal frequency. As the application of the GPS technique spreads rapidly in our country, radio frequency interference (RFI) should be considered a serious threat. The new geodetic control network (OGPSH) in Hungary is based on GPS measurements and incorporates more than 1100 sites. The paper reports the experiences gained during the establishment of the network. Interference sources were tracked to ground-based digital data transmissions for telecommunications, which operate mostly in the Western part of Hungary. Telecommunication regulations exceptionally allow such transmissions in specified countries. In order to warn potential GPS users, the interference sources are being mapped.

1. Introduction

Radio frequency interference (RFI) can seriously affect GPS applications, particularly in geodetic field work. In such cases, the unsuspecting operators of geodetic GPS receivers may experience lock-on failures, nonsense error messages or total malfunctioning of their receivers. Several authors have already reported GPS interference in different countries (e.g. Butsch 1997, Haagmans 1994). In this paper we report such disturbances from Hungary. Identification and mapping of the interference sources is of national interest in order to inform GPS users about potential dangers.

In 1995 and 1997 a GPS-based geodetic control network (OGPSH) was established, consisting of more than 1100 sites well distributed in the territory of Hungary. The average distance between the sites is about 10 km. During the field measurements, our engineers experienced "inexplicable" receiver failures at some locations, which later proved to be caused by RFI. In this paper we report our preliminary investigations and identify some of the interference sources. Clearly more detailed study of the GPS interference problem is necessary to map RFI disturbing GPS applications in the country.

2. Spectral Measurements

We received 10 reports of receiver failures, which we decided to investigate in more detail. In these cases the receivers were failing in certain locations, whereas

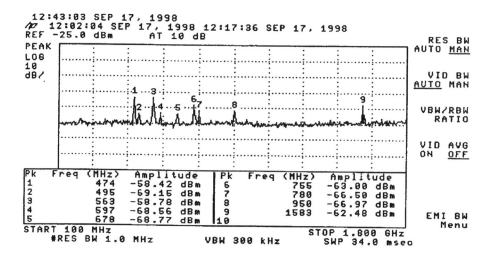

Figure 1. 1700 MHz wide-band spectrum obtained near Lentikápolna using the horn antenna without preamplifier. Interference peaks are numbered.

on moving them to another place, the receiver operations were normal. This was a clear indication of RFI. In the interval March - September 1998 we inspected the reported locations using Trimble 4000 SSE, Trimble 4000 SST, Trimble Pathfinder ProXL and handheld Trimble Scout type GPS receivers. We found serious receiver failures at two locations: near Zalalövő and near Lentikápolna. According to these tests the most sophisticated geodetic receiver (SSE) was the most sensitive to interference while the handheld navigation receiver (Scout) was the least affected in operation. Susceptibility of GPS receivers to RFI has been investigated e.g. by Sluiter and Haagmans (1995).

In September 1998, in collaboration with the Technical University of Budapest, spectral measurements were carried out at Zalalövő and Lentikápolna locations using a Hewlett Packard 8790L type spectrum analyser. Two different antennae were applied. The first one was a wide-band horn antenna without preamplifier. The second was a SEL Alcatel Globos M2000 type GPS antenna equipped with narrow-band filter and preamplifier.

Figure 1. shows a wide-band spectrum in the range 100–1800 MHz, obtained with the first antenna at Lentikápolna. Several peaks can be identified. The peaks below 1 GHz were not pursued further, since their harmonics were not considered to be harmful for GPS signal reception. The peak No. 9 at 1583 MHz, however, falls very close to the GPS L1 signal at 1575.42 MHz. Therefore it deserved a closer look.

Figure 2. shows a 200 MHz wide spectrum centred on 1584 MHz. Peaks of transmissions at 1562.5, 1576.5, 1582.5, 1587.0, 1595.0 and at 1607.0 MHz, with levels in the range -24.54 to -58.74 dBm, are clearly identified. The peaks are superposed on a more than 50 MHz wide-band signal at about -60 dBm. These signals overlap the GPS L1 frequency range.

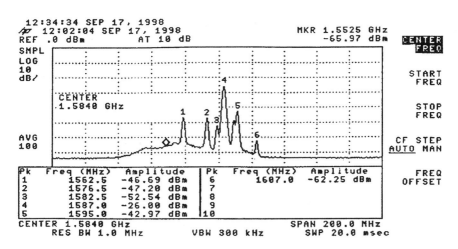

Figure 2.   RFI spectrum overlapping the GPS L1 (1575.42 MHz) frequency near Lentikápolna. Interference peaks are numbered.

Figure 3. shows a 710 MHz wide spectrum obtained near Zalalövő, centred on 1584 MHz. Peaks of transmissions at 1566.4, 1582.3 and at 1591.2 MHz, with levels in the range -34.41 to 30.67 dBm, are clearly identified. Here again the peaks are superposed on a more than 50 MHz wide-band signal at about -60 dBm level. These signals overlap the GPS L1 frequency range. The signal structure is characteristic of digital data transmission. Using the directional sensitivity of the GPS antenna we identified a nearby fixed telecommunication tower as the source of the signal. The disturbing transmission was clearly directional, since, on moving the receivers by a few hundred meters, the interference disappeared.

3.   Are Ground-to-Ground Transmissions Legal at the L1 Frequency?

According to the Radio Regulations of the International Telecommunication Union (ITU RR) the 1559–1610 MHz frequency band is primarily allocated to Aeronautical Radionavigation and Space-Earth Radionavigation-Satellite Services. But this regulation is complemented with a footnote, which adds:

> "*Additional allocation:* in the Federal Republic of Germany, Austria, Bulgaria, Cameroon, Guinea, Hungary, Indonesia, Libya, Mali, Mongolia, Nigeria, Poland, the German Democratic Republic, Romania, Senegal, Czechoslovakia and U.S.S.R., the bands 1550–1645.5 MHz and 1646.5–1660 MHz are also allocated to the fixed service on a primary basis."

So MATÁV uses these frequencies legally and according to the ITU regulations. This means that the L1 GPS signal frequency is not protected in the listed countries, including Hungary.

We have contacted the Hungarian Communication Authority (HCA) asking for more information about this matter. The HCA kindly provided a detailed

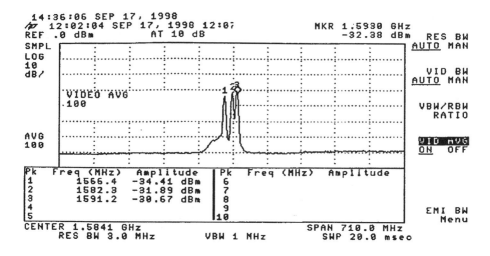

Figure 3.   RFI spectrum overlapping the GPS L1 (1575.42 MHz) frequency near Zalalövő. Interference peaks are numbered.

list of transmitters operating in this frequency band. The HCA also informed us that Hungary is preparing steps to protect the band 1574.5–1576.5 MHz for the sake of GPS users. For the fixed services, new licences will not be issued, and existing licences will not be extended in Hungary. A considerable decrease in the number of cases of interference originating from fixed services on the L1 frequency is projected for the coming years.

4.   Mapping GPS Interference Sources in Hungary

According to the list provided by the HCA, we have found 60 fixed stations operating near the GPS L1 frequency. Figure 4. shows the geographic distribution of these stations in Hungary. The stations dominantly cluster in the Western part of the country, but some can be found South of Pécs and also near the river Tisza in Eastern Hungary.

The precise geographic extension of the RFI around the towers has not yet been mapped. In the course of OGPSH measurements we had to change the position of 3 preplanned sites out of 1150 due to interference. This seems a surprisingly small proportion. Nevertheless, we plan to carry out more detailed investigation of this problem and continue to collect information on GPS interference sources in Hungary.

5.   Planned Activities

We intend to collect countrywide information on the RFI experienced by Hungarian GPS users. More detailed mapping of the extent of interference near the listed transmitters is necessary in order to assess the dangers for GPS users.

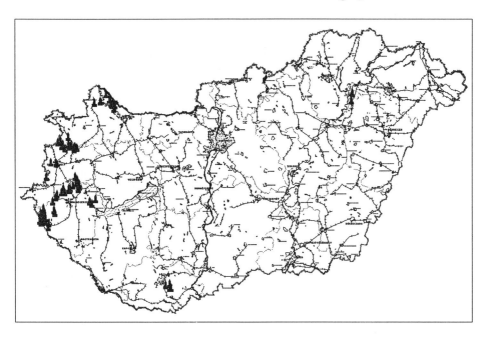

Figure 4.   Fixed transmitters operating in Hungary in the frequency
band 1560–1610 MHz.

Therefore further systematic spectral measurements are necessary. The sensitivity of different GPS receiver types to RFI will be another topic for study. We are aware of the fact that fighting interference is a continuing struggle and new RFI sources can appear unexpectedly anytime, anywhere.

Acknowledgments. The authors thank G. Ijjas (TU Budapest) for contributions to the spectral measurements and P. Tomka (HCA) for providing helpful information on fixed transmitters. This work was supported by the Ministry of Transport, Communication and Water Management under contract HEKI 288/A.

References

Butsch, F. 1997, GPS interference problems in Germany. Proc. ION Annual Meeting, Abuquerque, N.M., USA.

Haagmans, M.E.E. 1994, GPS signal reception problems: the situation in the Netherlands. GPS Niusbrief, May 1994, p. 67–69

Sluiter, P.G., Haagmans, M.E.E. 1995, Comparative test between geodetic Y-code GPS receivers. Susceptibility to radio frequency interference. GPS Niewsbrief, May 1995, p. 11–17

Preserving the Astronomical Sky
IAU Symposium, Vol. 196, 2001
R. J. Cohen and W. T. Sullivan, III, eds.

Protecting Space-Based Radio Astronomy

V. Altunin

*Jet Propulsion Laboratory, California Institute of Technology,
Pasadena, CA 91109*

Abstract. This paper outlines some of the radio frequency interference issues related to radio astronomy performed with space-based radio telescopes. Radio frequency interference that threatens radio astronomy observations from the surface of Earth will also degrade observations with space-based radio telescopes. However, any resulting interference could be different than for ground-based telescopes due to several factors. Space radio astronomy observations significantly enhance studies in different areas of astronomy. Several space radio astronomy experiments for studies in low-frequency radio astronomy, space VLBI, the cosmic microwave background and the submillimetre wavelengths have flown already. The first results from these missions have provided significant breakthroughs in our understanding of the nature of celestial radio radiation. Radio astronomers plan to deploy more radio telescopes in Earth orbit, in the vicinity of the L_2 Sun-Earth Lagrangian point, and, in the more distant future, in the shielded zone of the Moon.

1. Introduction

The explosive development of radio techniques in response to various commercial and scientific applications has resulted in a wealth of data about the Universe obtained through radio astronomy. But it has also created "radio frequency interference" (RFI) that threatens radio astronomy observations from the surface of Earth. This man-made radio noise will also degrade radio astronomy observations taken with space-based radio telescopes. However, any resulting interference could be different than for ground-based telescopes, because of different factors, such as the location of space radio telescopes at relatively large distances from the Earth (the source of man-made radio noise) or even (in the future) on the far side of the Moon, as well as the telescope's orientation relative to Earth.

The frequency sharing and protection considerations associated with space radio astronomy are, in general, more complex than in ground-based radio astronomy because of the requirements not only for observing bands but also for communication links needed to support spacecraft and space radio telescope operations. Moreover, unlike ground-based radio telescopes, space-based radio telescopes are located in close proximity to transmitters and receivers used for spacecraft operations and for data transmission.

The subject of this paper is radio interference to radio astronomy observations originating from man-made systems transmitting radio waves for broadcasting, communications, navigation or other radio "active" services. From the point of view of an astronomer observing with a radio telescope, this man-made RFI is often sporadic in position, intensity, or frequency. This makes it difficult to distinguish sources of natural radio emissions from RFI, and may lead to inaccurate interpretation of an observation. Even more threatening, the powerful interfering radio signals from "active" services may damage the sensitive receivers of a radio telescope. As a result of such harmful interference, loss of radio astronomy data may occur.

Similar to ground-based radio astronomy, radio frequency interference from other services must be taken into account when the space radio astronomy missions and experiments are designed and operated. However, the basic principles of frequency sharing and protection commonly used in ground-based radio astronomy are not always applicable to space-based radio astronomy. In addition, in order to realize the full scientific potential of space radio telescopes, their designers tend to build space receivers with the receiving bands much wider than assigned to radio astronomy by radio regulations. Because of this, the need to minimize a loss of data due to interfering radio signals may have an impact on the mission design (e.g., orbit selection, antenna type).

Space radio astronomy, by its nature, requires international spectrum coordination. Space radio astronomy developments are truly an international effort. International cooperation is driven by the need to share the cost of the mission. International spectrum coordination is also needed because of the orbital location of the space radio telescope as well as the need to provide ground operations support through a significant part of its orbit.

2. Radio Astronomy Observations with Space-Based Telescopes

Space radio astronomy observations are defined as those astronomical measurements conducted in the radio band $f < 3 \times 10^3$ GHz (Radio Regulations 1994, S1.4, S1.5) by coherent (radio) detection techniques (Kitchin 1991) with a space-based radio telescope or network of radio telescopes at least one of which is located in space. Space radio astronomy observations already significantly enhance such radio astronomy fields as:

(i) studies of natural radio sources with the Space Very Long Baseline Interferometry (SVLBI) technique with an angular resolution not achievable with ground-based interferometry,

(ii) studies of the cosmic microwave background radiation with satellite-based observations achieving unprecedented sensitivity by avoiding atmospheric noise and terrestrial RFI,

(iii) studies of natural radiation below approximately 10-30 MHz that are difficult or impossible with ground-based radio astronomy due to the Earth's ionosphere,

(iv) astronomical studies in millimetre and submillimetre wavebands where the Earth's atmosphere significantly attenuates (or completely blocks) the radiation of astronomical sources from reaching the surface.

A few dozen space radio astronomy experiments and dedicated telescopes for studies in low-frequency radio astronomy, space VLBI, and the cosmic microwave background have already flown, while the first submillimetre mission, SWAS, was successfully launched in December 1998. The results from these missions have provided significant breakthroughs in our understanding of the nature of celestial radio radiation. Among them, for example, is the discovery of the cosmic microwave background radiation anisotropy by the COBE satellite in 1993, which provided insights on the Universe as it was about 1 million years after the Big Bang. These first successes of space radio astronomy led to significant efforts to develop the successors to these experiments.

The following subsections summarize some important features of space radio astronomy experiments of the past, present and future. This list does not include early experiments performed mainly in the field of low-frequency radio astronomy, or the solar system radio astronomy experiments performed with interplanetary spacecraft.

2.1. Space VLBI

Space Very Long Baseline Interferometry (SVLBI) potentially can provide a microarcsecond or better angular resolution at radio wavelengths. This can be achieved if the radio interferometer consists of antennas separated by a distance exceeding the Earth's diameter. Today's first generation of SVLBI missions use only one antenna in space, located in relatively low orbits (apogee 20,000 - 100,000 km) and operate simultaneously with a ground-based network of VLBI telescopes. Radio astronomers envision a future where networks of space radio telescopes are located in high-Earth orbit or at the Sun-Earth L_2 point, as well as VLBI radio telescopes located on the Moon.

Table 1.   Space VLBI missions / experiments.

Mission / Experiment	Dates	Orbit	Frequency bands (GHz)	References
TDRSS	1986-1988	Geosynchronous 38,000 km	2.271 - 2.285 15.35 - 15.43	Levy et al. 1986 Linfield et al. 1990
VSOP	12 Feb 1997 (launch)	Elliptical Apogee =20,000 km Perigee = 500 km	1.6-1.722 4.8-5.0 22.2-22.3	Hirabayashi 1998(a)
Radioastron	2002 - 2006	Elliptical Apogee = 78,000 km Perigee = 2000 km	0.32-0.328 1.633-1.697 4.8-5.0 22.2-22.3	Kardashev 1997
VSOP - 2	2005 - 2008	Elliptical Distance from Earth up to 100,000km	4.8-5.0 22.2-22.3 42-44	Hirabayashi 1998(b)
ARISE	2010 - 2015	Elliptical Distance from Earth up to 80,000km	8.0-9.0 21-23 42-44 84-88	Ulvestad et al. 1998
Millimetron	2010 - 2015	Earth-Sun system L_2 point, 1.5×10^6 km from Earth	18-26 45-53 104-112 217-225 266-274	Kardashev 1995

SVLBI astrophysical objectives include studies of the physics of the most compact and remote objects known in the Universe, associated with such energetic events as the nuclear activity in galaxies (compact continuum extragalactic radio sources, megamasers, the compact radio source in the centre of the Milky Way). The SVLBI technique is also useful for studies of the birth, life and death of stars (masers in protostar disks, flaring radio emissions from stars, neutron stars (pulsars), and X-ray binaries associated with the final stages of a star's evolution).

The list of space VLBI missions already flown or in the preparation/planning stage is given in Table 1. The current generation of Space VLBI missions and future missions will also rely heavily on ground support. Radio interference in SVLBI observing bands as well as in SVLBI communication and data channels (phase reference signal transfer, data transfer, and spacecraft operations) can jeopardize SVLBI observations.

2.2. Microwave Studies of the Early Universe

The microwave background radiation discovered in the mid 1960s is crucial to understanding the origin and evolution of the Universe. In fact, it is the only way so far to see what the Universe "looked like" when it was about a million years old (the current age of the Universe is estimated to be about 15 billion years). No electromagnetic radiation can come from the earliest stage of the Universe's evolution because the Universe was so dense and ionized at this stage that all electromagnetic radiation was absorbed by matter.

Table 2.   Missions to study the microwave background

Mission / Experiment	Dates	Orbit	Frequency bands (GHz)	References
Prognoz-9 / Relict 1	07/01/1983- 01/1984	Highly elliptical Apogee = 750,000 km Perigee = 1000 km	37 -37.4	Strukov et al. 1984 Klypin et al. 1992
COBE / DMR	11/18/1989- 12/23/1993	Circular distance from Earth = 900 km	31.25-31.775 52.631-53.458 90.909-91.777	Smoot et al. 1990 Mather et al. 1991
MAP	2000 - 2003	Earth-Sun system L_2 point, 1.5×10^6 km from Earth	18-96	Bennett 1997
PLANCK	2005 - 2010	Earth-Sun system L_2 point, 1.5×10^6 km from Earth	27-33 39.6-48.4 63-77 90-110 122-178 177-257 287.5-418.5 444-646 698.5-1015.5	Bersanelli et al. 1996

This relic radiation has almost an isotropic character and the spectrum of a blackbody with a temperature very close to 2.7 K, with the maximum intensity at millimetre wavelengths. Extremely small variations in the temperature of this radiation (on the order of $\delta T/T = 10^{-5}$), with characteristic angular sizes in degrees to tens of degrees, are believed to contain information on the "primordial seeds" from which all the galaxies in the Universe evolved. Observations of this phenomenon require extraordinary sensitivity which is achieved through the use of wideband cryogenic receivers and hours of integration time for measurement of one data point. Though microwave background studies are also performed from the ground, ground-based observations are ultimately constrained by amplitude and polarization fluctuations due to propagation in the atmosphere. Also, a full-sky mapping of the microwave backround radiation is required for these studies. Ground-based observations with different instruments observing different parts of the sky incorporate calibration and other systematic errors. Only observations from space are free from these constraints. The most spectacular results in this area of astronomy research have been obtained with space-based radio telescopes.

A list of space radio astronomy missions for the study of the cosmic microwave background emissions which have already flown or are under development is given in Table 2.

2.3. Low-Frequency Radio Astronomy from Space

Observations at frequencies below approximately 10-30 MHz are difficult or impossible to conduct from the ground due to the Earth's ionosphere. Information on radio emissions from astronomical objects at frequencies below the ionospheric cutoff (about 6 MHz) has come entirely from spacecraft observations. The first space radio astronomy low-frequency experiments have given us just a glimpse of the wealth of astronomical phenomena expected to be manifested in this band. Key astronomical information has already been obtained on the Sun and the planetary radio emissions, as well as the background radio emission of our Galaxy. It is expected that the next generation of low-frequency space radio astronomy experiments/missions will allow us to better forecast the impact of solar activity on the Earth, to understand the late stages of the evolution of radio galaxies (to discover galaxy "fossils"), and perhaps to discover radio emissions from Jupiter-like planets in extrasolar planetary systems.

Table 3.   Low-frequency radio astronomy missions

Mission / Experiment	Dates	Orbit	Frequency bands (MHz)	References
RAE-1	Jul.4, 1968- Jul. 1972	Earth orbit Circular, 5800 km	0.45 - 9.18	Weber et al. 1971
RAE-2	Jun.15, 1973- Feb. 1976	Lunar orbit 360,000 km from Earth	0.25 - 13.1	Alexander et al. 1975
WIND/WAVES	Jan.11, 1994 (launch)	Halo orbit around L_1 point, 1.5×10^6 km from Earth	0.02-13.85	Bougeret et al. 1995
ALFA	2003 - 2006	10^6 km from Earth	0.03 - 30	Jones et al. 1998

Numerous low-frequency radio astronomy experiments have been performed with spacecraft since 1960. A few of these experiments are listed below (see Table 3). A new mission, Astronomical Low Frequency Array (ALFA), consisting of 16 spacecraft, is being studied by NASA. It will study low-frequency natural radio emissions with a sensitivity and angular resolution orders of magnitude greater than previous experiments.

2.4. Millimetre and Submillimetre Radio Astronomy from Space

Millimetre and submillimetre astronomy observations from the ground are restricted to a few atmospheric windows due to absorption by water vapor and oxygen. Even in these "atmospheric windows" the sensitivity of observations is degraded by the influence of the atmosphere, forcing astronomers to build millimetre and submillimetre telescopes at high altitudes. A space location entirely eliminates these atmospheric effects and also offers access to the frequency bands where the ground-based observations are not at all possible.

The millimetre and submillimetre wavebands are essential to the study of the "cold matter" in the Universe -
(i) cold molecular and dust clouds in our Galaxy, which are the site of the origin of stars,
(ii) dust and molecules in young galaxies at high redshifts in the early Universe, and
(iii) the microwave background radiation, which at the equivalent temperature 2.7 K has its maximum intensity at 150 GHz.
Space-based millimetre and submillimetre radio telescopes will significantly enhance the ability of astronomers to conduct research in these bands and may lead to fundamental astronomical discoveries. This is recognized by the space agencies, which are developing an impressive set of space missions for millimetre-submillimetre studies of the Universe. Table 4 contains a full list of such missions that have been or are expected to be launched in the next two decades.

Table 4.   Millimetre and submillimetre radio astronomy missions

Mission	Dates	Orbit	Frequency bands (GHz)	References
SWAS	1998 - 2000	Earth orbit, Circular, 600 km from Earth	487-493 547- 557	Melnick 1993
ODIN	1999 - 2001	Earth orbit, Circular, 600 km from Earth	118.25-119.25 486-502 541-579	Scheele 1996
FIRST	2005 - 2010	Highly elliptical Earth orbit / 70,600 km, or halo orbit about Sun - Earth L_2 point / 1.5×10^6 km	490 - 642 640 - 802 800 - 962 960 - 1122 1120 - 1250 1600 - 1800 2400 - 2600	Pilbratt 1997
ARISE (single dish mode)	2005 - 2010	Elliptical, Apogee = 100,000 km	50-70	Ulvestad 1998

Most of the past and current experiments have been flown on Earth-, Moon- or Sun-orbiting spacecraft. It is planned that the new generation of space radio telescopes are to be deployed in the vicinity of the L_2 Sun-Earth Lagrangian point. In the more distant future, radio astronomers plan to place radio telescopes in the shielded zone of the Moon.

2.5. Radio Astronomy from the Sun-Earth Lagrangian Point

Quasi-stable (halo) orbits can be established for spacecraft around five special libration (Lagrangian) points in the gravitational field of the Sun-Earth system.

Two of them, the L_1 and L_2 points, are located along the Sun-Earth line at distances of about 1.5×10^6 km from each side of the Earth - the L_1 point is between the Sun and the Earth and the L_2 point is on the other side, furthest from the Sun. Halo orbits having radii up to about 250 000 km are possible in the vicinity of the L_2 point. Favourable conditions for maintaining and operating space telescopes near the L_2 point (e.g., efficient radiative cooling of a telescope and receivers) have led to proposals for a number of such astronomical missions (see Tables 1-4). Because of the great distance from the Earth, the orbits around the L_2 point are expected to be "quiet" in terms of radio interference.

2.6.   Radio Astronomy from the Shielded Zone of the Moon

The far side of the Moon offers excellent conditions for the location of astronomical observatories including radio observatories. The advantages and opportunities for radio astronomy on the Moon have been discussed in numerous studies describing a wide range of projects from low-frequency arrays (Landecker et al. 1991), to highly-sensitive radio astronomy observations at centimetre wavelengths and SETI (Heidmann 1998), to submillimetre radio interferometers (Mahoney 1991). One of the most important advantages to radio astronomy is an environment relatively free from terrestrial radio transmissions provided by the natural shielding of the Moon.

> "The shielded zone of the Moon comprises the area of the Moon's surface and an adjacent volume of space which are shielded from emissions originating within a distance of 100 000 km from the centre of the Earth" (Radio Regulations 1994, S22.22.1).

In recognition of the great scientific potential of the location of radio telescopes on the shielded zone of the Moon, international radio regulations prohibit any emissions in this zone causing harmful interference to radio astronomy observations in the *entire* frequency spectrum, except in the bands allocated to space research and required for the support of space research services (Radio Regulations 1994, S22.22 Section V).

3.   Radio Frequency Interference in Space Radio Astronomy

At the present time, the radio frequency spectrum from 9 kHz to 275 GHz is completely allocated to one or more radio services. Because of the fast progress in utilizing even higher radio frequencies, efforts have begun for allocation of the frequencies from 275 GHz to 1000 GHz. The radio spectrum is allocated by the International Telecommunication Union in blocks of frequencies to provide the necessary frequency bands for operations of different radio services. There are currently two dozen bands below 275 GHz, with relative bandwidth $\delta f / f$ between 0.2 to 10 percent, allocated for radio astronomy observations on a primary basis. Three of them are effective only in one of three ITU geographical areas. Additionally, a few allocations exist on a secondary basis. Ground-based radio astronomy observations are conducted actively in all of these bands. Moreover, radio astronomers are in the forefront of developments at even higher frequencies, conducting observations with ground-based radio telescopes at frequencies as high as 900 GHz. Thus, ground-based radio astronomy utilizes practically the entire frequency spectrum available at which cosmic radio waves are not absorbed or reflected by the Earth's atmosphere and ionosphere. For observations with space radio telescopes, radio astronomers intend to use the existing radio astronomy frequency allocations plus the parts of the spectrum at which the atmosphere is opaque. Moreover, they hope to be able to observe in the entire radio frequency spectrum, not just in the frequency bands designated by the radio regulations, to fully realize the potential of the space missions and to explore the frequency bands which it is impossible to study on Earth because of emissions from other radio users.

3.1. RFI in Space-Based vs. Ground-Based Radio Astronomy

There are important factors regarding radio frequency sharing and protection for space radio astronomy missions that are different from such considerations for ground-based radio astronomy.

Firstly, the space telescope's location away from the Earth means that RFI sources, which are contained mainly within the distance of 100,000 km from the centre of the Earth (Radio Regulations S22.22.1, 1994), will occupy only a portion of the sky. A space radio telescope equiped with a high-gain antenna can avoid observations in the direction of Earth. The main disadvantage is that such limitations will constrain the scientific operations by removing from any potential observations a significant portion of the sky. Evidently, such a constraint is more severe for Earth-orbiting space telescopes with apogees smaller than the distance to the Moon. From the L_2 point, however, the angular size of the noisy area will be just a few degrees across. Also, radio astronomy observations with a low-gain space antenna (e.g. the dipole type used for low-frequency radio astronomy) will be more affected by RFI since they have low selectivity to the direction of the upcoming signals.

In addition to the RFI from the vicinity of the Earth, transmissions from deep space probes as well as transmissions required to support space research operations on the Moon and other planets can affect space radio astronomy observations. Careful coordination between radio astronomy, space research, and space operation services is required to protect space radio astronomy.

Secondly, geographical spacing between ground-based radio observatories and ground stations supporting active radio services allows simultaneous operations by both of these services in same the same frequency band. This has allowed the co-allocation of bands to both radio astronomy and active radio services transmitting from the Earth towards space-based receivers. Geographical sharing may work for a ground-based telescope, but an orbiting telescope may go right above the source of the RFI and be completely unprotected. Also, since the space-based radio telescope is likely to observe at any point in its orbit, different criteria of sharing and protection may apply while it is above the different ITU frequency allocation regions.

Thirdly, such radio services as inter-satellite communications are in general compatible with ground-based radio astronomy because they tend to use bands which ground-based telescopes cannot use due to the atmosphere's opacity. In turn, however, such intersatellite links may interfere with space-based radio astronomy experiments which will be designed to observe also in the bands not accessible to ground-based telescopes.

Finally, it is well known that because of the extremely high sensitivity of the radio telescopes, sharing of the radio astronomy bands with an active service with a transmitter located within line-of-site of the ground-based radio telescope at a distance even as far as that to the geostationary orbit is practically impossible (Handbook 1995). Estimates show that the level of detrimental interference from such transmissions in radio astronomy bands exceeds the threshold of the interference required to protect the radio astronomy observations. But the remote location of space radio telescope will help to reduce line-of-sight RFI from the emissions of these active services.

3.2. Protecting Space Radio Astronomy from Man-Made Radio Interference

The impact of RFI on space radio astronomy missions will depend upon the mission configuration (e.g., location of telescopes, space antenna type) and the type of radio astronomy observations to be performed (e.g. VLBI, extended source, continuum, spectroscopy).

Although the sensitivity of VLBI observations to RFI, in general, is lower than for other types of radio astronomy observations (Handbook 1995), the effect of RFI on SVLBI observations may be different in a few aspects. Recent experience with SVLBI observations in L-band with the Japanese VSOP satellite HALCA has shown a high vulnerability to RFI entering through the space radio telescope sidelobes (Lioubtchenko et al. 2001). Particularly since sensitivity on the baseline between the space radio telescope and ground telescope is not very good (the space radio telescope has a modest size of 8-m diameter), RFI in the observing channels of both the ground and space telescopes makes the initial detection of interferometric fringes difficult. This may lead to the loss of precious observing time.

Space radio telescopes for cosmic microwave backround studies are also very sensitive to RFI. They have
(i) extremely high sensitivity (a few tenths of mK for an integration time of only 1 second),
(ii) very wide bandwidth of the radiometers ($\delta f/f = 0.1$-0.2),
(iii) a need for extremely high stability of the receiver's gain and calibration accuracy, and
(iv) a need for continuous data collection during a long duration (about 1 yr).
Locating space radio observatories for such observations at the L_2 point or on the shielded side of the Moon is practically a mandatory requirement.

Low-frequency radio astronomy observations from space at frequencies below the ionospheric cut-off are partially protected from ground-based RFI by ionospheric shielding. However, observations from space at frequencies between 6 and 30 MHz, where ground-based observations are still very difficult due to the ionosphere, may be significantly affected by terrestrial RFI because
(i) these observations are usually conducted with low-gain antennas,
(ii) this part of the radio spectrum is heavily used by the broadcasting services,
(iii) a high dynamic range is required (the signal power of observing targets, ranging from solar radio phenomena to extragalactic radio sources, may differ

by more than 90 dB), and

(iv) these measurements require accurate calibrations.

It has been reported that man-made radio emissions introduce significant interference to low-frequency observations in the vicinity of Earth at distances up to 1.5×10^6 km (the L_1 Sun-Earth system Lagrange point) (Alexander et al. 1975; Kaiser et al. 1996). The future space observatories operating in these bands will need to be located on the shielded side of the Moon or far enough from the Earth that other means of suppression of the RFI can be used (Jones et al. 1998).

The RFI in the observing bands of a millimetre-submillimetre radio telescope can lead to at least three potential reasons for the loss of data. First, since the front-end devices based on SIS (Superconductor-Insulator-Superconductor) junctions have a very small level of destruction power, about 10 mW, these front-ends can be easily destroyed if powerful radar or intersatellite link signals are accidentally intercepted by the telescope antenna beam. Secondly, such interfering signals at the level of -90 dBW picked up, for example, in antenna sidelobes, can easily saturate the receiver. Thirdly, narrow bandwidth RFI appearing in the observing band by leaking through the sidelobes can lead to the spurious identification of a non-existent molecular line.

The ITU frequency allocations for radio services are currently limited to frequencies below 275 GHz. It is important for future allocations above 275 GHz to take into consideration the new developments in millimetre-submillimetre space radio astronomy.

Finally, preliminary studies show that the levels of radio emissions on the surface of the Moon even from the deep space probes (deep space is defined as the region beyond 2×10^6 km from the Earth) may exceed the thresholds for harmful interference established for total power continuum radio astronomy measurements (Gutierrez-Luaces 1997). This indicates that protection of Moon-based radio astronomy will require careful coordination of the frequency allocations for all permitted active services.

4. Conclusions

Space radio astronomy is an integral part of future developments in radio astronomy. Consideration of frequency protection and sharing for this field is urgent and timely because of the rapid growth. Initial steps to protect space radio astronomy have been made by establishing a quiet zone on the back side of the Moon. Efforts have also been made to establish a coordination zone to protect the radio observatories in the vicinity of the Sun-Earth L_2 point. Much more work needs to be done.

Acknowledgments. The research described in this paper was carried out by the Jet Propulsion Laboratory, California Institute of Technology, under a contract with the National Aeronautics and Space Administration. The author is grateful to his colleagues from the Jet Propulsion Laboratory and from the ITU Working Party 7D for discussions and useful suggestions on the subject of this paper.

334 *Altunin*

References

Alexander, J.K., Kaiser, M.L., Novaco, et al. 1975, A&A, 40, 365

Bennettt, C.L., Halpern M., Hinshaw, G. et al. 1997, AAS, 191, 87.01

Bersanelli, M., Bouchet, F.R., Estathiou, G., et al. 1996, ESA Report, D/SCI(96)3

Bougeret, J.-L., Kaiser, M.L., Kellogg, P.J., et al. 1995, Space Sci.Rev., 71, 231

Jones, D.L., Weiler, K.W., Allen, R.J., et al. 1998, PASP, 144, 393

Handbook on Radio Astronomy 1995, ITU-R, Geneva, Ch.5

Heidmann J. 1998, AdvSpaceRes, 22, 347

Hirabayashi, H., (a) 1998, PASP, 144,11

Hirabayashi, H., (b) 1998, in Proceedings of the COSPAR-98, Japan, July 10-19, 1998 (to be published)

Gutierrez-Luaces, B.O. 1997, TDA Progress Report, 42-129, 1-9

Kaiser , M.L., et al. 1996, GeophysRes. (Letters), 23, 1287

Kardashev, N.S., et al. 1995, Acta Astronautica, 37, 271

Kardashev, N. S. 1997, Experimental Astronomy, 7, 329

Kitchin, C.R. 1991 Astrophysical techniques, Bristol, Philadelphia and New York: Adam Hilger

Klypin, A.A., Strukov, I.A., Skulachev, D.P. 1992, MNRAS, 258, 71

Landecker, P.B., Choi, D.U., Drean, R.J., et al. 1991, 42nd Int. Astron. Congr., IAF, Oct. 5-11, 1991

Lioubtchenko, S., Popov, M. V., Hirabayashi, H., Kobayashi, H. 2001, Proceedings of this conference

Levy, G.S., Linfield, R.P., Ulvestad, J.S. et al. 1986, Science, 234, 187

Linfield, R.P., Levy, G.S., Edwards, C.D., et al. 1990, ApJ, 358, 350

Mahoney, M.J., Marsh, K.A. 1991 SPIE, 1494, 182-193

Mather, J.C., Hauser, M.G., Bennet, C.L., et al. 1991, AdvSpaceRes, 11, 181

Melnick, G.J. 1993, AdvSpaceRes, 13, 535

Pilbratt, G. 1997, ESA SP-401, 7

Radio Regulations 1994, ITU, Geneva

Scheele, F. 1996, 47th Int. Astron. Congr., IAF, Oct 7-11, 1996

Smoot, G., Bennet, R., Weber, J., et al. 1990, ApJ, 360, 685

Strukov, I.A., Skulachev, D.P. 1984, Soviet Ast.(Letters), 10, 1

Ulvestad J. S., Linfield, R.P. 1998, PASP, 144, 397

Weber, R.R., Alexander, J.K., Stone, R.G. 1971, Radio Sci., 6, 1085

Preserving the Astronomical Sky
IAU Symposium, Vol. 196, 2001
R. J. Cohen and W. T. Sullivan, III, eds.

Origin of Major L-Band Interference Received by the HALCA Space Radio Telescope

S. Yu. Lioubtchenko and M. V. Popov

Astro Space Center FIAN, Profsoyuznaya str. 84/32, Moscow 117810, RUSSIA

H. Hirabayashi and H. Kobayashi

Institute of Space and Astronautical Science 3-1-1 Yoshinodai Sagamihara Kanagawa 229-8510 JAPAN

Abstract. About 40 hours of observing data received by the space radio telescope HALCA at L-band (1.6 GHz) were analyzed in order to investigate interference received by the space radio telescope. Autocorrelation spectra for this study were specially prepared at the DRAO S2-correlator with a 7.8125 kHz frequency resolution in each 16 MHz channel. It was found that during 20% of the observing time the interfering signal was above the tolerable level of 1% of total receiver noise in a 16 MHz channel. The major source of interference is identified with uplink communication from ships to geostationary satellites in the International Maritime Satellite service (INMARSAT). The frequency range allocated for INMARSAT is 1636.5–1645.0 MHz. INMARSAT uses four geostationary satellites, two of which are located above the Atlantic Ocean where the strongest interference was observed. To avoid this interference it is recommended to move the HALCA observing frequency range from the currently used 1634–1666 MHz to 1645–1677 MHz. A simple criterion is proposed to predict harmful interference from INMARSAT. This criterion may be used in scheduling of future HALCA observations at L-band.

1. Introduction

In February 1997 the first space-VLBI radio telescope HALCA was launched into elliptical orbit around the Earth by the Japanese M-V rocket from the Kagoshima Space Center of the Institute of Space and Astronautical Science. Thus, space radio astronomy entered a new era. The satellite's orbit has a revolution period of 6.3 hours, an inclination of 31.4° and an eccentricity of 0.60. The 8 m radio telescope changes its height relative to Earth's surface from 560 to 21400 km. HALCA's very sensitive receiving systems allow radio interferometric observations at 1.60–1.73 GHz (L-band) and 4.7–5.0 GHz (C-band).

From the first observations it became clear that the space radio telescope quite often received significant interfering signals at L-band. However, interference with the same frequency distribution and time behavior was never reported as being observed at ground radio observatories. The purpose of this study was to identify the origin of the main interference received by HALCA at L-band and to develop recommendations for the planning of future interferometric observations with space-ground radio interferometers.

2. Observations

The autocorrelation spectra for our analysis were prepared by Brent Carlson at the Dominion Radio Astrophysical Observatory (DRAO) Correlator using 2048-channel frequency resolution in each main 16 MHz channel (7.8125 kHz resolution) and an integration time of 500 s. The total observing time was about 40 hours with the spacecraft being at quite different positions in its orbit relative to the Earth's continents. Therefore, we can consider the results of our study statistically meaningful. In our analysis we paid attention only to major interfering signals which might be harmful for interferometric observations. Only harmonics with amplitudes greater than 50σ were taken into account (where σ is the root-mean square deviation in those 7.8125 kHz channels free from interference).

Figure 1 shows a spectrum damaged by strong interference whose integrated power constitutes 30% of the total noise power received from the sky in the lower of HALCA's two 16 MHz channels; the calibration tones, at 1 MHz intervals, are just visible.

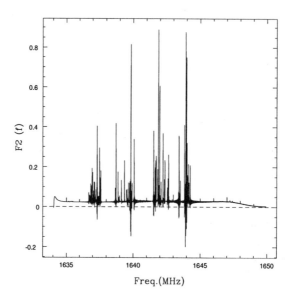

Figure 1.   Example of spectrum with strong harmful interference.

In the following sections we will analyze the main parameters of the observed interference in order to identify its source(s).

3. Parameters of the Observed Interference

3.1. Frequency Distribution

The frequency distribution of the observed interference is shown in Figure 2. One can see that the major portion is located in the range 1634–1645 MHz. In this report we will consider only this interference, since interference observed in other ranges was never found to be harmful for continuum interferometric observations.

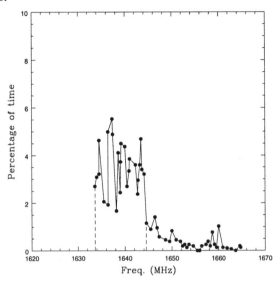

Figure 2.   Frequency distribution of the observed interference.

3.2. Power Distribution

Figure 3 presents the percentage of time when the interference was greater than a given percentage of the total receiver noise power in a 16 MHz channel. For Very Long Baseline Interferometry (VLBI) observations the tolerable interference level is determined by the requirement that the power level of the interfering signal should not exceed 1% of the receiver noise power, as set forth in Recommendation ITU-R RA 769. One can see from Figure 3 that in about 80% of the observing time the interfering signal did not exceed the 1% limit. To keep the figure compact, the rightmost box (of 5% height located between 10 and 11% of interference power) includes in fact interfering signals as strong as 40%. This whole box corresponds to a single occasion when the interfering signal was that high: October 11, 1997, from 19:40 till 21:20 UT (experiment code v101e). For the upper channel (1650–1666 MHz) the interfering signal never exceeded the tolerable level of 1%.

3.3. Geographical Coverage

To understand the origin of the interference we calculated the position of the HALCA satellite in its orbit for the experiments under consideration. For this

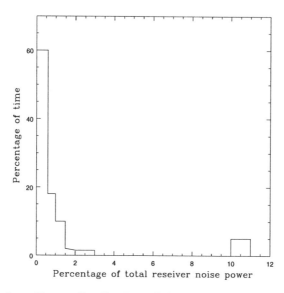

Figure 3.   Power distribution of the observed interference.

purpose we used a Keplerian orbit with the osculating elements corresponding to the date of the observation. We found the accuracy of this orbit approximation quite sufficient for our purposes by comparing our calculation of HALCA tracking passes with those used for mission operations. We have not found any correlation between the strength of the interference and the spacecraft position above a particular continent. There was also no correlation between the power of the interfering signal and the height of the satellite above the Earth's surface.

3.4.   Relative Velocities of HALCA and the Interfering Sources

The frequency resolution of the autocorrelation spectra was 7.8125 kHz, which corresponds to a Doppler shift of 1.43 km s^{-1}. This resolution was sufficient to distinguish between two possible origins of interference: Earth stations or space stations. It was found by visual inspection of the time-frequency diagrams produced with this frequency resolution that the interfering signals appear as a regular sequence of harmonics with a spacing of 50 kHz. Such regular structures in the spectra permit us to investigate the time behavior of the relative velocity between the source of interference and the spacecraft. Because of the large number of harmonics involved, the accuracy of the velocity measurements was as high as 0.1 km s^{-1}. We compared the calculated radial velocity of HALCA as seen from the Earth's centre and the velocity behavior of the interfering signal. It was evident that the source of interference was located at the Earth's surface. For a geostationary satellite, and even more so a low orbit satellite, the velocity curves would have been very different!

As a result it was not difficult to find the communication system responsible for the interfering signals received by HALCA at L-band: it is the INMARSAT International Maritime Satellite service which provides communication for the maritime industry.

4. General Characteristics of INMARSAT Communication Service

INMARSAT provides a complete range of ship-to-shore and shore-to-ship communications via geostationary satellites. At present, INMARSAT serves about 50000 Mobile Users (MU) with voice, fax, E-mail and telex services.

INMARSAT's geostationary satellites (Space Stations - SS) are located at the following ranges of longitude: 178°E, 179°E (Pacific Ocean Region); 54°W, 55°W (Atlantic Ocean, Region-West); 15.5°W, 17°W (Atlantic Ocean, Region-East); 64.5°E, 65.5°E (Indian Ocean Region). There are about 20 Ground Stations (GS) communicating with the SS and connecting calls from ships to the ground network. Uplink communications from ships to satellites (MU to SS) produce interfering signals for HALCA at L-band. There are 339 frequency channels in the range 1636.5–1645 MHz; the bandwidth of a single channel is about 12 kHz and the central frequency of any channel is given by

$$f_n = 1636.5 + 0.025 * n \text{ MHz.} \qquad (1)$$

There are many kinds of MU stations, providing an EIRP (effective isotropic radiated power) of 37.0 dBW in an 8.5 MHz band. Antenna diameters of MU stations vary from 0.6 m to 2.0 m. For example, the 1.3 m antenna for the Russian MU station "Volna" has the following beam pattern (Zhilin, 1988):

$HBWD = 8°;$ $\quad G_t = 24$ dB at $\theta = 0°;$ $\quad G_t = 8$ dB at $16° < \theta < 21°;$
$G_t = 41 - 25 * \log \theta$ dB at $21° < \theta < 57°;$ $\quad G_t < 3$ dB at $\theta > 57°.$

Using the simple relation for free space transmission loss

$$dP = P_t G_t G_r (\lambda/4\pi D)^2 \qquad (2)$$

one can estimate the necessary separation distance D between a single MU and HALCA when the interfering signal from this single MU would not be harmful. The HALCA radio telescope at L-band has a maximum gain of 39.7 dB, its zero gain is at 23°, and the gain is reduced to -28 dB for angles greater than 90°. Let us consider that the angle between the HALCA beam and the direction to the MU is always greater than 90°, and calculate the separation distance for three different positions of the HALCA spacecraft relative to the MU communicating with the SS:

Angle between HALCA and SS as seen from MU	Distance beyond which interference tolerable (km)
$\theta = 0°$	15000
$16° < \theta < 21°$	2700
$\theta > 45°$	1000

In these calculations we assumed the tolerable level of interference for continuum VLBI observations in a 16 MHz bandwidth at L-band to be equal to -156.5 dBW, as determined in our previous research (Popov 1996).

It is evident that when HALCA is in the main beam of a MU it would receive harmful interference at nearly any portion of its orbit even from the single MU, while it is safe from harmful interference at angles greater than 30°. Of course, the real situation is even worse when we take into account the density of mobile users and the operational load of the INMARSAT service.

5.   Conclusion

The frequency allocations in the HALCA L-band observing range are shown in Figure 4, where "RA" denotes the frequency bands allocated for radio astronomy.

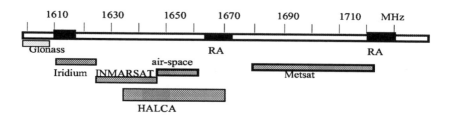

Figure 4.   Frequency allocations in the HALCA L-band observing range.

The best frequency allocation for HALCA continuum measurements seems to be 1645–1677 MHz. This frequency range overlaps with the air-fleet to space communication band (1646.5–1660 MHz), which was not found to be harmful for HALCA. Overlap with the Metsat band (1675–1710 MHz) may be harmful for ground observatories. In fact, there were reports of interference around 1680 MHz and 1690 MHz at Narrabri (Australia) which was strong enough to double the system temperature in a 64 MHz band. At the VLBA significant interference has been occasionally observed in the 1674–1679 MHz frequency range. Also at the VLA frequencies lower than 1660 MHz suffer less from interference.

If the current frequency allocation for continuum observations with HALCA at L-band (1634–1666 MHz) remains the same, then a prediction of the expected interference from INMARSAT should be included in the scheduling of observations.

We would like to emphasize that space close to the Earth (especially the volume inside the geostationary radius of 36000 km) is very dangerous for radio astronomy because of numerous telecommunication systems and services, and careful selection of the bands for radio astronomy must be underaken in developing such space radio astronomy projects.

Acknowledgments.   We gratefully acknowledge the VSOP Project, which is led by the Japanese Institute of Space and Astronautical Science in cooperation with many organizations and radio telescopes around the world. We are also grateful to Brent Carlson, who prepared data for our analysis at the DRAO S2 Correlator in Penticton, and to Phil Edwards for his remarks and corrections.

References

Zhilin V.A. "International Maritime Satellite Service INMARSAT", Handbook, pp.1-159, Sudostroenie, Leningrad, 1988 (in Russian)

Popov M.V. "Estimates of Potential Radio Interference for VSOP from Space Related Systems", Memo ASC FIAN 1439/1, ISAS/VSOP 003/1, 1996

Part 5
Outreach

Preserving the Astronomical Sky
IAU Symposium, Vol. 196, 2001
R. J. Cohen and W. T. Sullivan, III, eds.

Saving Our Skies: Communicating the Issues to the Media

Richard West and Claus Madsen

European Southern Observatory (ESO)
Karl-Schwarzschild-Strasse 2, D-85748 Garching, Germany
E-mail: rwest@eso.org, cmadsen@eso.org

Abstract. We discuss possible mechanisms for setting up a global outreach campaign centred on the main theme of this meeting: **save our skies!** Effective communication of this message to the world's media and the wide public is a prerequisite for successful sensitisation of decision-makers in different countries to the crucial issues at stake. We emphasise the need for careful planning of such a programme, especially in terms of definition of the key issues, the way they are presented, as well as the communication channels to be employed. It is important to differentiate the arguments used in connection with different types of pollution (light, radio, space debris). It will be necessary to identify clear and forceful messages that convincingly stress that these problems are of ultimate concern, not just a small group of astronomers, but to all of humanity. With their extremely sensitive instruments, astronomers constitute an avant-garde that is the first to detect the adverse effects, but as these intensify, increasingly broader sectors of society will be affected. It appears feasible, within the limited means available to the IAU and IDA, to initiate such an outreach effort with a comprehensive web-based campaign that highlights astronomical "pollution". This may also serve as a useful test-bench for subsequent campaigns based on more communication vehicles and with a wider spectrum of associated activities.

1. Introduction

Astronomers all over the world have become seriously concerned about increasing "pollution" in space and in the atmosphere. They sense a global deterioration of observing conditions that may become a real dilemma in the future. Today, few locations on the Earth enjoy the privileged situation of a truly unperturbed view of the sky. ESO's observatories at La Silla and Paranal in the Chilean Atacama desert, as well as the high-altitude Chajnantor site for the future ALMA facility in the same region, are far from inhabited places and belong to this rare class of 'last resorts' for astronomical observations from Earth. However, even though these near-pristine sites are still reasonably well protected against ground-based pollution, they are of course exposed to space-based effects like all others.

Most causes of *sky pollution* that adversely influence astronomical observations have been identified and analysed and various technical remedies are known. However, it is not possible for astronomers alone to implement these and to change the situation. They need the help and support from other agencies and individuals. There are many examples of interactions with local authorities with a happy outcome, but this is not so in all cases. At the global level, scientists act primarily through the immediately concerned inter-governmental committees and commissions, e.g. COPUOS and IUCAF, so far with a reasonable degree of success.

But astronomers are worried about the future, and rightly so. The members of these international committees are delegates from different countries and organisations and rarely have a background in astronomy or space science. Not all will show understanding for astronomers' problems and many will mostly follow instructions from their bases. They are also, to a greater or lesser extent, subject to pressures and lobbying from influential interest groups.

There is thus a need for astronomers to exert some pressure. They have to inform others about their problem in such a way and on such a scale that the odds increase that they will obtain the support they ask for. In short, *they must call attention to their problems and do this in a way that is sufficiently convincing and persuasive that they will be heard by the decision-makers. Only then is there some hope that effective solutions to the mentioned pollution problems will be earnestly considered, perhaps even implemented.*

That is the current situation in a nutshell. It is not unlike what is experienced by other groups, in other contexts - it is in fact a recurrent scenario in a complex society where a multitude of often-conflicting interests confront each other. It is therefore reasonable to assume that communicative means and methods similar to those used by other groups may be employed to achieve the desired results, with the ultimate goal of halting or even reversing the present deterioration of the natural conditions of astronomical observations.

The astronomical community is geographically dispersed and it is not very large, especially when compared to the big environmentally oriented groupings. Nevertheless, it has certain prerogatives that, if used properly, may achieve effects that are quite powerful. The world's astronomers constitute a rather tight-knit, well-organised group of individuals with an inspiring mission and a generally positive image in the public mind, an advantage that cannot be overestimated in today's global village.

This paper will analyse the current situation in a somewhat unorthodox way and propose some specific remedies that may be applied at relatively short notice and which are compatible with the means of astronomers and their representative organs, in particular IDA and IAU and the associated commissions and committees. It goes without saying that a positive outcome can be expected only if proper collaboration is established from the outset with the national bodies and if vertical lines of communication to local communities are kept open and active.

Although activities at the local and national levels are both necessary and valuable, **we will focus on more global aspects** in what follows. The problems of sky pollution have now reached a stage where it is desirable to develop a global view, both of the issues themselves and of their possible solutions. However, some of these considerations may of course also have local applications.

2. Laying the Ground for a Media Campaign

Modern society has furthered the development of very powerful media on which we are all dependent, whether as citizens, scientists or decision-makers. There is no doubt that a major road towards alleviation of the current environmental problems in astronomy passes through them. To achieve the desired results, certain rules must be followed.

As in other media campaigns, it is necessary to identify the main issues in advance and to perform an evaluation of how the different instruments shall be played to achieve the best possible effect. Standard procedures can then be brought into action.

The first phase is the definition of the problem. This may seem trivial for astronomers, but not so for bystanders. What are the real *key issues* and who are the main players? Can you explain to somebody who is not necessarily interested in astronomy, what this science really is? Can you describe the types of pollution in a simple way? Can you condense the problems into easily comprehensible issues that can be transmitted to the public, the media, the politicians and other relevant groups?

What are the *consequences of the current situation* and the projections into the future? Not just for astronomy as a science, but for the individual, for other branches of human endeavours, for the individual town and country, for the world and for mankind as a whole?

Is it clear that something must be done at all? Will the *nuisance of this problem ultimately be greater than the expense of alleviating or eliminating it?* (If not, why bother at all?) Can you convincingly explain why the free view of the skies is so important that this must be guaranteed for future generations? Not just in the usual, idealistic or philosophical-cultural sense, but also from a more "applied", not to say "business" point-of-view?

Whilst bringing the problem to the fore is indeed the goal of a communication plan, there are associated risks which must not be overlooked. Thus a possible confrontation between the needs of science and commercial interests, if not well controlled, could in principle lead to the conclusion that short-term commercial considerations are more important than long-term scientific investigations. Such a situation must clearly be avoided.

Assuming that it is possible to analyse these questions in depth and to come up with satisfying, not to say promising answers and formulations, the next step will then follow - that of preparing the media campaign.

3. Understanding the Media and How They Work

The increasing sky pollution is closely linked to the development of our modern society on which we all depend and the progress of which is considered basically beneficial by most citizens. Why, then, should the common citizen be worried about side effects like light or radio pollution?

Any attempt to disseminate an answer that is supportive of the astronomers' point of view means getting access to the media by presenting it in the proper way. The issue must be dealt with, not only from the scientist's point of view, but also primarily from the perspective of the public that may not even realise that there is a problem. Of course, this does not mean that the scientific concerns should remain in the background. What it does imply, however, is that the presentation must be geared to take the specific requirements of the media into account.

It is in this context useful to consider some of key features of 'science items that are likely to 'succeed' as news stories', as listed by BBC producer Jana Bennett (1997):

- Science for the human race

- Could affect us all

- Viewer already knows enough to be able to integrate new information

- Focuses on one clear issue

- Shows awareness of viewers' concerns

While this short list is certainly not exhaustive, it already contains the seeds of success of a large-scale media campaign. Much thought should therefore be given to the integration of these and related considerations into the planning.

Successful communication in public mass media requires the use of clear language, striking analogies and/or good metaphors. It is important, though, only to use comparisons that can stand close scrutiny. At the same time, it must never be forgotten that there will be large differences in cultural background and perception within a global audience - what may be a great way of presenting the problem in one country could utterly fail in another.

4.   A Media-Campaign, Step-by-Step

In practical terms, the steps towards a successful media campaign are:

- Define problem and relate to media requirements

- Set communication goal(s)

- Consider possible risks

- Develop communication strategy:
    - identify core issue (from a media perspective)
    - consider terminology/key expressions of language (including analogies, metaphors)
    - identify communication interfaces
        * scientists/audiences
        * communication/media

- Determine operational approach
    - top-down ('big' events)

- bottom-up ('grass roots')
- an organised network (?)

• Select communication channels

  - the World-Wide Web
  - public mass media (TV, radio, newspapers) meetings/conferences/ public events
  - performances (public lectures, planetaria, science theatre)
  - special opportunities (science festivals, others)

• Produce communication materials

  - texts (press releases, exhibition panels, leaflets)
  - video (video news reels for TV, information video for public)
  - images (photos of natural phenomena, explanatory graphics)
  - "gadgets" (buttons, stickers, T-shirts, memorabilia)

5. Developing Key Messages

Human-induced pollution of the astronomical sky can be divided into at least four distinct areas:

• Light pollution (from the ground, e.g. cities; from space, e.g. satellites)

• Radio interference (from the ground and from space; in the band and spill-over)

• Space debris (light effects in case of extremely sensitive equipment; risk of collision with astronomical and other spacecraft)

• Increasing cloud coverage and amount of absorbing gases and dust in the atmosphere (especially water), due to short-term (e.g. El Niño) and long-term (general greenhouse effect) climatic changes (to be realistic, there may of course both be natural and human causes to these effects)

Each problem obviously requires its own communication approach and specific language use. However, the overall notion of 'pollution' must be kept in mind.

It should be noted that considerable work in the area of *light pollution* has already been done by groups (Mizon 1998) in particular in the USA and in the UK (above all by the International Dark Sky Association and the British Astronomical Association).

The problem about the detrimental effects of man-made *radio noise* to astronomy may be less known to the public.

The other two effects, *space debris* and *climatic effects* are already generally recognised by the media and the public and have much wider impact than what is considered here. Their negative effects on astronomy are in fact only marginal when compared to their overall implications for mankind. It is not obvious how astronomers may contribute significantly in these areas to further increase the otherwise high level of awareness among the media, the public and the decision-makers.

In all of these contexts, it is important to stress **the central role of astronomy as a pathfinder**. Thanks to front-line technology and extreme sensitivity demands, astronomical instruments are the first to read the warning signals and to provide quantitative estimates of the deteriorating situation. Simple extrapolation then shows how an increased segment of mankind and a variety of its activities will be progressively influenced.

6.   Differentiation of Arguments

We now discuss some of the individual characteristics of the current threats to astronomy and provide a few, central considerations in connection with a related, future media campaign.

6.1.   Light Pollution

Light pollution can be likened, e.g., to chemical pollution. At low levels, it may not pose a threat to marine life or to surface vegetation, but as the concentration increases, it will produce progressively stronger effects. In the early phase, local street lights and advertisements may only be a nuisance for amateur astronomers in that geographical area, but if global urbanisation continues unabated and without stricter lightning regulations, a growing percentage of mankind will be denied access to a fundamental piece of Nature - the night sky. It concerns us all, the man in the street as well as the scientist.

The night sky is a very basic, natural as well as cultural resource. Throughout the ages, it is referred to in innumerable texts. It is the home of the gods, a display of beauty and serenity, a window to eternity. There are recent examples of its protection being elevated to a national goal, e.g. in Chile, not just to please astronomers, but most certainly also with profitable tourism in mind. Where will future generations go when they want to experience the real night sky? Obviously to those places where it as unaffected as possible!

Sacrificing the dark night sky for commercial interests may be compared to not producing any real books anymore in order to utilise all the paper in the world for supermarket leaflets. We can switch off the advertisements on the TV-set, but is it acceptable that we cannot see the Moon because it is hidden behind an orbiting soft-drink billboard?

It is important to remain truthful to the subject matter and analogies must not be taken too far. On the other hand, messages that are directed towards large sections of the public must be simple and easy to associate with. Thus, while it may be advisable to use softer formulations in front of a highly educated audience and in general, informed members of the public, a more dramatic metaphor may have a better impact and meet with an excellent response elsewhere. Here is one example of a simple message that may hit hard (too hard?) in many places. The ultimate result of global light pollution may be likened by locking up a person in a cell in eternal light, inhibiting the view of the darker surroundings outside and thus depriving him of any sense of orientation beyond the narrow confines of the cell. Shall this be mankind's future?

6.2. Radio Interference

Astronomers need all available channels to study the Universe. It is not enough to make pretty pictures in the visual part of the spectrum. It is exactly the addition of all the other wavebands during the past decades that has resulted in an information explosion with the associated fundamental discoveries of completely new classes of celestial objects and phenomena.

Radio astronomy is comparatively easy to portray to the public - we "listen" to the Universe, rather than seeing it (visual) or sensing it (infrared). Radio noise is well known to many people, certainly to those living in remote places and especially to short-wave amateurs, and also to the homesick tourist that tries to tune in to a favourite station in his far-away home country.

However, while he or she may accept that there are natural limits to the propagation of radio waves, it really ought to be more difficult to understand and accept that man-made signals need to flood those specific radio bands in which we may receive crucial information from the distant Universe. Let the satellite-based communications organisations carry the load of explaining in a simple and satisfactory way why they cannot manage within the other bands! They would be hard pressed to do so convincingly in front of an inquisitive public!

To further dramatise this effect, we may speak about the pitiable person above, now also being deprived of means of communication with the outside world - all messages from there are drowned in emissions from noise senders. And yet, this is exactly what large populations in countries with totalitarian regimes are being (were) subjected to. Would you then like to live, together with the rest of mankind, under such deplorable circumstances?

And then the tragicomic approach. Imagine that a SETI signal is on its way towards Earth, speeding towards a well-tuned radio telescope pointing in the right direction at the right time. Now is the moment of the greatest discovery mankind will ever make - we are not alone! And exactly at that time, somebody in the neighbourhood calls-a-pizza via his global mobile telephone... Wouldn't that really be too bad!

6.3. Space Debris

The issue of space debris is of course first of all a question of safety. There are clear signs that the space agencies and with them the media, are taking this new form of Russian roulette very seriously - many reports are appearing, dealing with the risk of collisions. For earthlings, the situation can be likened to living in a house surrounded by swarms of killer bees - perhaps a rather strong, but not necessarily misleading, metaphor. In any case, based on recent estimates of the rapidly growing number of orbiting objects and with permanent habitation on the International Space Station now imminent, it is fair to expect that the perception of this problem will soon take on an entirely new dimension.

It is safe to predict that a great media (and political and public) uproar will occur sooner or later, at the latest when an exceedingly costly communications satellite is destroyed by a 5-cent bolt left in space or a peanut-sized part from a long-abandoned spacecraft. However, in this context, astronomers must concede that their space observatories are only a subgroup of the vast orbital population, dominated by commercial and military satellites. There are equal risks for all, but research spacecraft are of course **unmanned**. Don't ask the public what is perceived as the lesser evil: whether Chandra or the ISS are hit!

The real problem that is specific to astronomers is the pollution caused by reflected sunlight from an increasing number of orbiting objects of all sizes, moving randomly through the fields-of-view of the world's large, expensive telescopes. Already now, a very substantial fraction of all exposures with large-field cameras are contaminated by visible trails from such objects. There are numerous examples, also demonstrated during this conference, of bright trails that overlap faint celestial objects for which the exposure was made, resulting in loss of valuable observing time - sometimes even of epoch-dependent, unique scientific data.

In all known cases, such a loss has been of no great consequence beyond the circle of astronomical observers. But what if sometime in the future, when the density of space debris presumably has become much larger than now, a crucial observation fails that may have wider implications? In a dramatic scenario, it could be an asteroid or comet that happens to be on collision course with the Earth and therefore remains undetected (although this event is of course most unlikely).

6.4. Increasing Cloud Coverage

The current global warming is receiving wide exposure in the media. While doubts have been expressed about the true extent of this effect and its speed of progression, it appears that the media (contrary to some governments) are now quite unison in presenting this as a major challenge to mankind in the next century. Unless something is done, it is said, we will be in for dramatic changes, on a global as well as a local level. The public is being constantly sensitised to these issues. It is thus reasonable to predict that at some moment - perhaps only some decades from now when the related effects have become even more obvious, especially in terms of changing climatic conditions and increased frequency of natural disasters - public opinion will force the enactment of stricter environmental laws in many more regions of the earth than is now the case.

Astronomers suffer from this as all other citizens do, and it is almost certain that most sites of astronomical facilities will ultimately be influenced. In the Pacific area, where some of the prime observatory sites are located, including those of ESO, it is above all the effects of El Niño that will cause concern. If, for instance, this phenomenon began to reappear more frequently than hitherto - or perhaps on a larger scale- with increased temperature excesses and impacting a larger geographical area, astronomical observations in these areas would become increasingly difficult.

Astronomers may contribute to raising the public awareness of these changes, since they have long-term objective records that may provide clear illustrations of what is going on. This is fundamental issue that concerns us all and we will float or sink with the rest of humanity.

7. Communication Tools

Giving the fact that the issue is (or rather should be) of wide concern, mass-media such as TV and the printed press are good channels for raising public awareness about astronomical pollution. Attracting the attention of the press, however, is not easy and will require a variety of approaches, including:

- Events - can be high-level meetings (such as this IAU Symposium), with highly respected scientists and politicians pronouncing statements on the issue - or local events, organised by science centres, societies, astronomy clubs, planetaria , etc. (i.e. 'grass roots' movements). They can be staged around the issue or can be integrated into other events, in particular astronomical events (eclipses, comets, meteor streams, spacecraft fly-bys etc.). Integration into the national science week(s) or day(s), now taking place in an increasing number of countries, may also be considered.

- Well-prepared information material, e.g., press releases on the right occasions, good illustrations, possibly video material in broadcast standard - and quality - is likely to open many media doors.

- Feature articles and letters in leading newspapers and magazines by scientists and lay people

- Using dedicated science programmes on radio and TV as a stepping stone

- Using dedicated youth programmes and youth magazines as a stepping stone to reach the next generation

- Scientists and science agencies may use their media connections to press the issue and possibly undertake co-ordinated efforts in that direction

- International bodies, including the IAU and IDA, are well placed to undertake co-ordinated efforts on a world scale

8. The World-Wide Web

In view of its global reach and ease of use, web-based activities can be extremely efficient in reaching well-defined communities and groups in society. However, although 'the Web' is clearly undergoing a dramatic expansion - and therefore gains in importance for communication - the enormous amount of information now on the Web tends to make it a victim of its own success. Furthermore, obtaining information from the Web requires a dedicated effort by the individual, much more so than being fed by the conventional mass media.

From the point of view of a *producer* of communication, linking with search engines can help, as can any other activity that directs Web users towards the appropriate Web pages. In this context, it should be mentioned that even the most successful internet-based educational activities by ESO, e.g. the 'Astronomy On-Line' programme, were always 'announced' in parallel by means of conventional printed matter. Hence a combination of the tools listed above is called for.

Nevertheless, there is an obvious potential for a dedicated effort to spread the word about a 'Save our Skies' global project via the Web. It should not be too difficult or costly to set up in a relatively short time a concerted action by involving major observatories, planetaria and amateur groups to achieve an efficient media and public sensitisation. This could be done within the IAU framework or via the IDA or, better, in a collaborative effort. It would involve the preparation of convincing, interesting and educational material, including such that is specially prepared for schools, but the main efforts would be in the coordination of the campaign.

It is also important to understand that what is required for such a dedicated, global campaign, as indeed in any other attempt to spread the word about sky pollution, is a *sustained effort*. The public mass media can play a crucial role in moving the issue out of the 'dark' corner, where it resides now (as seen from most countries), and into the public arena. Media attention, though, is short and must be won again and again.

9. Conclusions

From a communication point of view, *sky pollution* is similar to many 'standard' issues of contention that occur in our complex, modern society. Solving/removing the problems for science that arise from sky pollution requires a broader understanding and support than can be generated within the scientific community alone. Hence a broad public discussion about the related issues is highly desirable.

In moving the issue into the public arena, a clear and carefully prepared communication strategy is necessary, both with respect to the formulation of the key texts and the choice and interplay of communication tools.

It appears feasible, within the limited means available to the IAU and IDA, to initiate such an effort at the global level with a comprehensive web-based campaign that highlights astronomical "pollution". This may also serve as a useful test-bench for subsequent campaigns based on more communication vehicles and with a wider spectrum of associated activities.

A close collaboration between local, national and international groupings and organisations over an extended period of time is needed to achieve substantial positive results and an impact that is sufficient to reach decision-makers on a wide scale. This will take some hard work, but we have no doubts that the world's astronomical community is in possession of the dedication, the means and, not least, those structures that are necessary to ensure a successful outcome.

References

Bennett, J. 1997, 'Science on television: A coming of age?' in Farmelo, G. and Carding, J. (eds) *Here and Now, Contemporary Science and Technology in Museums and Science Centres*, (c) Trustees of the Science Museum.

Mizon, R 1998, 'The British Astronomical Association's Campaign for Dark Skies Achievements after 10 years'
(http://www.u-net.com/ph/cfds/info/last10yr.htm)

Preserving the Astronomical Sky
IAU Symposium, Vol. 196, 2001
R. J. Cohen and W. T. Sullivan, III, eds.

Light Pollution: Education of Students, Teachers and the Public

John R. Percy

Erindale Campus, University of Toronto
Mississauga ON Canada L5L 1C6

Abstract. The preservation of the astronomical environment is intimately connected to society's understanding and appreciation of astronomy. This requires effective education of students, teachers and the general public. We know how this can be done. It remains for us to become education-active and to convince our colleagues and students to do likewise.

1. Introduction

Many speakers at this Symposium have pointed out that education is the first step in preserving the astronomical sky. In fact, education is essential for the health of *all* of astronomy, both to attract and train the next generation of astronomers and to promote awareness, understanding and appreciation of astronomy on the part of the taxpayers who support us. The IAU has sponsored over 375 Symposia and Colloquia; only two of these have dealt with education!

The process of astronomical education is made more complicated (and interesting) by the fact that education takes place in a variety of situations - not just in the classroom (Fraknoi 1996). Furthermore, there are vast differences in the systems of education in different countries and even within different countries. IAU Commission 46 (Teaching of Astronomy), with its system of National Representatives, may be able to help. But astronomy education should not be left to educators alone: *if every astronomer spent an additional 1 per cent of their time on education and public outreach, the effect could be profound.*

Light pollution is obviously undesirable because it is a symptom of inefficient, ineffective lighting; major cities waste millions of dollars each year in lighting the night sky. This hinders the work of professional and amateur astronomers and robs everyone of dark skies. Is this a problem? Only if astronomy, and dark skies, are in some way "useful".

2. Why Is Astronomy Useful?

Astronomy is deeply rooted in history and culture, as a result of its practical applications and its religious and philosophical implications. Among the scientific revolutions of history, astronomy stands out. Recent lists of "the hundred most important people of the millenium" invariably include a few astronomers. Astronomy still governs the cycle of day and night, the seasons, and many aspects of long-term climate change; impacts of asteroids and comets have been implicated in mass biological extinctions. It contributes to the advancement of mathematics and computer science, and science and technology. It is a dynamic science in its own right; each year, many of the most important scientific discoveries are in astronomy and related fields. Astronomy deals with our place in time and space, and with our cosmic roots. It promotes environmental consciousness, both through images of our fragile planet taken from space, and through the knowledge that we may be alone in the universe. It also has aesthetic and emotional dimensions: it reveals a universe which is vast and beautiful; it has inspired artists and poets for centuries; it harnesses curiosity, imagination, and a sense of shared exploration and discovery. In school, it can be used to teach concepts such as light and gravitation, to give students a more meaningful appreciation of scales of distance and time, and to illustrate the *observational* approach to the scientific method. It is the ultimate interdisciplinary subject, and "cross-curricular connections" are highly-valued in modern curriculum development. It attracts young people to science and technology; it increases public awareness and interest in science; and it is an enjoyable hobby for millions of people worldwide. In the words of Henri Poincaré: "Astronomy is useful because it shows how small our bodies, how large our minds".

3.   Astronomy Education - Where and How Does It Occur?

Education can be divided into: (1) formal or school education, and (2) informal or public education. In fact, these two overlap: the public derive their basic knowledge of (and attitudes toward) astronomy in the schools, and students learn as much about astronomy out of school as within. As Fraknoi (1996) has eloquently pointed out, astronomy education "happens in hundreds of planetaria and museums around the country; it happens at meetings of amateur astronomy groups; it happens when someone reads a newspaper or in front of television and radio sets; it happens while someone is engrossed in a popular book on astronomy, or leafs through a magazine like *Sky & Telescope*; it happens in youth groups taking an overnight hike and learning about the stars; and it happens when someone surfs the astronomy resources on the internet. When we consider astronomy education, its triumphs and tribulations, we must be sure that we don't focus too narrowly on academia, and omit the many places that it can and does happen outside the classroom".

It follows that education about light pollution will be most effective when it comes through a coalition or partnership - the local "astronomical community". This includes professional astronomers; scientists, engineers, and other academics in related fields; students at every level; educators at every level, in every setting; journalists; amateur astronomers; environmentalists; and interested members of the general public. This provides a "critical mass" of concerned citizens, providing a broad base of support.

4.   Light Pollution and Formal Education

First - a brief introduction to formal science and astronomy education. We know many things about effective science teaching and learning: young people (and adults) have deep-seated misconceptions about scientific topics, which teachers must identify and deal with; scientific concepts must be introduced in a logical sequence, and at an age when students can absorb them; and science teaching is most effective when an inquiry-based or activity-based approach is taken (see various papers in Percy 1996). This suggests that sky-watching ("eyes-on astronomy") should be an important part of astronomy education. Even though day-time astronomy is more convenient than night-time astronomy ("the stars come out at night, the students don't"), *observing the night sky should be part of every astronomy course*. This can be achieved, for instance, by giving each student a star map, or by allowing every student to make their own planisphere from a template. They can be taught how to use this in the classroom, so that they can use it in the evening at their convenience.

Unfortunately, there are many barriers to the effective teaching of astronomy: lack of appreciation of its value; lack of understanding of astronomy and astronomy education by teachers; and the ingrained "classical" methods of teaching by memorization and regurgitation. This is true at all levels, including university; *effective professional development for teachers is essential*.

We should all work to get more and better astronomy in the school curriculum. In many countries, astronomy *is* part of the school science curriculum; it is part of the US National Science Education Standards (NRC 1996) and the Canadian equivalent. Simple observable changes (such as day and night) are covered in the lower grades of elementary school, followed by a more detailed discussion of solar system topics (such as moon phases) in the upper grades. In the lower grades of secondary school, the curriculum may include material on the planets, sun, stars and galaxies. In the upper grades, astronomy may be used to illustrate physics topics such as gravity, light and spectra.

Ironically, the study of light pollution makes a very good "cloudy night activity" for students. It can be done in an urban or suburban setting. It does not require clear skies. It helps students to understand scientific, technological and societal issues. See Percy (1998a) for a review of some educational activities and projects on light pollution. Light pollution may not be an explicit "topic" in the science curriculum, but the science curriculum includes processes, skills and applications, which can be taught effectively through the study of light pollution and transferred to other science topics.

Because light pollution is such a promising topic for promoting science education, and because Metaxa (this Symposium) was directing a successful light pollution project in the schools in Greece, Crawford, Metaxa and Percy (1998) recently produced a special teachers' newsletter on the topic - an issue of the Astronomical Society of the Pacific's *The Universe in your Classroom*. This newsletter is sent to thousands of classrooms across the world; it is available on the internet (www.aspsky.org/html/tnl/44/lightpoll.html); and it is translated into a dozen languages. In addition to background information and resource lists on light pollution, it includes three kinds of activities:

- Astronomy activities in which students investigate the effect of light pollution and other factors on the limiting magnitude of stars which they can see.

- Physics activities in which students use simple transmission diffraction gratings to observe the spectra of various natural and artificial light sources in their local environment. Sets of gratings for an entire class can be obtained at low cost from various science supply companies.

- Science-and-society activities in which students investigate the types of lighting in their community, the effectiveness of the lighting, and the channels which they can use to promote better lighting.

In my own province of Ontario, Canada, the school science curriculum has recently been revised, and astronomy and space are now a part of the curriculum in grade 6 (age 11 years) and grade 9 (age 14 years). One of the new textbooks (Plumb et al. 1999) includes a two-page section by Alan Hirsch on the issue of light pollution. Several hundred thousand students will be exposed to this material over the life of the textbook. If the astronomical community could provide appropriate information, activities and resources on light pollution to textbook authors and publishers, there would be a greater chance that this material would be presented to students. To illustrate the impact of textbook authors: Professor Jay Pasachoff, an astronomer at Williams College USA, not only writes textbooks for the 200,000 post-secondary students who take introductory astronomy courses in North America each year, but also co-authors textbooks for high school, junior high school and elementary school; he also writes astronomy guidebooks for the general public. This is a powerful contribution to astronomy education, which should not go unnoticed or unused. Many other professional astronomers also author excellent textbooks for these purposes.

Perhaps the best way to generate effective educational material would be through a formal science education project, tied to documents such as the US National Science Education Standards (NRC 1996) and carried out with input and assistance from educators. In the US, the National Science Foundation funds a variety of science education projects and programmes. I urge organizations like the International Dark-Sky Association, and astronomers throughout the world, to find and use such funding to develop the best possible education material. Astronomers and school-teachers in France have been especially successful in developing effective educational materials.

Other threats to the astronomical environment - pollution of the electromagnetic spectrum, and space debris - also fit naturally in the school science curriculum. The electromagnetic spectrum is introduced in almost every secondary school and university astronomy course, but there is seldom any specific discussion of which radio wavelengths are used for astronomy and which for everyday practical and commercial applications. The specific contributions of radio astronomy should be highlighted. Likewise, almost every astronomy textbook discusses orbital motion in the earth's gravitational field and shows a diagram of near-earth space. A discussion of space debris - where it is and how it moves - would provide a natural application.

5.   Light Pollution and Informal Education

Since other speakers at this Symposium have discussed public education about light pollution, I will mention a few specific topics only.

The Mass Media. The mass media have a profound effect on public understanding of, and attitudes to, science. Scientists often complain about media coverage (or lack thereof) of science topics. The situation can be helped if scientists learn why and how the media operate, and work with them to improve science coverage. In my experience, media coverage of light pollution has been accurate and sympathetic - perhaps because it is an issue to which everyone can relate. During a recent (July 1999) joint meeting of the Astronomical Society of the Pacific, the Royal Astronomical Society of Canada and the American Association of Variable Star Observers, in Toronto, there were three major articles about Ontario's new "dark sky reserve", including a front-page article in the *Globe & Mail*, which styles itself "Canada's national newspaper". (The Torrance Barrens Conservation Reserve is a provincial park; dark skies will be one of its aims and one of its attractions.)

Specialized Media. As well as the "mass media", there are scientific magazines which have a very large readership and influence - *National Geographic* and *Scientific American*, for instance. Preservation of the astronomical environment would be an interesting and appropriate topic for either of these publications and would have tremendous impact; priority should be given to getting articles into such publications.

Amateur Astronomers. There are various definitions of *amateur astronomer*. The simplest is "someone who enjoys astronomy and cultivates it as a hobby". They may be content to read about astronomy, attend lectures, or watch TV documentaries. They may enjoy sky-gazing and be adept in using star charts and telescopes. Williams (1988) has proposed a more stringent definition: "someone who does astronomy with a high degree of skill, but not for pay". Such individuals can make important contributions to astronomical research, through the measurement of variable stars, for instance, or the discovery of comets, asteroids, novae and supernovae. They can also make important contributions to astronomical education (Percy 1998b): they produce radio and TV programmes; they write newspaper and magazine articles and books; they give astronomy courses and lectures; they volunteer in planetaria, science centres and public observatories; they organize star parties (an excellent means of educating the public about light pollution); they are the driving force behind International Astronomy Day. They can even contribute to astronomy education in the schools, through programmes such as the Astronomical Society of the Pacific's *Project ASTRO* (Bennett et al. 1998). The number of "master" amateur astronomers in North America is comparable with the number of professional astronomers; the total number of astronomy hobbyists is 10-50 times greater (Gada et al. 1999). They are certainly our "grass-roots allies" in the battle against light pollution; see Percy (1998a) for specific examples.

Other Means of Informal Education. Planetaria, of course, are the last bastions of darks skies in the cities. They are especially appreciated by young people. When Toronto's McLaughlin Planetarium closed, several years ago, the most eloquent lament was written by a student, in the University newspaper. Astronomy programmes in parks, camps, youth groups (Boy Scouts and Girl Scouts, for instance) and outdoor education centres are also an ideal way of introducing people to the beauty of dark skies. We should remember, though, that not everyone has the opportunity to travel to these dark-sky locations.

Education should begin with astronomers themselves: *every astronomy library and ideally every astronomer's bookshelf, should include a complete set of IDA Factsheets and reprints of key articles on light pollution.*

Educating the world about astronomy and about light pollution is a major challenge which can be met if we work together with like-minded organizations. In all parts of the world, this concept of "partnership" can lead to "astronomy education initiatives" which can provide both a rationale and a possible source of funding for education about the preservation of the astronomical environment. IAU Commission 50 (for the professional astronomy community) and the International Dark-Sky Association (for the "astronomical community" in general) can play a key role in these partnerships; *we should all support their work.*

Acknowledgements. My participation in IAU Symposium 196 was made possible by a research grant from the Natural Sciences and Engineering Research Council of Canada. I thank David Crawford, Margarita Metaxa and Syuzo Isobe for many interesting and useful discussions.

References

Bennett, M., Fraknoi, A., & Richter, J. 1998, in *New Trends in Teaching Astronomy*, ed. L. Gouguenheim *et al.*, Cambridge University Press, 249

Crawford, D., Metaxa, M., & Percy, J.R. 1998, *"Light Pollution"*, special issue of The Universe in your Classroom, ASP, San Francisco CA

Fraknoi, A., 1996, in *Astronomy Education: Current Developments, Future Coordination*, ed. J.R. Percy, ASP Conf. Series. 89, 9

Gada, A., Stern, A., & Williams, T.R. 1999, paper presented at Amateur-Professional Partnership in Astronomical Research and Education, Toronto, Canada, 5 July 1999

National Research Council 1996, "National Science Education Standards", National Academy Press, Washington DC

Percy, J.R., (ed.) 1996, Astronomy Education: Current Developments, Future Coordination, ASP Conf. Series. 89

Percy, J.R. 1998a, in *Preserving the Astronomical Windows*, ed. S. Isobe & T. Hirayama, ASP Conf. Series 139, 7

Percy, J.R., 1998b, in *New Trends in Teaching Astronomy*, ed. L. Gouguenheim *et al.*, Cambridge University Press, 205

Plumb, D., Ritter, B., James, E., & Hirsch, A. 1999, Science 9, ITP Nelson, Toronto, Canada

Williams, T.R., 1988, in *Stargazers: The Contribution of Amateurs to Astronomy*, ed. S. Dunlop & M. Gerbaldi, Springer-Verlag, 24

Preserving the Astronomical Sky
IAU Symposium, Vol. 196, 2001
R. J. Cohen and W. T. Sullivan, III, eds.

The Light Pollution Programme in Greece

M. Metaxa

Arsakeio High School 63 Eth.Antistaseos, 15231 Athens, Greece
email:mmetaxa@compulink.gr

Abstract. The problem of light pollution exists almost everywhere, and is still growing rapidly. The preservation of the astronomical environment requires effective education. In this paper we present our educational programme, based in Greece and also including other countries. For full information see our web pages:

http://www.uoi.gr/english/EPL/LP/lp.htm

1. Introduction

The Greek "light pollution" educational programme has been arranged through the Greek Ministry of Education and Religion with support and finance from the EPEAEK Action III initiative. The two year programme has run from 1997 to 1999. The programme was a proposal of the Astrolaboratory of the Second Lyceum of the Arsakeio of Athens. Two schools were acting as partners in the programme, the EPL from Ioannina, Greece and the Grammar School of Manchester, UK.

It was the first time in Greece that almost all the scientific organizations related to a topic cooperated for an educational programme. To be more precise, this programme involved observatories (Athens, Crete), universities (Athens, Crete, Thessaloniki, Ioannina), environmental organizations (Sea Turtle, Mio), foundations (Eugenides, Goulandri), municipalities (Athens, Pireus, P. Faliro, Lemnos, Axioupoli, Ag. Nicolaos), lighting companies (Philips, Siemens, iGuzzini), and schools from all over the world (40 schools with 720 students)

2. Description of the Programme

Besides the three main partners, a great interest was expressed from other Greek and foreign schools. As a result in total 40 schools (32 from Greece and 8 from abroad) with 76 teachers and 720 students collaborated with us.

The programme's purpose was:

- to familiarize students with the problems of light pollution through astronomy, physics and computer science

- to make students consider the cultural and social dimensions of the impact of light pollution, and

- to help students to appreciate the effects of light pollution on heritage and environment throughout their country.

In order for this programme to be effective, the students were divided into four groups, according to their preferences and abilities:

1. The Astronomical Group - which studied the astronomical dimension of the problem.

2. The Lighting Group - which studied the design of lighting, the different types of lamp and their effects on the problem.

3. The Social Group - which studied the social dimension of light pollution such as its effects on the ecosystem, the psychological dimension and the laws that must be enacted to minimise these problems.

4. The Public Relations Group - which consisted of members from the other groups. Their task was to inform the local authorities, media and society about the problem by organising special events. Their main target was to attempt to influence planning authorities to produce efficient and effective lighting schemes.

2.1. Structure of the Programme

The programme had three main parts. The first was related to the background knowledge that the students and teachers should learn and the activities they should be involved in. In the second part of the programme the participants communicated their activities and results to the local mayors and other authorities. In that way the programme was opened to the public and to the media, and thus given a different dimension. The third part consisted of meetings for the participants to get know each other, communicate their results, make presentations, and learn from experts. Exhibitions of photographs as examples of good and bad lighting were also held.

In order to accomplish all these, a very tight organization was needed. We thus established in 18 cities light pollution centres, whose main duties were:
(a) to coordinate and inform the schools situated in their local area about the programme;
(b) to inform and communicate about the light pollution problem with their local authorities, local environmental associations and scientific societies, and
(c) to inform local media about the light pollution problem.

2.2. Students' Activities

Through the various activities that were part of the programme, we wanted our students and the general public to become familiar with the night sky. Additionally we wanted to increase awareness of the effects of light and air pollution and to attempt to influence planning authorities to produce efficient and effective lighting schemes. Well designed lights not only cut down light pollution but also save energy. The activities proposed were similar to those that have been run in other countries for a number of years: United Kingdom, Japan, United States and Canada. Student activities included:

1. A visual project, i.e. students reported which stars they could see in the Pleiades cluster and/or in a selected part of the constellation of Ursa Minor.

2. A photographic project, i.e. students took standardized photographs of the sky.

In addition, students were asked to take photos of good and bad lighting examples in their local area. After familiarizing themeselves with the CIE zoning system, they were then asked to implement the system and check it within their local area.

3. Theoretical Framework

Contemporary teaching requires a connection with events in our everyday lives. This is very succesfully provided by the UNESCO model for environmental studies. This model requires that each project should be placed in the following environment: natural, historical, social, and technological. The goal of this model is "the development of citizens/people with knowledge, sensitivity, imagination and an understanding of their relationship with their physical and human environment, ready to suggest solutions and participate in decision making and implementation".

This model fits excellently with our light pollution project and so we followed it. In our case, the natural environment is provided by the night sky, or the loss of the night sky! The historical environment is provided by the historical activities, buildings and persons who were related to the night sky, e.g. the ancient Greeks. The social environment is provided by the social activities concerned with communication of the problem to the local authorities, the media and professionals. The technological environment is provided by the means that students use in order to communicate, get information, etc. These means are mainly computers and the world wide web. Use of technology in the classroom can foster an environment that more closely reflects the methods scientists use in doing research.

4. Key Elements of our Programme

(a) Meetings
During the programme four meetings have been held: one in Manchester, two in Athens and one in Agios Nicolaos of Crete, Greece. All these meetings were successful and were attended in total by 500 students and 30 teachers. Bad and good lighting examples were presented in students' posters (by photographs), while they also talked about their local light pollution problems and the actions they had taken to overcome them. Scientists also gave lectures during the meetings. The final event was a Light Pollution Symposium held in Athens. Here we would like to express our deepest thanks to, Dr. D. Crawford (USA), Prof. J. Percy (Canada), Prof. J. Pasachoff (USA), Dr. S. Isobe (Japan) and Prof. Osorio (Portugal) for the honour of their participation in our Symposium and for giving our students the opportunity to better understand many aspects of the

problem. Furthermore, we hosted teachers and students from Italy, Switzerland and our partner school from Manchester, UK. From Greece, 30 schools participated with oral or poster papers.

(b) Publications
The students and teachers that participated in our programme often opened their results to the public. They communicated regularly with their local authorities, environmental associations and scientific societies on the light pollution problem. In this way they gained a real sense of participation.

5.   Results and Conclusions

Through this programme many interesting things came to the surface. Most of the teachers and students were totally unaware of the problem at first, and after hearing about it were surprized and became strong supporters of finding solutions. After all, it is an important and "different" environmental issue.

After the meeting that was held at Ag. Nicolaos of Crete, a pilot programme of a good lighting environment was implemented for a local road. Also, in Tarrega, Spain ordinances were implemented by the local authorities after the communication of the problem through our participating school.

Goulandri Museum, in Athens, will host a diorama constructed by the students of Arsakeio School, with material from our programme, concerning light pollution. This Museum is visited by 1,000,000 people annually. Thus a different dimension of the programme will reach the public.

We conclude that our educational programme, over two years of implementation, reached society in a most effective way. Let me here just repeat the words of Prof. Percy: "The only good thing about light pollution is that it provides students with an excellent way of learning about science, technology and society." The result was the establishment of a Light Pollution Office in collaboration with IDA. This Office will continue the programme's efforts by
(a) exchanging views with Greek and foreign organizations;
(b) promoting standards that have been established;
(c) contacting municipalities about the problem in order to take actions, and
(d) educating the public, having as a final target regulations that will reduce light pollution.

Acknowledgments.   I am grateful to the organizers of this Symposium for the opportunity to deliver this paper, and to thereby become more involved in this important topic. I especially thank Dr. D. Crawford and the IDA for all the help and material given to us, as well as Dr. D. Schreuder of the CIE. I would like to acknowledge the funding of the programme through the Greek Ministry of Education, and to thank Philekpaideutiki Etereia and all of our foreign collaborators. I thank all the members of the scientific committee and especially Prof. J. Percy, who has actively supported and helped the programme. I would like especially to thank IDA, the lighting companies and especially iGuzzini Illuminazione SA, and the Greek Municipalities that cooperated with us.

Preserving the Astronomical Sky
IA U Symposium, Vol. 196, 2001
R. J. Cohen and W. T. Sullivan, III, eds.

Educating the Public about Light Pollution

Syuzo Isobe and Shiomi Hamamura

National Astronomical Observatory, Mitaka, Tokyo, Japan

Christopher D. Elvidge

NOAA National Geophysical Data Center, Boulder, Colorado, U.S.A.

Abstract. Using low-gain DMSP data, we obtained absolute values of light energy loss ejected to space from different cities. Showing these data to local people in each city encourages them to try to reduce light energy loss. This educational approach is very effective in reducing light pollution.

1. Introduction

It is true that the public need and enjoy illuminating light. However, outdoor lighting is producing light pollution. We professional and amateur astronomers have a tendency to request light reduction from the public with arguments that star fields are beautiful and that scientific outputs are important for human-beings. Sometimes we succeed in making the public turn down the lights, especially for astronomically interesting events such as Comet Hyakutake, Comet Hale-Bopp and the Leonid showers.

It is also true that the fraction of people enjoying astronomical observations and star-watching is not large, that is, only one tenth or one hundredth of the total population. In order to get the support of the majority of people, we have to develop a clear way to educate the public. Since we have the DMSP (Defense Meteorological Satellite Program) data, we are trying to develop a new way to reduce light pollution.

2. Energy Loss

Electrical engineering has been much developed in the 20th century, especially in these last decades, during which people have enjoyed a much brighter night-time environment. However, the public and also lighting designers want to have lighting fixtures which are well decorated, especially in day time. Then they most easily use lighting fixtures which eject a large fraction of light towards the sky (Kawakami & Isobe 1998), and this light becomes energy loss.

Light pollution of astronomical observations is mainly produced by light ejected in directions with elevation angles of 0° to 45°. As an example, such an effect was studied by Osman et al. (2001) for the case of Kottamia Observatory in Egypt. City light ejected directly in near-zenith directions is only energy loss but does not create much light pollution of nearby observatories.

3. DMSP Data

The US Air Force started the DMSP in 1972 and made continuous observations using a series of satellites. However, after the US Air Force had used those data, they had no interest in keeping the data in digital form and just kept photographic prints. Using those prints Sullivan (1991) produced a famous map of Earth at Night and later Nakayama (1992) read the brightness distributions of all the photographic prints into a computer, corrected the projection effects and obtained a better map of the Earth at Night.

Fortunately, since 1993 the National Geographic Data Center receive digital data from the US Air Force and keep all the data in the form of 8-mm tapes. Through some communications with a leader on this programme, Dr. Elvidge of NGDC, we also get the data interesting for us as a collaborative project with NGDC.

The DMSP data from 1993 to 1996 were obtained using a high-gain mode which can detect light levels within the range $8 \times 10^{-11} \mathrm{W/cm^2/sr/\mu m}$ to 7×10^{-9} $\mathrm{W/cm^2/sr/\mu m}$ (Elvidge et al. 1999), but a large fraction of the area of big cities was usually saturated. Therefore, we cannot measure absolute values of light energy detected. However, we can still obtain a reasonable result for small cities and we have demonstrated increases of light energy loss from 1993 to 1997 in the cities of Akita, Shizuoka, Hiroshima, Matsuyama, and Tokushima in Japan (Isobe & Hamamura 2000).

By a strong request of Dr. C. Elvidge, the US Air Force made several observations with low gain. Gain number is different from time to time, but in most cases we could get non-saturated data, except for the central parts of very big cites such as Tokyo.

Table 1 shows all the results that we could reduce to date. Although we have a much larger amount of data, we have reduced only a small part of them because of a shortage of manpower. However, we will extend our efforts. Figures 1 and 2 show two maps of the brightness distribution. Further results of this ongoing work, tables and a picture gallery, can be viewed at the following web site, http://neowg.mtk.nao.ac.jp/ .

Fortunately, in 1999 the US Air Force plan to observe with low-gain much more frequently and then we will be able to compare values of light energy loss at different years.

4. Discussion

In Table 1, column 2, the observed value is just for observed total intensity in each area. Light energy loss per year is estimated assuming that the amount of energy lost to space is constant for 10 hours per night for each night of one year. Then light energy loss/km^2/year is calculated. Since the dynamic range of the detector is not so large, the resulted light energy loss/km^2/year estimated for most of the cities has values spanning only a factor of 20.

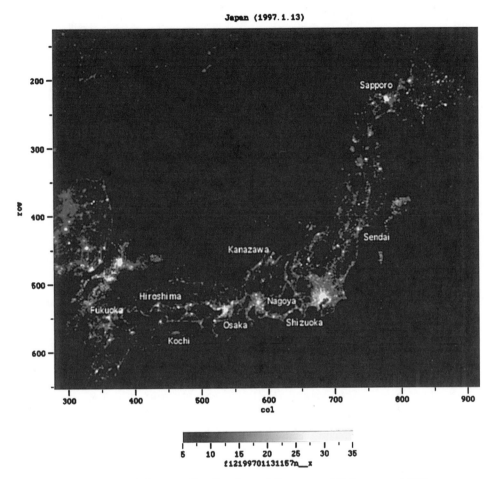

Figure 1. Brightness distribution of Japan on 13 January 1997.

There are fairly high values for cities in Canada, which may be caused by light reflection from snow on the roads. We see a similar effect for the Japanese city of Sapporo. To escape this snow effect, we should use low-gain data obtained in a season other than winter. However, only winter data are available in 1997 and therefore we have to wait to make a definite conclusion.

Although our data have still a problem to be resolved, we have now a fairly large number of light energy loss values for cities. In Japan, the Environmental Agency set a guideline to protect light pollution and assigned six cities to work on this. Then, if some city tries to reduce upward light, we can detect a decrease of light energy loss. Local people can see that their effort is directly linked to reduced energy loss and get a triggering motivation to consider the conservation of energy for future generations.

This kind of work is not a direct method to reduce a light pollution, but a certain method for making it happen.

Table 1.   Energy detected by the DMSP and values estimated from it.

	Observed Value $(10^{-8}\text{W/cm}^2/\text{st}/\mu\text{m})$	Light Energy Loss/Year (10^6kWh)	Area (km^2)	Light Energy Loss/Area/Year (10^6kWh/km^2)
Japan (1997.1.13)				
Sapporo	2.47×10^3	14.8	1046	1.41×10^{-2}
Sendai	7.40×10^2	4.43	463	9.57×10^{-3}
Kanazawa	5.18×10^2	3.10	543	5.71×10^{-3}
Shizuoka	4.56×10^2	2.73	528	5.17×10^{-3}
Nagoya	3.83×10^3	22.9	1519	1.51×10^{-2}
Osaka	5.85×10^3	35.1	1896	1.85×10^{-2}
Hiroshima	8.72×10^2	5.22	1001	5.21×10^{-3}
Kochi	2.39×10^2	1.43	729	1.96×10^{-3}
Fukuoka	1.56×10^3	9.35	1026	9.11×10^{-3}
Korea (1997.2.27)				
Seoul	7.07×10^3	42.4	2266	1.87×10^{-2}
Pusan	1.49×10^3	8.96	910	9.85×10^{-3}
Pyongyang	2.38	0.0143	133	1.08×10^{-4}
Europe (1997.1.13)				
London	4.84×10^3	29.0	2030	1.43×10^{-2}
Amsterdam	1.07×10^3	6.43	367	1.75×10^{-2}
Leiden	2.16×10^2	1.29	138	9.35×10^{-3}
Bruxelles	9.64×10^2	5.78	536	1.08×10^{-2}
Paris	6.33×10^3	37.9	2091	1.81×10^{-2}
Europe (1997.2.3)				
Wein	1.20×10^3	7.19	1080	6.66×10^{-3}
Budapest	1.58×10^3	9.44	1331	7.09×10^{-3}
Praha	1.26×10^3	7.55	1020	7.40×10^{-3}
Bratislava	4.25×10^3	2.55	389	6.56×10^{-3}
Warszawa	1.47×10^3	8.81	950	9.27×10^{-3}
Dresden	9.23×10^2	5.53	1162	4.76×10^{-3}
Brno	4.02×10^2	2.41	384	6.28×10^{-3}
Krakow	7.35×10^2	4.40	592	7.43×10^{-3}
Milano	2.32×10^3	13.9	1434	9.69×10^{-3}
Zagreb	4.78×10^2	2.86	380	7.53×10^{-3}
Greece (1997.2.5)				
Athinai	2.49×10^3	14.9	1837	8.11×10^{-3}
Tessaloniki	6.67×10^2	4.00	711	5.63×10^{-3}
Larisa	1.13×10^2	0.674	219	3.08×10^{-3}
Volos	1.25×10^2	0.749	210	3.57×10^{-3}
Lamia	65.6	0.393	148	2.66×10^{-3}
Iraklion	1.06×10^2	0.637	273	2.33×10^{-3}
Middle East (1997.1.9)				
Tel Aviv-Yafo	1.72×10^3	10.3	813	1.27×10^{-2}
Jerusalem	7.40×10^2	4.43	511	8.67×10^{-3}
Amman	8.77×10^2	5.25	478	1.10×10^{-2}
Haifa	5.53×10^2	3.31	253	1.31×10^{-2}
Damascus	4.98×10^2	2.98	320	9.31×10^{-3}
Beirut	6.48×10^2	3.88	464	8.36×10^{-3}
Baghdad	9.39×10^2	5.62	1510	3.72×10^{-3}
Egypt (1997.2.5)				
Cairo	4.51×10^3	27.0	1968	1.37×10^{-2}
Alexandria	6.52×10^2	3.90	818	4.77×10^{-3}
Ismailiya	2.88×10^2	1.73	273	6.34×10^{-3}
Suez	3.38×10^2	2.02	264	7.65×10^{-3}

Table 1.   continued

	Observed Value $(10^{-8}\text{W/cm}^2/\text{st}/\mu\text{m})$	Light Energy Loss/Year (10^6kWh)	Area (km^2)	Light Energy Loss/Area/Year (10^6kWh/km^2)
Canada (1997.1.12)				
Quebec	6.13×10^3	36.7	1767	2.08×10^{-2}
Trois Riviere	1.23×10^3	7.37	36	0.205
Montreal	2.32×10^4	139	4039	3.44×10^{-2}
Ottawa	5.44×10^3	32.6	1612	2.02×10^{-2}
Toronto	2.29×10^4	137	4330	3.16×10^{-2}
Sudbury	1.41×10^3	8.45	603	1.40×10^{-2}
Chicoutimi	1.28×10^3	7.65	400	1.91×10^{-2}
Calgary	1.39×10^4	83.4	1901	4.39×10^{-2}
Edmonton	9.83×10^3	58.9	1819	3.24×10^{-2}
U.S.A. (1997.2.4)				
New York (Long Is.)	2.26×10^4	136	9095	1.50×10^{-2}
Philadelphia	8.10×10^3	148.5	2690	1.80×10^{-2}
Boston	2.51×10^3	15.0	1122	1.34×10^{-2}
Baltimore	4.88×10^3	29.2	1854	1.57×10^{-2}
Washington D.C.	6.98×10^3	41.8	3087	1.35×10^{-2}
Buffalo	3.34×10^3	20.0	1250	1.60×10^{-2}
U.S.A. (1997.1.12)				
Mineapolis	2.04×10^4	122	4329	2.82×10^{-2}
St. Louis	1.55×10^4	93.0	4061	2.29×10^{-2}
Kansas City	1.19×10^4	71.5	4611	1.55×10^{-2}
Las Vegas	6.35×10^3	38.0	1552	2.45×10^{-2}
Phoenix	9.18×10^3	55.0	4782	1.15×10^{-2}
Tuscon	2.20×10^3	13.2	1804	7.32×10^{-3}
Middle America (1997.2.8)				
Mexico City	9.82×10^3	58.8	4015	1.46×10^{-2}
Monterrey	1.63×10^3	9.79	1701	5.76×10^{-3}
Guadalajara	2.56×10^3	15.3	1260	1.21×10^{-2}
Guatemala	7.23×10^2	4.33	1184	3.66×10^{-3}
San Salvador	4.57×10^2	2.74	1038	2.64×10^{-3}
Tegucigalpa	3.23×10^2	1.93	489	3.95×10^{-3}
Managua	2.75×10^2	1.65	630	2.62×10^{-3}
San Jose	8.64×10^2	5.17	1141	4.53×10^{-2}
Panama	5.35×10^2	3.21	891	3.60×10^{-3}
Habana	3.61×10^2	2.16	706	3.06×10^{-3}
Kingston	7.33×10^2	4.39	891	4.93×10^{-3}

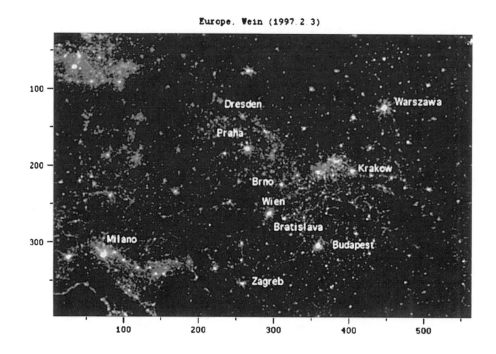

Figure 2.   Brightness distribution of Europe on 3 February 1997.

References

Elvidge, C. D., Baugh, K. E., Dietz, J. B., Bland, T., Sutton, P. C., and Kroehl, H. W. 1999, Radiance Calibration of DMSP-OLS Low-Light Imaging Data of Human Settlements, Remote Sensing and Environment, Vol. 68, 77-88

Isobe, S. and Hamamura, S. 1998, Ejected City Light of Japan Observed by a Defense Meteorological Satellite Program, ASP Conf. Ser., 139, 191-199

Isobe, S. and Hamamura, S. 2000, Monitoring Light Energy Loss Estimated by the DMSP Satellites, Mem. Soc. Astr. Italiana, 71, 131-138

Kawakami, K. and Isobe, S. 1998, A Study of Luminous Intensity Distribution of Lighting Lamps, ASP Conf. Ser., 139, 161-165

Nakayama, Y. 1992, Poster of Earth at Night

Osman, A., Isobe, S., Nawar, S. and Morcos, A. 2001, this symposium

Sullivan III, W. T. 1991, Earth at Night-An Image of The Nighttime Earth Based on Cloud-free Satellite Photographs, ASP Conf. Ser., 17, 11-17

Preserving the Astronomical Sky
IAU Symposium, Vol. 196, 2001
R. J. Cohen and W. T. Sullivan, III, eds.

The Cultural Value of Radio Astronomy

Woodruff T. Sullivan, III

*Dept. of Astronomy Box 351580 Univ. of Washington
Seattle, WA 98195 USA*
woody@astro.washington.edu

Abstract. In order to compete successfully in the marketplace of the radio spectrum radio astronomers must appeal not to economic gain, but to the cultural value of their enterprise. In the real world this can be problematic, but it is not hopeless. This paper gives arguments why radio astronomy, no less than astronomy as a whole, has great cultural value whether considered from an environmental or an intellectual point of view.

1. Introduction

Protecting and preserving radio astronomy takes place in the political arena, where the larger body politic, society as a whole, makes the vital decisions. The political process leading to these decisions considers economic, social and cultural values. Therefore, if we are to be effective in protecting radio astronomy, we must marshal strong and persuasive arguments that address as many of these factors as possible.

In this paper I argue that our strongest hand in demonstrating the worth of radio astronomy is in the *cultural* realm. The economic and social factors are, unfortunately, woefully and powerfully against us. When dealing with telecommunications firms, we bring only bad economic news to their managers, their satellite designers and their stockholders. What we are asking will cost them money and time and be a hindrance to providing the services eagerly sought by their public (and by many of us, too!). We must thus appeal to other countervailing values if radio astronomy signals are to survive amidst the cacophony of modern communications systems[1].

There are two primary aspects of the cultural value of radio astronomy. The first is *environmental* and has the potential of becoming a powerful approach, but at this point requires more sensitization and education of the public and

---

[1] As discussed by Finley in this volume, the situation is vastly different with regard to light pollution. In addition to the cultural arguments about the value of (optical) astronomy, the International Dark-Sky Association and its allies have established that making the changes desired by astronomers also yields economic, social, aesthetic and environmental benefits to the general public - it is a win-win situation (see the article by Crawford in this volume). Progress is happening and at an accelerating pace, although there remains the need for a huge educational and political effort, and at optical wavelengths there are no large regulatory bodies with which to work.

of decision-makers. The second is *intellectual* and is our ultimate strength. If we cannot persuade people that we are asking profound questions and finding exciting and important answers, then the battle is lost.

2.   Radio Astronomy as an Environmental Issue

The history of the past forty years has seen a steady enlargement in what are considered "environmental" issues. Initially attention was focused on the more obvious dangers to the fundamental natural resources of air, water, soil, and flora and fauna. Environmentalists today, however, realize that other, more subtle, entities also affect the well-being of planet Earth and its inhabitants. Examples include the stratospheric layers (ozone depletion), general burning of hydrocarbons (global warming), the sonic environment (noise pollution) and the night-time environment (light pollution). For its environmental welfare the public is now accustomed to considering new phenomena and new ways of looking at old situations.

So it is with radio astronomy, an abstract and arcane field of whose value we must persuade the average person. Radio astronomers must start talking about the electromagnetic spectrum as a finite resource, not unlike virgin forests. Just as with the forests, the great bulk of the spectrum resource has gone to serve the material, everyday needs of society. The question is what to do with the remaining small portion? Adding it to the other "useful" spectrum will minimally enhance overall economic prosperity, but will leave us with zero for radio astronomy. Is the trade-off worth it? Once again, this is a political issue, but it is winnable - with proper education the public on the whole is supportive of preservationist stances.

The task then is to convince the public that the value of radio astronomy is not unlike the value of a forest of 500-year-old trees. In the case of the forest, our emotions respond to its magnificence and our intellect craves to study its unique and fragile ecosystem. In the case of radio astronomy, can we likewise show that fundamental human needs are satisfied? First, we must show the sustaining and exciting nature of astronomy as a whole, and then demonstrate that radio astronomy is one of astronomy's key components. People must value scientists probing the Universe to better determine what's there and how it works, searching for clues about where we fit in the Big Picture. Then they must be made aware of how our society is endangering our ability to view this Universe.

We need not start from scratch. In a recent article on the concept of the *commons*, environmentalists Snider and Warshall (1998) have discussed a wide variety of common resources, ranging from local fisheries to public roads to sacred heritages to humankind's legacy of ethics. And amongst their list of thirty entities is the electromagnetic spectrum, or "spectral commons"! Once this principle becomes established, namely that the electromagnetic spectrum should be treated as a public commons, then the way is greatly eased for preservation of a portion of it for other than practical uses. We can also appeal to the familiar warning of the "tragedy of the commons". Ecologist Garrett Hardin coined this phrase for the erosion and eventual destruction of any commons that can occur if each person acts only in his or her own self-interest, without regard to the interests of the community.

Morimoto (1993) has called the beautiful view of the Universe afforded by a dark sky, whether optical or radio, a "rare and precious treasure of Nature". Paul Vanden Bout (1994) has nicely developed the environmental analogy in his article "Preserving wilderness areas in the radio spectrum". He points out that 1 MHz of bandwidth in the cellular telephone band may have a value of one billion dollars, but does that mean it should all be sold to the highest bidder? What is the value of the acreage of the Grand Canyon, or of the Great Barrier Reef, or of Fujiyama? Vanden Bout points out that spectrum is like public lands:

> It is fixed in overall amount, controlled by the government, serving many varied constituencies, with important roles in the economy and safety of the nation. In this view of the spectrum, to give a few examples, the protected radio astronomy bands are analogous to wilderness areas and national parks, the navigation bands to ... defense installations and the broadcast bands ... to lands leased for cattle grazing or mineral extraction. Market forces are applied *after* recognizing the importance of some services beyond their purely economic value.

Continuing the analogy, he points out that the lack of wisdom in siting heavy industrial activities adjacent to a national park is the same as that of assigning frequencies for ubiquitous and powerful satellite transmitters next to those for radio astronomy.

Reserving slices of spectrum as wilderness is of course an abstract concept. But we can also more directly profit from the concept of wilderness areas by seeking actual zones or regions on the Earth free of humanmade interference. The largest and earliest such zone is the National Radio Quiet Zone, a region of ∼160 km×160 km established in 1956 for the benefit of the National Radio Astronomy Observatory at Green Bank, West Virginia, USA. At this symposium Butcher has argued for the creation of one or more International Radio Quiet Zones to protect the next generation of radio telescopes, such as the Square Kilometre Array. The ultimate such zone would be the far side of the Moon and indeed some are working to that end, but this is only a partial (and very expensive) solution and only for the distant future. Radio astronomy from the Moon, if and when it happens, will no more eliminate the need for ground-based radio observatories than the orbiting Hubble Space Telescope has made ground-based optical observatories unnecessary.

3. Radio Astronomy as an Intellectual Value

To make our case one must first establish the intellectual value of all of astronomy, then focus on radio astronomy and show that it is integral to modern astronomy. The cultural value and impact of astronomy as a whole, however, is a huge topic and beyond the scope of this paper. Plato even said in the *Timaeus* (47a) that philosophy itself began with astronomy:

> For had we never seen the stars and the sun and the heaven, none of
> the words which we have spoken about the universe would ever have
> been uttered ... [These] have created number and have given us a
> conception of time and the power of inquiring about the nature of
> the universe. And from this source we have derived philosophy.

If one were to write a book on "astronomy and culture", some of the possible chapter titles reveal how intimately astronomy is woven into who we are and how we see ourselves: time and calendrics, navigation, astrology, eclipses, art and music, extraterrestrial life, physics, etc. It might be thought that most of these connections apply only to the past, but this is wrong on almost every count. A few examples: (1) regarding time and navigation, the Global Positioning Satellite (GPS) system (which was largely designed by astronomers and is still fine-tuned by the US Naval Observatory) is transforming not just navigation, but everything from ornithology to police work; (2) regarding the arts, cosmic themes are still frequent in modern art (and certainly in popular culture); and (3) regarding astrology and extraterrestrial life, widespread beliefs not unlike those of the past are rampant (and astronomy needs to exist to weigh in with its scientific input); and (4) regarding new physics, the "astronomical laboratory" still supplies unmatched testing grounds for nuclear physics, high-energy physics and gravitation.

Astronomy as a whole thus continues as a vital part of culture, as it always has. But what about radio astronomy, which has been with us for only a half century? An examination of its short history reveals its critical role in our present understanding of the cosmos. Radio waves were the first non-optical portion of the spectrum to be exploited in a major way. The astronomy of the 1940s and 1950s was revolutionized by discoveries of radio sources that allowed probing of the farthest reaches of the Universe, relativistic electrons threading the Milky Way, compact sources of unprecedented energy, dark cool clouds that traced our Galaxy's rotation, and huge bursts of energy in the solar corona. As radio astronomy matured in the 1960s and 1970s it led to the discovery of quasars, the cosmic microwave background radiation, pulsars and complex interstellar molecules. The last two decades have seen the first evidence for gravitational waves (emitted by a pulsar system), pulsar clocks as accurate as anything on Earth, excellent evidence for the presence of huge black holes at the centres of galaxies, and mapping of the early Universe's structure through subtle corrugations in the cosmic background.

4.   A Few Examples of Radio Astronomy's Importance

Rather than try to cover all of the major contributions of radio astronomy, this section reviews three of the most important in more detail. These examples have been chosen because of the profound effect they have had on how we view our Universe. As it turns out, these have also led to no less than three Nobel Prizes.

4.1.   Cosmic Background Radiation

Using a strange, sugar-scoop-shaped antenna at Bell Labs in New Jersey, Penzias and Wilson (1965) deployed a sensitive and stable receiver system for a study of the Milky Way's background synchrotron radiation. The experiment was exquisitely designed and allowed an *absolute* measurement of radio intensity from any direction in the sky[2] [3].

The surprising finding was that there was in fact no sector of the sky that was radio-quiet: rather, the entire sky emitted as if at a uniform temperature of 3 K. This radiation soon came to be accepted in the context of the Big Bang theory as the cooled, redshifted remnant of an originally hot, dense Universe. Three decades of subsequent measurements, most accurately by the COBE satellite in the mid 1990s, have confirmed this basic idea to the point where this cosmic background radiation is still looked upon as one of the two or three key pieces of observational evidence underpinning Big Bang cosmology.

4.2.   Pulsars

Once again, a serendipitous observation, doggedly tracked down, led to a basic discovery. This time it was an array of 2,048 dipole antennas at Cambridge University in England, designed to study fast intensity scintillations that radio sources exhibit because of variable propagation conditions (primarily changing electron density in the interplanetary medium). But amongst the expected scintillating sources were discovered four that changed not in the usual irregular fashion, but with incredible regularity. Already in their first article Hewish, Bell, Pilkington, Scott and Collins (1968) mentioned the possibility that these pulsating sources might be neutron stars, which is still the best interpretation. Neutron stars had been talked about as a theoretical construct since the 1930s, but here was the first observational evidence for their existence. They are the fast-rotating, collapsed-core remnants of supernova explosions, superdense objects with the mass of a star but the size of a mountain. Besides providing a powerful laboratory for testing ideas of the physics of matter under such extreme conditions, the pulsars (of which hundreds are now known) are beautiful clocks. The most accurate are those with the fastest periods (approaching one millisecond) and their long-term stability is as good as any of the manmade time standards on Earth.

One pulsar was found by Hulse and Taylor (1975) to be in orbit about an unseen companion object (itself soon inferred to be a neutron star, but radio quiet). A monitoring program using the 305-m diameter Arecibo dish soon revealed that the regular orbital motion of the pulsar about its companion every 7.75 hours was slowly shifting in a complex manner. The stability of the pulsar's 59.03 msec periodicity (about 1 part in 10^{17}) and the accumulation of many measurements of it, each accurate to 14 decimal places, allowed the subtle orbital changes to be detected and ascribed to a variety of relativistic effects. While most of these effects had been measured in other contexts, one provided wholly new evidence for Einstein's general theory of relativity, which predicts that oscillating masses should emit gravitational waves roughly analogous to

---

[2] In contrast, almost all radio astronomical measurements, then as now, are not absolute, but relative to a nearby patch of sky assumed to have no signal.

[3] It is ironic that this fundamental discovery was made with technology available as an offshoot of the then nascent development at Bell Labs of the first earth-orbiting communication satellites (Telstar, Early Bird, etc.).

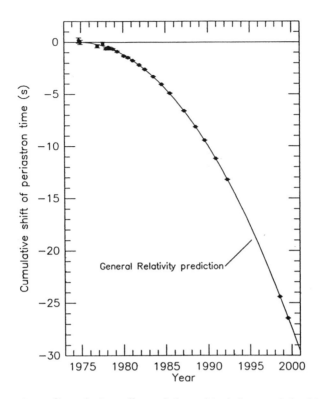

Figure 1.   Cumulative effect of the orbital decay of the binary pulsar
PSR B1913+16 over the past 25 years. The points are measured values
based on timing of the pulsar at Arecibo Observatory, while the line
is the effect expected from general relativistic emission of gravitational
waves. (Courtesy J.H. Taylor)

how oscillating electrons emit electromagnetic waves. Despite many previous
attempts to detect such waves directly, no one had succeeded. Within a decade
of monitoring, however, Taylor and his colleagues showed that the pulsar's orbit
was decaying exactly as would be expected if the two, orbiting neutron stars
were losing energy via emission of gravitational waves. The radio technique's
marvellous sensitivity uncovers the tiny decrease in every orbital period, two
parts in 10^{12}, or 76 μsec per year (Fig. 1). Our best theory of gravity became
even better.

4.3.   Microwave and Millimetre Spectroscopy of Molecules

Astronomical spectroscopy is an essential technique. By measuring radiation
at distinctive frequencies (spectral lines) characteristic of each element and
molecule, radio astronomers can determine not only the density and location
of these species, but also their velocity, temperature, pressure, ambient magnetic field, etc. The first radio spectral line to be detected (in 1951) and still

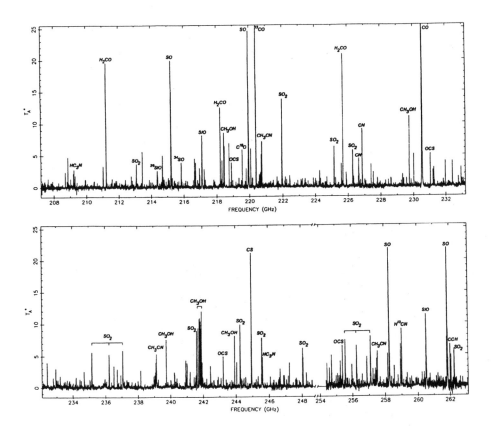

Figure 2. The spectrum of the Orion Molecular Cloud over the range 208 to 262 GHz, as measured at the Owens Valley Radio Observatory. Molecular identifications are given for the strongest of more than 800 spectral lines detected in this range. (Adapted from Blake et al. 1987)

the most important, was that at 1.420 GHz from the hydrogen atom, the most abundant element in the Universe. But the real floodgates opened in the late 1960s when a varied suite of molecules was found to be present in many interstellar regions of our galaxy the Milky Way. First spectral lines to be found were hydroxyl (OH), water vapor (H_2O), ammonia (NH3), carbon monoxide (CO), cyanide (CN), hydrogen cyanide (HCN) and formaldehyde (H_2CO). By now over 140 different molecular species, some with up to 13 atoms, have been detected. They include ethyl alcohol (!, CH_3CH_2OH), the sugar glycoaldehyde ($C_2H_4O_2$) and perhaps even the amino acid glycine (H_2NCH_2COOH). They reveal a previously unsuspected, rich carbon-based chemistry not unlike what may have occurred in the early solar system long before life appeared on Earth. Through the study of these molecules our ideas about the formation of stars and planets have been revolutionized.

With few exceptions these spectral lines happen to occur at frequencies above 20 GHz, most above 100 GHz. Figure 2 shows a spectrum over a range from 208 to 262 GHz, with some of the rich array of spectral lines identified. These all originate from a single direction in the sky, a particularly dense and dusty interstellar cloud and region of star formation.

5.   Conclusion

Whether considered from an environmental or a cultural point of view, radio astronomy is a resource vital to society. It is the duty of astronomers to demonstrate this to the public at large and to those who make decisions regarding usage of the radio spectrum. How foolish it would be if, after half a century of exciting radio discoveries that have profoundly affected how we view our Universe, society were to squelch this activity by wrapping the Earth in an impenetrable electromagnetic fog.

References

Blake G. A., Sutton E. C., Masson C. R. & Phillips T. G. 1987, Ap. J. 315, 621

Hewish, A., Bell, S.J., Pilkington, J.D.H., Scott, P.F.& Collins, R.A. 1968, Nature 217,709.

Hulse, R. & Taylor, J.H. 1975, Ap. J. L. 195, L51.

Morimoto. M. 1993, Modern Radio Science 1993, 213.

Penzias, A.A. & Wilson, R.W. 1965, Ap J 142, 419.

Snider, G. & Warshall, P. 1998, Whole Earth, Fall issue, 4-7 + 16-23 + 50.

Vanden Bout, P.A. 1994, in The Vanishing Universe, Ed. D. McNally (Cambridge: Cambridge Univ. Press), p. 97.

Preserving the Astronomical Sky
IAU Symposium, Vol. 196, 2001
R. J. Cohen and W. T. Sullivan, III, eds.

Educating the Public About Interference to Radio Observatories

David G. Finley[1]

National Radio Astronomy Observatory, P.O. Box O, Socorro, New Mexico 87801 USA

Abstract. Educating the public about interference to radio observatories is a different and more difficult task than educating the public about light pollution. Convincing and successful arguments against light pollution can be based on aesthetic, economic, cultural, safety and security considerations without relying solely on the need to preserve the environment for astronomy. In contrast, it is necessary to first convince members of the public of the value of radio astronomical research before making the case for interference protection. Once this is done, arguments about interference must be presented in ways understandable to a public that is, by and large, woefully uninformed about the technology involved. Successful approaches often borrow from the language of environmental protection and draw parallels to such issues as air and water pollution in justifying the expense of engineering measures to protect radio astronomy.

1. Introduction

Public support is a vital element in any effort to protect radio observatories from harmful interference that jeopardizes their ability to advance our knowledge of the Universe. While the support of the general public will not, by itself, ensure success in our efforts to protect the radio astronomical spectrum, lack of public support would certainly doom such efforts. The battle for spectrum protection is at its heart a political battle, and thus must be fought, at least in part, in the arena of public opinion.

This fact has been recognized by the optical-astronomy community in their efforts to reduce light pollution. The numerous successes at various governmental levels in implementing regulations to reduce light pollution have resulted from campaigns that convinced diverse segments of the population to support such measures.

While radio astronomers can follow such an example in building public support, they face significant obstacles not faced in the light-pollution battle. First, light pollution can be opposed on a number of grounds that resonate with people who have little or no interest in astronomy. The economic, aesthetic, cultural, safety and security benefits of reducing light pollution attract broad

---

[1]The National Radio Astronomy Observatory is a facility of the National Science Foundation, operated under cooperative agreement by Associated Universities, Inc.

377

support without the need to convince people to protect astronomical research. On the other hand, measures to mitigate interference to radio observatories do not produce a similar set of non-astronomical benefits. In addition, radio interference is a highly technical issue that is beyond the understanding of much of the public, making the task of gaining public support more difficult.

Successful efforts to build public support for protecting the radio astronomical spectrum thus must be built on a foundation of public support for astronomy itself. The arguments for spectrum protection then must be made in non-technical terms understandable to the public.

2. Public Involvement in Policy Issues

The public faces a wide variety of policy issues, and most individuals devote the time to become informed or active about only a small number of these. For any specific, specialized issue, the level of public interest and involvement has been described as a pyramidal structure (Almond 1950). In this pyramid (Figure 1), officials with the power to enact policy measures sit at the pinnacle. Below them is a group of experts who actively lead public opinion on the issue. Among the general public, those who are interested in the issue are divided among the *attentive* public and the *interested* public. The attentive public is generally more active in support of their viewpoint on the issue, while members of the interested public are not active. At the bottom of the pyramid, with the largest numbers, are members of the *residual* public, who have little or no interest in the issue.

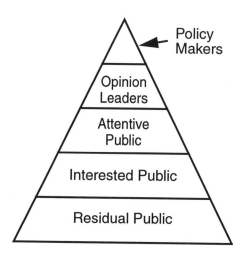

Figure 1.   Public Involvement in Policy Issues.

In the United States, only 14 percent of adults are included in the attentive public for all science and technology policy issues (National Science Board 1998). Much of the success of the optical-astronomy community in gaining support for measures to control light pollution comes from the fact that their arguments effectively motivate many people from the large residual public for science policy issues. A campaign against light pollution can, in effect, build its own policy pyramid that includes many people with no interest in astronomical research. Unfortunately, that is not the case with the issue of interference to radio observatories.

Radio astronomers must assume the role of opinion leaders and work to educate the public about the interference issue. Our objective must be to increase the numbers of our attentive and interested publics for this issue, and to move members of the interested public up into the ranks of the attentive public, and thus of active supporters. Since the principal difference between an active member of the attentive public and a passive member of the interested public is a self-perception of being adequately informed on the issue (Miller 1999), our educational efforts should be directed toward giving members of the public confidence that they understand the interference problem and also our suggested remedies.

3. The Value of Astronomy and of Radio Observations

Astronomy has an excellent record of contributing to human progress from ancient times to the present. It has inherent cultural value in providing humans with a sense of their place in the Universe. It has produced numerous practical advances, and will continue to do so in the future. It has proved to be a powerful tool for science education and attracting young people to technical careers.

The inherent appeal of the night sky dates to prehistory and is part of cultures worldwide. Today, the basics of astronomical knowledge provide a conceptual framework for the universal human desire to understand our place in space and time.

The practical contributions of astronomy range from the calendar that helped early humans to progress beyond hunting and gathering to an agriculture-based society, to advances in modern medical imaging. The world-wide commerce we take for granted began because astronomers made it possible for mariners to navigate accurately across the oceans. Today, the communications and weather satellites upon which we depend represent a technology made possible by Kepler, Newton and the celestial mechanicians who followed them. Mathematics, computer science, telecommunications and other fields owe large debts to astronomy.

The visual appeal of astronomy and the excitement of new discoveries about the Universe make astronomy a powerful educational tool. People of all ages are drawn to science by the latest news from the frontiers of astrophysics. Astronomy serves as an entry point for a wide variety of scientific and technical careers, as young people first drawn to astronomy branch out into other technical fields that power today's high-tech economies. Non-scientists who maintain their interest in astronomy increase their scientific literacy, becoming more valuable as citizens, leaders and managers. Professional astronomers are aided greatly in their educational efforts by hundreds of thousands of amateur astronomers who volunteer their time to educate the public.

The Universe provides us with a "cosmic laboratory" of extreme conditions that cannot be replicated on Earth. This laboratory represents an invaluable resource for all humanity. From it, we continue to gain fundamental new knowledge about physics – knowledge that in the future may spawn entirely new technologies.

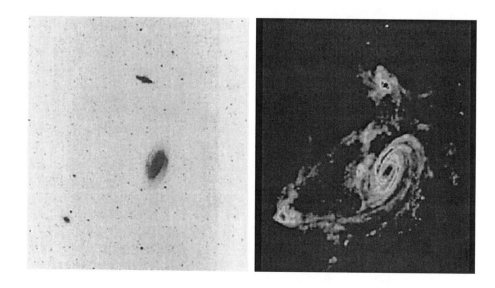

Figure 2.   Radio astronomy helps show the "whole picture". Optical image (left) of the M81 Group of galaxies fails to show the interactions revealed by a VLA image at 21 cm wavelength (right) (Yun et al. 1994).

Radio astronomy, in a mere six decades of existence, has revolutionized our understanding of the Universe. From revealing the remnant radiation of the Big Bang to showing the afterglows of Gamma Ray Bursters, radio observers have provided unique insights. Six of the ten Nobel laureates cited for work in astronomy did that work with radio telescopes. For a wide variety of astronomical phenomena, radio observations provide essential pieces of the "whole picture" required for full understanding (Figure 2). Astronomy has become a multi-wavelength enterprise, and radio observations are an integral part of this effort.

Convincing people of the value astronomy – and radio observations – provide to society is the essential first step in educating the public about interference to radio astronomy. We have a good case and we must make it vigorously. Every astronomer should become familiar with the contributions our science has made and its promise for the future. In public lectures, classrooms, and in casual conversations, we must not be shy or defensive about the value of astronomy. Indeed, we must convey this message at every opportunity. Doing so will not only help in the fight to preserve the radio astronomical spectrum but, in most countries, will also help maintain support for public financing of astronomical institutions and efforts.

4. Obstacles to Public Education

In the pyramidal structure of public involvement, our attentive public for the radio interference issue is very small, because few non-astronomers are aware of the problem or sufficiently knowledgeable about its technical aspects. Our interested public could include a majority of amateur astronomers and others who follow astronomical discoveries through the news media. Our rather formidable challenge is to move people upwards in the pyramid by gaining their attention for this issue and making them feel well informed about it.

The principal obstacles to effective public education are the public's lack of understanding of the technical issues surrounding interference mitigation and our own tendency to lapse into jargon when discussing the issue.

In an objective study of public knowledge, only 15 percent of U.S. adults were categorized as "civic scientifically literate", meaning that they understood a basic scientific vocabulary, understood the process of scientific research, and understood the impact of science on society (Miller 1999). Less than half of U.S. adults correctly answered that the Earth goes around the Sun once a year (National Science Board 1998).

Widespread anecdotal evidence suggests that many people fail to realize that "wireless" technologies such as mobile telephones, pagers, etc., rely on radio transmissions that could affect astronomy. In a survey of tourists at the Very Large Array (VLA) Visitor Center in New Mexico, 73 percent rated their own astronomical knowledge as "basic" or better. However, only 11 percent rated their radio astronomy knowledge above "basic", while 55 percent claimed no knowledge whatever of radio astronomy. These were tourists who had chosen to visit a radio observatory in a remote location. A majority of these tourists held college degrees and 78 percent had some college training.

A disturbingly large number of people on guided tours of the VLA are unaware of the difference between electromagnetic radiation and sound, and in fact seem quite surprised when disabused of the common misconception that radio astronomers "listen to sounds".

Clearly we are challenged to present information about radio astronomy and the interference issue in clear and accurate, but non-technical, terms understandable to a public largely unaware of the basics of our science. This requires serious effort to keep the language understandable and avoid lapsing into technical jargon. Jargon will intimidate our audience and raise their resistance to our message.

5. Successful Approaches

5.1. Strategy

First, we must take every opportunity to educate the public about radio astronomy and its contributions, through the news media, museums and science centers, observatory visitor centers, public lectures and other outlets. We need to do a better job of publicizing the results of radio astronomical research. While we rarely can compete with the dramatic visual impact of images from optical telescopes, radio astronomers continuously produce exciting new knowledge that, conveyed properly, captures the attention of the public and reinforces their support.

The importance of introductory college-level astronomy courses aimed at non-science majors cannot be overemphasized. The U.S. National Science Board's surveys have shown that those who have taken college science courses are much more likely to retain an interest in science policy issues and to support science. Thus, it is vital that radio astronomy and its contributions be well represented in astronomy courses aimed at non-science majors. This can be achieved by having radio astronomers teach such courses and by working with textbook and educational media publishers to assure that the work of radio observatories is included in the curriculum for these courses.

In specifically addressing the radio-interference problem, we need to define the issue in our own terms – terms that are both understandable to non-scientists and favorable to our cause. By thus defining the issue, we gain the high ground in the debate. Here we can follow the example of the optical-astronomy community and borrow from the language of environmental protection.

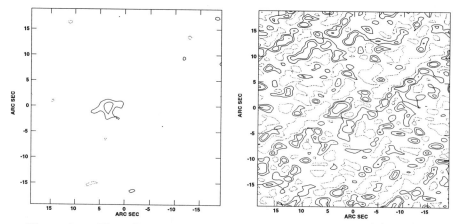

Figure 3.   The effect of interference on radio images. A VLA image of an OH/IR star observed at 1612 MHz with no satellite signal present (left) and with an Iridium satellite approximately 22 degrees from the star (right). The increased noise in the image with the satellite present is clearly obvious, even to a non-technical audience. Courtesy G.B. Taylor, NRAO/AUI.

Our campaign against radio interference should use phrases such as "polluting" interference, "radio waste", "spillover" and "dirty" transmitters. Harmonics and spurious emissions should be portrayed as trespassers and polluters into territory where they do not belong. Simple illustrations of spillover into radio-astronomy bands and of the effect of interference on radio images (Figure 3) can effectively convey our message.

As reported elsewhere in this volume, radio astronomy may not remain the lone victim of interference problems. We were the first because of the extreme sensitivity of our systems, but other users of the radio-frequency spectrum are becoming vulnerable. This "canary in the mine" status of radio astronomy can strike a responsive chord in many minds.

In pointing out the value of astronomical research within the unique "cosmic laboratory", we must point out that our use of this invaluable resource can be degraded or prevented by radio interference. Those who indiscriminately pollute the radio spectrum are thus cutting off the rest of humankind from an important source of scientific knowledge and technological progress. No one, we must argue, has the right to do so.

We are, of course, in favour of efficient use of the radio spectrum, but only in a responsible manner that also allows astronomical research to continue. This position is similar to the successful arguments that convinced public policy makers that industry should not only make profits but also protect the air, water and soil from pollution. Counterarguments that measures to mitigate pollution would cost too much have largely been unsuccessful. In the U.S., public-opinion polls for the past decade have shown strong support (by a margin of more than 2-1) for protecting the environment even at the risk of curbing the economy (Gillespie 1999). We should tap this strong source of public support.

5.2. Tactics

When addressing a specific audience on this issue, one must carefully gauge the appropriate technical level of the discussion to avoid intimidating the listeners. It is better to keep the discussion simple, and keep the audience confident in their understanding, than to try covering all technical aspects and turn off the listeners. Questions from a more sophisticated member of the audience can be answered at the level of the question, but then the discussion should be returned to a level appropriate to the majority of the audience.

In keeping the public discussion simple, we once again can follow the successful example of campaigns for environmental protection. Preventing and remediating air, water and soil pollution are, in fact, complex and sophisticated technical disciplines. However, public discussion about protecting these resources rarely reaches the level of complexity facing professionals in those fields. The public simply wants the resources protected and most people feel little need to learn the details of how that is done.

In dealing with news media, it is important to recognize the differences among various types of media. In a television or radio news show, two or three minutes of air time would constitute significant coverage for that show, but would mean we must make our case very quickly in accurate but simple language. In discussing the issue with reporters for print media, we can be more complete and, depending on the publication and the reporter's own technical background, delve into some of the more technical aspects. Because of the great differences in the level of discussion possible in different media, separate press releases, tailored to these differences, may be advisable.

The more informed members of the public tend to get more of their information from newspapers and magazines than from broadcast outlets, so, at least initially, coverage in these media may prove more productive.

Finally, we must not overlook the growing importance of the World Wide Web as a means of information distribution, particularly to the more technically-sophisticated public. Radio observatories and other astronomical institutions should provide information on the interference issue aimed at an interested, but non-technical, public audience in their Web sites.

6. Summary

Radio astronomers must take an active role as leaders of public opinion to build broad support for protecting radio observatories from interference. The starting point for this effort is to aggressively promote the value and contributions to society as a whole of astronomy in general and radio observations in particular. We must all familiarize ourselves with astronomy's solid record of contributing to human progress and the promise for continuing to do so in the future. Then, we must take every opportunity to make this case to our fellow citizens.

In addressing the specifics of interference and its effects on radio observations, we must recognize that we are dealing with a public that is not well-educated technically and that will not respond well to intricate, jargon-laced technical discussions. Instead, we can convince a generally receptive public by using the language of protecting from pollution a valuable resource for all humanity. By defining the issue in these terms and placing the burden on those who for economic gain would deprive all humans of the benefits of knowledge to be gained from the "cosmic laboratory", we can build a broad base of public support.

References

Almond, G. A. 1950, *The American People and Foreign Policy,* New York: Harcourt, Brace

Gillespie, M. 1999, *U.S. Public Worries About Toxic Waste, Air and Water Pollution as Key Environmental Threats,* Princeton, N.J.: The Gallup Organization

Miller, J. D. 1999, "Scientific Literacy, Issue Attentiveness, and Attitudes toward Science and Space Exploration", paper presented to AAS meeting, May 1999

National Science Board 1998, *Science & Engineering Indicators,* Washington: U.S. Government Printing Office

Yun, M.S., Ho, P.T.P. and Lo, K.Y. 1994, Nature, 372, 530

Part 6
Outcomes

Preserving the Astronomical Sky
IAU Symposium, Vol. 196, 2001
R. J. Cohen and W. T. Sullivan, III, eds.

Optical Workshop Report: Statements Relative to Environmental Protection for Optical Astronomy

1. We reaffirm that all IAU Resolutions relative to the preservation of the skies are still valid.

2. We recommend that the IAU establish a working group with no more than 10 members within Commission 50 to specify the specific needed items for protecting the dark sky for optical astronomy, to include at least:

   a. monitoring and quantifying the level of night sky brightness, from space observations and from the ground (at observatories and at amateur observatories).

   b. educational projects and to select ones of special short term application.

   c. developing fiscal support for the cause.

   d. establishing and maintaining liaison with other organizations involved in the issues.

   e. including the particular sensitivity to the environments in developing countries.

Work to begin via email as soon as established, with a formal meeting to take place at the IAU General Assembly in Manchester in 2000. The first agenda item is to produce a Policy Statement that all people can use in support of the cause. Members to include observatory directors, astronomers, amateur astronomers, educators and lighting engineers.

3. We recommend that an inform/formal working group be formed to discuss media outreach for the cause, members to be from the allied organizations involved, such as the IAU, IDA, BAA, RAS, AAS, ASP, etc. (both internationally and nationally). Most contact among members can be by Internet, with a formal meeting to be held at the IAU General Assembly in Manchester in 2000.

4. Relative to the lighting industry, we recommend that:

   a. All luminaire manufacturers measure and publish, for fixed angle luminaires, their ULOR and DLOR.

   b. All luminaire photometric data make clear which intensity figures have been measured as zero and which have not been measured, but defaulted to zero.

    c. All lamp manufacturers print both on the lamp and its packaging the initial (100hr) lumen output figure.

5. Astronomical support for the cause: We urge that astronomers and observatories work with and support those individuals and organizations involved in the protection of the dark skies. The new Commission 50 working group can be one of the mechanisms. Each observatory should have a light pollution officer as a point of contact.

6. We recommend that Heritage Sites of Dark Sky Preserves be created an maintained to preserve dark skies for all of humankind. The IAU welcomes and supports the initiative taken near Toronto for establishing the first such Dark Sky Preserve.

7. Education: We recommend that the IAU and other scientific and educational organizations work together to promote public awareness, understanding and appreciation of the problem of light pollution and other threats to the astronomical environment through school systems and other appropriate means.

8. We urge all observatories to measure and monitor the sky brightness and other adverse environmental impacts at their observatories and we urge the IAU to set up a mechanism to coordinate such measurements.

Preserving the Astronomical Sky
IAU Symposium, Vol. 196, 2001
R. J. Cohen and W. T. Sullivan, III, eds.

Radio Workshop Report: Technical Methods and Strategies for Mitigating Radio Frequency Interference (RFI)

J. R. Fisher

NRAO, P.O. Box 2, Green Bank, WV, USA

1. Introduction

This discussion session opened the Radio Workshop on Thursday 15th July 1999. There were four agenda items proposed:

1. The role of the IAU in controlling RFI

2. Support of spectrum management

3. Characterization of the RFI environment

4. Research and development.

Discussion of the first item continued throughout the day, but it was pointed out at the beginning that the IAU is only one of a number of international and national organizations that play a role in spectrum management and RFI control for radio astronomy, e.g. IUCAF, ITU-R (Working Party 7D), URSI, OECD, CRAF, CORF/NAS, PTT's/FCC, and the AAS.

This session also opened the discussion of suggestions for revisions to the General Secretary's draft report to the Technical Forum of UNISPACE III.

2. Action Items

Four major action items emerged from this workshop session:

- First, Sullivan, Gergely, Butcher and Fisher were assigned the task of drafting a letter to radio observatory directors asking for their support of spectrum management activities and RFI research and development. The letter will remind them of the Kyoto Declaration and will make the following points:

  1. Ask directors to encourage staff involvement in RFI matters. A particular concern is that this encouragement include adequate credit towards career advancement for participation this activity.

  2. Encourage directors to ensure that their own observatory is in compliance with the ITU-R R.A.769-1 radiation limits. The key point here is that there is a moral imperative on us to ensure that our own activities do not corrupt the radio astronomy bands to any greater extent than the levels which we ask other users of the radio spectrum to respect. This

is particularly important to radio astronomers' bargaining position when they are negotiating in various spectrum management fora.

3. Offer the IAU's assistance in obtaining funds for research and development of RFI measurement and mitigation techniques.

- The second of the action items was a request for a new chapter for the ITU-R Handbook on Radio Astronomy, to cover RFI measurement techniques and standards. Fisher will compose the first draft and submit it to ITU-R Working Party 7D for revision and final disposition.

- The third item that resulted from both this discussion and the discussion on RFI databases was an agreement to organize a session on RFI Measurements, Standards, and Databases for the IAU General Assembly in Manchester in August 2000. Van Driel and Fisher undertook to organize this session.

- Finally, it was agreed to reactivate the IAU Commission 40 Working Group on Astrophysically Important Spectral Lines, under the Chairmanship of Ohishi.

Editor's Note: The following Internet site was created shortly after the end of Symposium 196:

http://www.atnf.CSIRO.AU/SKA/intmit/

"a meeting place for anyone interested in the technical problems of making radio astronomical measurements in the presence of other radio signals."

Preserving the Astronomical Sky
IAU Symposium, Vol. 196, 2001
R. J. Cohen and W. T. Sullivan, III, eds.

Radio Workshop Report: Public Awareness of Radio Interference

David G. Finley[1]

National Radio Astronomy Observatory, P.O. Box O, Socorro, New Mexico 87801 USA

This, the final radio workshop session, covered the vital topic of public outreach. A number of valuable ideas emerged, with at least one specific goal to be achieved in a relatively short amount of time.

First, an important point that perhaps should be expected to come out of such an international gathering is that there is no single solution or "one size fits all" outline for educational efforts aimed at the public and at policy-makers. There are differences in how the news media operate in different countries and cultures that must be taken into account. Also, regulatory agencies in different countries vary, sometimes dramatically, in their receptiveness to public input on policy and technical matters. In each country, therefore, the educational efforts must be tailored to the audiences and political structures and traditions prevailing there.

One audience that should prove receptive to our message is the "general" scientist and engineer – not an astronomer, but someone capable of readily grasping both the technical aspects of radio frequency interference and also the importance of the issue to radio astronomy.

As part of the effort to educate both the general public and the general scientist and engineer, we need to publicize radio astronomy research *and* the effects of radio interference on that research. Publicity materials such as press releases, brochures, videos, Web sites, etc., about research results from radio astronomy should mention the frequency at which the observation was made. If we show the public a beautiful image or significant new scientific discovery made with a radio telescope, the public should understand that producing it required use of radio spectrum.

In a number of countries there are annual events such as "Outreach Week" or "Astronomy Day," in which the research community is encouraged to present science to the public. Radio observatories and individual radio astronomers should participate in such events, raising the level of public awareness. These events often receive considerable publicity and we can use these opportunities to make the case for interference mitigation.

Many radio observatories have well-attended visitor centres. The audiences at these visitor centres already are mostly well disposed towards radio astronomy. We could make our point about radio interference to these audiences by making our visitor centres as RFI-compliant as possible. Having visitors enter a Faraday cage for example and explaining to them why it is necessary to prevent the

---

[1]The National Radio Astronomy Observatory is a facility of the National Science Foundation, operated under cooperative agreement by Associated Universities, Inc.

microprocessors in the exhibits from interfering with the radio telescopes outside, could quickly and dramatically sensitize them to the interference issue.

Finally, it was pointed out that, despite the importance of the World Wide Web as a tool for education and communication, there does not exist, to our knowledge, any material on the Web specifically designed to explain the interference issue to the general public. Individual radio observatories have often-extensive Web sites aimed at providing observers with information on their local RFI environments, but there are no "non-technical" Web sites on the topic.

To remedy this situation, we decided that NRAO will take the lead in producing a Web page or pages explaining the interference issue in terms understandable to the general public. This material will be either duplicated or linked to by any other observatory, by the IDA, the IAU and other astronomical institutions. In addition, we will seek to produce a Web repository of material and graphics that can be used for viewgraphs, providing a resource for anyone who wishes to talk about the interference issue in public lectures.

Preserving the Astronomical Sky
IAU Symposium, Vol. 196, 2001
R. J. Cohen and W. T. Sullivan, III, eds.

Postscript

R. J. Cohen

*University of Manchester, Jodrell Bank Observatory, Macclesfield,
Cheshire SK11 9DL, UK*

Abstract. A summary is given of the results of WRC-2000 for radio
astronomy and other developments since the end of Symposium 196 (up
to April 2001).

1. Apologia

The interval between the end of Symposium 196 and the final editing of these
proceedings saw great progress of the issues of our Symposium. I therefore took
the liberty of adding a summary of some major results, to complement Section 4
of the article by Johannes Andersen (these proceedings).

2. Results from WRC-2000

Radio astronomers made major gains at WRC-2000 in Istanbul. The month-long
meeting was attended by about 2,500 delegates, including 17 radio astronomers.
Our success was largely due to international coordination of the radio astronomy
position over several years. The articles by Ruf and Ohishi (these proceedings)
explain the background to the WRC.

2.1. New allocations above 71 GHz

Under WRC-2000 agenda item 1.16, new allocations above 71 GHz, guaranteed
access was secured to nearly all of the spectrum that is observable from the
ground through the three major atmospheric windows. Under the new allocations radio astronomy now has primary allocations to most of the spectrum
in the three atmospheric windows between 71 and 275 GHz. In return radio
astronomy has given up some of its exclusively passive spectrum.

The allocations before and after WRC-2000 are shown in Figure 1. In
essence, radio astronomers have gained access to almost the entire useable spectrum in this frequency range, while renouncing exclusive possession of a small
part of it (totalling 3.45 GHz). Furthermore, radio astronomy use of bands up
to 945 GHz is now officially acknowledged via footnote S5.565. The current limit
of ITU-R allocations is 275 GHz. Allocations above 275 GHz have been placed
on the agenda for WRC-2006.

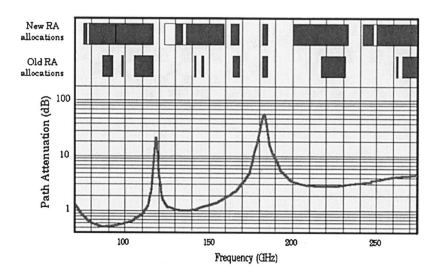

Figure 1.   New frequency allocations to radio astronomy in the frequency range 71 to 275 GHz, compared to previous allocations. The solid curve shows how atmospheric attenuation varies with frequency. (Figure courtesy of John Whiteoak, ATNF.)

2.2.   Unwanted emissions

WRC-2000 finalized several issues concerning spurious emission limits. The design objectives for spurious emissions from satellites are now hard limits which will apply to all new systems from 1 January 2003 and to all systems from 1 January 2012. This is an important step for the protection of radio astronomy. The actual limits introduced at WRC-2000 will not guarantee protection of radio astronomy in all frequency bands, but the possibility has been left open for WRC-03 to set tighter regulatory limits in specific bands for specific satellite services (either for spurious emissions or out-of-band emissions). A new ITU-R Task Group 1/7 is responsible for the technical studies.

2.3.   Regulatory protection from new satellites

Under agenda items 1.4, 1.14 and 1.15, radio astronomy had mixed fortunes. WRC-2000 made further allocations of satellite downlinks very close to radio astronomy bands. However regretable this may be to astronomers, the administrations have tried to help us. The affected radio astronomy bands have been given strong regulatory protection from the relevant satellite transmissions, via a new generation of tough footnotes. The new footnotes state that the satellite systems *shall* protect radio astronomy observatories to specified power flux-density levels calculated by radio astronomers, for a percentage of time agreed by radio astronomers. This is the first time that interference thresholds for radio astronomy have appeared in the Radio Regulations.

3. Developments within the UN

The seeds planted by Symposium 196 have started to bear fruit in the UN arena. In February 2000 the IAU contributed, by invitation, to the COSPAR/IAF Symposium on "Space Commercialization: An Era of New Opportunities and Challenges", held during the thirty-seventh session of the UN Committee for the Peaceful Uses of Outer Space (COPUOS). The following year, the IAU, ITU and OECD were invited to make presentations on the regulatory mechanisms for radio astronomy to the Scientific and Technical Subcommittee of COPUOS at its thirty-eighth session, in Vienna in February, 2001. The IAU also made an invited contribution to a COSPAR/IAF Symposium on "Terrestrial Hazards from Outer Space Objects and Phenomena", held in conjunction with the COPUOS meeting. Johannes Andersen was asked to summarise the Symposium on behalf of COSPAR, and also read an official statement on behalf of the IAU, which outlined the broader environmental perspective.

As an unexpected windfall, the US delegation announced that Congress had passed a law banning, in quite sweeping terms, all "Obtrusive space advertising" (i.e., visible from the ground with the naked eye), and had instructed the President to initiate negotiations towards international agreement to the same end.

4. Other Progress

On 1st October 1999, the Chilean Lighting Guideline (Norma Luminica) for controlling light pollution was established (Decreto Supremo No. 686/99, signed by President Frei); this guideline sets up a legal framework for protecting the skies of Northern Chile, based on environmental legislation. There is now an "Office for the Protection of the Chilean Skies" (Oficina Para la Protecciòn de los Cielos Chilenos, OPCC, with its own web site http://www.opcc.cl).

In August 2000 the IAU General Assembly in Manchester set up a new Commission 50 Working Group on Light Pollution, with Malcolm Smith as the first Chairman. Division X, Commission 40 (Radio Astronomy), also set up a new Working Group on Interference Mitigation, with Tassio Tzioumis (ATNF) as its first chairman.

The OECD Global Science Forum Task Force on Radio Astronomy held its first meeting in March 2001, under the chairmanship of Mike Goddard (Radiocommunications Agency, UK). The terms of reference of the group are given in Appendix 2.

Appendix 1: S196 Proposal to UNISPACE III [1]

Vienna
19-30 July 1999
Committee I
Agenda item 9
Benefits of basic space science and capacity-building

Conclusions and proposals of the International Astronomical Union/Committee on Space Research/United Nations Special Environmental Symposium "Preserving the Astronomical Sky"

Technical Forum

Recalling the paragraphs of the draft report of the Third United Nations Conference on the Exploration and Peaceful Uses of Outer Space (UNISPACE III) (A/CONF.184/3 and Corr.1 and 2) referenced in parentheses below, and noting that:

(a) Understanding the nature of the universe is one of humanity's oldest and strongest fascinations and has been of immense scientific, cultural and practical value for many centuries. Observations at all wavelengths of the electromagnetic spectrum, from the ground and from space, have been vital in the phenomenal progress in all areas of astronomy in the twentieth century, from the exploration of the solar system to discoveries of the echo of the big bang and the beginnings of structure in the universe (paras. 1, 2, 6 and 28);

(b) The space treaties adopted by the United Nations have defined outer space and the space environment as the province of all mankind, to be protected from harmful contamination and adverse changes of all kinds, the exploration and peaceful use of which should be carried on for the benefit and in the interests of all mankind (para. 313). This principle is also strongly supported by the International Astronomical Union and the Committee on Space Research;

(c) Nevertheless, continued scientific studies of the origin and evolution of the universe and mankind's place within it are being jeopardized worldwide by man-made environmental problems of rapidly growing severity. In space, interference in radio frequencies by telecommunications satellites and their ever-increasing demand for frequency space (para. 158) cloud the future of radio astronomy and the operation of scientific satellites for astronomy and remote sensing; space debris is a growing threat to scientific satellites and interferes with ground-based observations (para. 70); and projects to launch bright ob-

---

[1]Input document A/CONF.184/C.1/L.2, which was accepted by UNISPACE III and published in the final Report, UN Document A/CONF.184/6, as Annex III, item II, pp.111-112.

jects into space to illuminate the Earth or for artistic, celebratory or advertising purposes present a growing danger to observational astronomy against which no international protection at present exists (para. 73). On the ground, man-made light pollution has already made large areas of the world unsuitable for astronomical observations and is beginning to influence wildlife;

(d) Space is not just another place to do business (para. 273), but a finite natural resource common to all of humanity and already showing inexorable symptoms of over-exploitation (para. 70). The problems enumerated above are global in extent and some are long-term or irreversible in time. Owing to the extreme sensitivity of astronomical observations, science has been the first to detect and suffer from these effects, but it will not be alone for long;

It is recommended that:

(a) Member States should continue to cooperate, at the national and regional levels, and with industry and through the International Telecommunication Union, to implement suitable regulations to preserve quiet frequency bands for radio astronomy and remote sensing from space (para. 162), and to develop and implement, as a matter of urgency, practicable technical solutions to reduce unwanted radio emissions and other undesirable side-effects from telecommunications satellites;

(b) Member States should cooperate to explore new mechanisms to protect selected regions of Earth and space from radio emissions (radio quiet zones), and to develop innovative techniques that will optimize the conditions for scientific and other space activities to share the radio spectrum and coexist in space;

(c) Member States should cooperate, as a matter of urgency, to ensure that future space activities that would cause potentially harmful interference with the scientific research or natural, cultural and ethical values of other nations (para. 73) are subjected to an environmental impact assessment and international consultations before approval;

(d) Member States should cooperate to ensure that the implementation of measures, at the international level, to preserve all aspects of the space environment in the long term, are included in the work plan of the Committee on the Peaceful Uses of Outer Space and its Subcommittees (paras. 318-321). It is proposed that section III, subparagraph (b) of the draft Vienna declaration on space and human development be formulated more adequately as follows:

> "To improve the protection of the near and outer space environment through further research in, and implementation of, measures to control and reduce the amounts of space debris and unwanted emissions at all wavelengths of the electromagnetic spectrum";

(e) Member States should act to control pollution of the sky by light and other causes, for the benefit of energy conservation, the natural environment, nighttime safety and comfort and the national economy, as well as science.

Appendix 2: OECD Global Science Forum, Task Force on Radio Astronomy and the Radio Spectrum

Terms of Reference, March 2001

I. Background

The deployment of new low-orbiting telecommunications satellites will have a major impact on the future of radio astronomy. Radio telescopes are extremely sensitive, and signals from telecommunications satellites can overwhelm the signals from astronomical sources, thus preventing scientists from collecting data. For example, a simple portable telephone, if placed on the Moon, would be among the brightest astronomical objects, as seen from Earth. In the past, radio astronomers have sought protection from man-made signals by placing their telescopes in remote locations, These measures may no longer permit scientific observations, since the new satellites are designed to provide 100% global coverage. Also, radio astronomers made their observations in the frequency windows reserved for their exclusive use by the International Telecommunications Union. However, astronomers have now discovered that signals from the most distant and exotic objects in the sky are shifted in frequency due to the expansion of the Universe – and hence must be observed outside the bandwidths reserved for science.

To ensure that the science of radio astronomy can flourish along with a thriving telecommunications industry, new innovative and effective measures to permit astronomical signal detection will need to be developed before the next generation of radiotelescopes will be built. These instruments, which will be very few in number, will be a hundred times more sensitive than current telescopes - sensitive enough to permit observation of nearly the entire extent and history of the observable Universe.

The need to reconcile and support the interests of both the telecommunications industry and the radio astronomy community was highlighted in the report of a working group of the OECD Megascience Forum, now referred to as the OECD Global Science Forum, and was discussed by the Science Ministers of the OECD countries at their meeting in June 1999. In the communiqué of the meeting, the Ministers endorsed the establishment of an informal high-level Task Force to map out a strategy for ensuring the future of radio astronomy, while allowing for the continued vigorous growth of commercial space-based telecommunications.

The Task Force should have approximately ten members, among them senior executives of telecommunications companies, members of the scientific community, and persons who are active in national and international regulatory bodies. Task Force members should be leaders in their fields, with a strategic view of future developments, an appreciation of the value of astronomy and of telecommunications, and a desire to find win-win solutions for the scientific and business communities. They will participate in a personal capacity; however, they will be expected to maintain contacts with their organisations and their professional colleagues, to ensure openness and balance in the deliberations of the group. The Task Force is expected to complete its work in approximately one year, with two meetings convened during that time. Detailed studies by experts could be commissioned as needed.

II. Objectives

The Task Force should produce a brief, policy-level report that contains findings and action-oriented recommendations. While the Task Force is expected to formulate a long-term strategic plan, it should, whenever possible, propose specific activities that can be undertaken by existing national and international organisations and institutions that are already active in this area, or have the authority to act and generate concrete results. It is recognised that the ultimate resolution of this difficult problem could take many years, but the deliberations of the Task Force should lead to the initiation of appropriate activities in the near future. Examples of such activities could include: co-operative R&D projects involving satellite manufacturers and radio astronomy laboratories, analysis and discussion of regulatory and technical issues by ITU working parties, and feasibility studies by international bodies for new agreements or treaties. As it proceeds, the Task Force may wish to provide progress reports to the OECD Global Science Forum, the successor to the Megascience Forum. This committee consists of senior government science policy officials who will be available to provide a governmental perspective on the ongoing work of the Task Force.

III. Scope and Work Program

While the Task Force will be free to define its own work programme, it might choose to examine three general areas that were previously identified by the working group of the Megascience Forum:

A. **Technological solutions**: Radio astronomers and industry representatives could identify and jointly implement interference mitigation schemes.

B. **Regulation**: If new, innovative ways of sharing the radio spectrum are to be found, the appropriate discussions must begin soon within national agencies and the ITU, as the implementation of new regulations usually requires a considerable amount of effort over several years.

C. **Radio-quiet zones**: Remote areas on the Earth's surface could be designated where future radio observatories could be located, and where radio emissions, especially from spaceborne and airborne sources, would be restricted in frequency and time. The technical, regulatory, and legal dimensions of this concept would have to be carefully examined.

Accordingly, the Task Force could choose:

A. To investigate in detail the three-pronged approach indicated above, and to propose specific programs that will constitute a balanced, co-ordinated strategy.

B. To sound out members of their respective spheres of interest about the elements of acceptable regulatory and other solutions, and about their willingness to participate in achieving those solutions.

C. To develop a road map of long-term solutions, with a time schedule and milestones, for consideration by the relevant national and international bodies. Special attention should be given to taking advantage of (and strengthening) the procedures of the ITU.

Appendix 3: Abbreviations

1hT	One Hectare Telescope: original working title for the Allen Telescope Array (USA)
AIAA	American Institute of Aeronautics and Astronautics
ALCOR	Astronomical Lighting Control Regions (for Observatories)
ALFA	Astronomical Low Frequency Array
ALMA	Atacama Large Millimeter Array
ATN	Automatic Telescope Network
ATNF	Australia Telescope National Facility
AUASS	Arab Union for Astronomy and Space Science
AURA	Association of Universities for Research in Astronomy
BAO	Beijing Astronomical Observatory (China)
BRM	Bureau of Radio Management (China)
BSS	Broadcasting Satellite Service
CCD	Charge Coupled Device
CCIR	Comité Consultatif International des Radiocommunications (ITU)
CEN	European Committee for Standardization
CEPT	Conférence Européen des Postes et des Telecommunications
CIE	Commission Internationale d'Eclairage http://www.cie.co.at/cie/
COBE	Cosmic Background Explorer (NASA satellite)
CONAMA	Comision Nacional del Medio Ambiente (Chile)
CONICYT	Chilean Science Foundation
COPUOS	Committee on the Peaceful Uses of Outer Space (UN) http://www.oosa.unvienna.org/COPUOS/copuos.html
CORF	Committee on Radio Frequencies (USA) http://www.nas.edu/bpa/corf/
COSPAR	Committee on Space Research http://cospar.itodys.jussieu.fr/
CRAF	Committee on Radio Astronomy Frequencies (ESF) http://www.nfra.nl/craf/
CrAO	Crimean Astrophysical Observatory
CTIO	Cerro Tololo Inter-American Observatory (Chile)
DLOR	Downward Light Output Ratio
DLR	German Aerospace Centre (Deutsches Zentrum für Luft- und Raumfahrt)
DMSP	Defence Meteorological Satellite Program
DRAO	Dominion Radio Astrophysical Observatory (Canada)
EAAE	European Association for Astronomy Education
EC	European Commission
EESS	Earth Exploration-Satellite Service
EIRP	Effective Isotropic Radiated Power (radio)
EIS	Environmental Impact Statement
EISCAT	European Incoherent Scatter Scientific Association
EMI	Electro Magnetic Interference
ERC	European Radiocommunications Committee
ERO	European Radiocommunications Office

ESA	European Space Agency
ESF	European Science Foundation
ESO	European Southern Observatory
ETSI	European Telecommunications Standards Institute
FAAQ	Fédération des Astronomes Amateurs du Quebec
FAST	Five-hundred-meter Aperture Spherical Telescope (China)
FCC	Federal Communications Commission (USA)
FIRST	Far InfraRed and Submillimetre Telescope
FLWO	F. L. Whipple Observatory (USA)
FFT	Fast Fourier Transform
FSS	Fixed Satellite Service
GBT	Greenbank Telescope (USA)
GEO	Geostationary Orbit
GLONASS	Global Navigation Satellite System (Russia)
GMRT	Giant Metre-wave Radio Telescope (India)
GMS	Geostationary Meteorological Satellite
GMSK	Gaussian-filtered Minimum-Shift Keying (modulation scheme)
GNAT	Global Network of Astronomical Telescopes
GPS	Global Positioning System (USA)
GSM	Grating Scale Monitor
GSO	Geostationary Orbit
HAPS	High Altitude Platform System
HCA	Hungarian Communication Authority
HF	High Frequency (band) 3-30 MHz
HPS	High-Pressure Sodium (lamp)
HSOS	Huairou Solar Observing Station (China)
IAA	International Academy of Astronautics
IADC	Inter-Agency Space Debris Coordinating Committee
IAU	International Astronomical Union
	http://www.iau.org/
ICAO	International Civil Aviation Organization
ICSU	International Council for Science
	http://www.icsu.org/
IDA	International Dark-Sky Association
	http://www.darksky.org/ida/
IDMP	International Daylight Measurement Program (CIE, WMO)
IEC	International Electrotechnical Commission
	http://www.iec.ch
IF	Intermediate Frequency (in a radio receiver)
ILE	Institution of Lighting Engineers (UK)
INMARSAT	International Maritime Satellite service
IRA	Institute of Radio Astronomy (Ukraine)
IRAM	Institut de Radio Astronomie Millimetrique
IRQZ	International Radio Quiet Zone
ISO	International Organization for Standardization
	http://www.iso.ch
ITU	International Telecommunication Union
ITU-R	Radiocommunication Sector of ITU
IUCAF	Inter-Union Commission on Frequency Allocation for

	Radio Astronomy and Space Science
	http://www.mpifr-bonn.mpg.de/staff/kruf/iucaf/
IUGG	International Union of Geodesy and Geophysics
JAC	Joint Astronomy Centre (Hawaii, USA)
JIVE	Joint Institute for VLBI in Europe
KARST	Kilometer-square Area Radio Synthesis Telescope (China)
KPNO	Kitt Peak National Observatory (USA)
LDEF	Long Duration Exposure Facility (NASA)
LED	Light emitting diode
LEO	Low Earth Orbit
LO	Local Oscillator (in a radio receiver)
LOFAR	Low Frequency Array (Netherlands and USA)
LPS	Low-Pressure Sodium (lamp)
LSB	Lower Sideband (radio receiver)
MAO	Main Astronomical Observatory (Kiev, Ukraine)
MASTER	Millimetre-wave Acquisitions for Stratosphere-Troposphere Exchanges Research (ESA)
MH	Metal halide (lamp)
MLS	Microwave Landing System
MLS	Microwave Limb Sounder (EESS)
MMDS	Multichannel Microwave Distribution System (China)
MSRT	Miyun Synthesis Radio Telescope (China)
MSS	Mobile Satellite Service
MU	Mobile User (INMARSAT)
NASA	National Aeroanutics and Space Admininstration (USA)
NATO	North Atlantic Treaty Organization
NFRA	Netherlands Foundation for Research in Astronomy
NGDC	National Geophysics Data Center (USA)
NGSO	Non Geo Stationary Orbit (satellite system)
NOA	National Observatory of Athens (Greece)
NOAA	National Oceanographic and Atmospheric Administration (USA)
NOAO	National Optical Astronomy Observatories (USA)
NRAO	National Radio Astronomy Observatory (USA)
NRO	Nobeyama Radio Observatory (Japan)
NSA	Nançay Surveillance Antenna (France)
NSB	Night sky brightness
OCS	Observatory Control System
OECD	Organization for Economic Cooperation and Development
OLC	Outdoor Lighting Code (USA)
OLS	Operational Linescan System (of the DMSP satellites)
PEDAS	Potentially Environmentally Detrimental Activities in Space (COSPAR Panel)
PMO	Purple Mountain Observatory (China)
PMT	Photo Multiplier Tube
PVS	Polynesian Voyaging Society
RCT	Remotely Controlled Telescope (USA)
RFI	Radio Frequency Interference
RQZ	Radio Quiet Zone

RSA	Russian Space Agency
SAWF	Surface-Acoustic-Wave Filter (radio)
SAIt	Italian Astronomical Society
SCOPE	Standing Committee on the Problems of the Environment
SERNATUR	Servico Nacional de Turismo (Chile)
SEST	Swedish ESO Submillimetre Telescope (Chile)
SETI	Search for Extra-Terrestrial Intelligence
SHAO	Shanghai Astronomical Observatory (China)
SIS	Superconductor-Insulator-Superconductor (mixer device)
SKA	Square Kilometer Array
SOAR	Southern Observatory for Astrophysical Research
SOC	Scientific Organizing Committee
SOFIA	Stratospheric Observatory for Infrared Astronomy
SOPRANO	Sub-mm Observation of Processes in the Atmosphere Noteworthy for Ozone (ESA)
SS	Space Station (INMARSAT)
SSB	Single Sideband (radio receiver)
SVLBI	Space Very Long Baseline Interferometry
TC	Technical Committee (of the CIE)
TCS	Telescope Control System
TDRSS	Tracking and Data Relay Satellite System
TG	Task Group (of the ITU)
TÜBİTAK	Scientific and Technical Research Council of Turkey
UAO	Urumqi Astronomical Observatory (China)
ULOR	Upward Light Output Ratio
UN	United Nations
UNESCO	United Nations Educational, Scientific and Cultural Organization
UN-OOSA	UN Office for Outer Space Affairs
URSI	Union Radio-Scientifique International http://www.intec.rug.ac.be/ursi/
ULOR	Upward Light Output Ratio
USB	Upper Sideband (radio receiver)
UWLR	Upward Waste Light Ratio
VHF	Very High Frequency (band) 30-300 MHz
VLA	Very Large Array (USA)
VLBI	Very Long Baseline Interferometry
VLT	Very Large Telescope (European Southern Observatory)
VMM	Vlaamse Mileu Maatschappij (Belgium)
VSOP	VLBI Space Observatory Programme
VST	VLT Survey Telescope (European Southern Observatory)
WARC	World Administrative Radio Conference (ITU)
WEBT	Whole Earth Blazar Telescope
WMO	World Meteorological Union
WRC	World Radio Conference (ITU)
YAO	Yunnan Astronomical Observatory (China)

Author Index

Subject Index

ASTRONOMICAL SOCIETY OF THE PACIFIC CONFERENCE SERIES VOLUMES

and

INTERNATIONAL ASTRONOMICAL UNION VOLUMES

Published
by

The Astronomical Society of the Pacific
(ASP)

INTERNATIONAL ASTRONOMICAL UNION (IAU) VOLUMES
Published by the Astronomical Society of the Pacific

PUBLISHED: 1999

Vol. No. 190   NEW VIEWS OF THE MAGELLANIC CLOUDS
eds. You-Hua Chu, Nicholas B. Suntzeff, James E. Hesser, and David A. Bohlender
ISBN: 1-58381-021-8

Vol. No. 191   ASYMPTOTIC GIANT BRANCH STARS
eds. T. Le Bertre, A. Lèbre, and C. Waelkens
ISBN: 1-886733-90-2

Vol. No. 192   THE STELLAR CONTENT OF LOCAL GROUP GALAXIES
eds. Patricia Whitelock and Russell Cannon
ISBN: 1-886733-82-1

Vol. No. 193   WOLF-RAYET PHENOMENA IN MASSIVE STARS AND STARBURST GALAXIES
eds. Karel A. van der Hucht, Gloria Koenigsberger, and Philippe R. J. Eenens
ISBN: 1-58381-004-8

Vol. No. 194   ACTIVE GALACTIC NUCLEI AND RELATED PHENOMENA
eds. Yervant Terzian, Daniel Weedman, and Edward Khachikian
ISBN: 1-58381-008-0

PUBLISHED: 2000

Vol. XXIVA   TRANSACTIONS OF THE INTERNATIONAL ASTRONOMICAL UNION
REPORTS ON ASTRONOMY 1996-1999
ed. Johannes Andersen
ISBN: 1-58381-035-8

Vol. No. 195   HIGHLY ENERGETIC PHYSICAL PROCESSES AND MECHANISMS F OR EMISSION FROM ASTROPHYSICAL PLASMAS
eds. P. C. H. Martens, S. Tsuruta, and M. A. Weber
ISBN: 1-58381-038-2

Vol. No. 197   ASTROCHEMISTRY: FROM MOLECULAR CLOUDS TO PLANETARY SYSTEMS
eds. Y. C. Minh and E. F. van Dishoeck
ISBN: 1-58381-034-X

Vol. No. 198   THE LIGHT ELEMENTS AND THEIR EVOLUTION
eds. L. da Silva, M. Spite, and J. R. de Medeiros
ISBN: 1-58381-048-X

PUBLISHED: 2001

IAU SPS   ASTRONOMY FOR DEVELOPING COUNTRIES
Special Session of the XXIV General Assembly of the IAU
ed. Alan H. Batten
ISBN: 1-58381-067-6

Vol. No. 196   PRESERVING THE ASTRONOMICAL SKY
eds. R. J. Cohen and W. T. Sullivan, III
ISBN: 1-58381-078-1

Vol. No. 200   THE FORMATION OF BINARY STARS
eds. Hans Zinnecker and Robert D. Mathieu
ISBN: 1-58381-068-4

Vol. No. 203   RECENT INSIGHTS INTO THE PHYSICS 0F THE SUN AND HELIOSPHERE: HIGHLIGHTS FROM SOHO AND OTHER SPACE MISSIONS
eds. Pål Brekke, Bernhard Fleck, and Joseph B. Gurman
ISBN: 1-58381-069-2

INTERNATIONAL ASTRONOMICAL UNION (IAU) VOLUMES

Published by the Astronomical Society of the Pacific

PUBLISHED: 2001

Vol. No. 204   THE EXTRAGALACTIC INFRARED BACKGROUND AND ITS COSMOLOGICAL
IMPLICATIONS
eds. Martin Harwit and Michael G. Hauser
ISBN: 1-58381-062-5

Vol. No. 205   GALAXIES AND THEIR CONSTITUENTS AT THE HIGHEST ANGULAR
RESOLUTIONS
eds. Richard T. Schilizzi, Stuart N. Vogel, Francesco Paresce,
and Martin S. Elvis
ISBN: 1-58381-66-8

Complete lists of proceedings of past IAU Meetings are maintained at the
IAU Web site at the URL: http://www.iau.org/publicat.html

Volumes 32 - 189 in the IAU Symposia Series may be ordered from
Kluwer Academic Publishers
P. O. Box 117
NL 3300 AA Dordrecht
The Netherlands

All other book orders or inquiries concerning volumes listed should be directed to the:

Astronomical Society of the Pacific Conference Series
390 Ashton Avenue
San Francisco CA 94112-1722 USA

Phone: 415-337-2126
Fax: 415-337-5205
E-mail: catalog@aspsky.org
Web Site: http://www.aspsky.org